Advances in Computer Vision and Pattern Recognition

For further volumes:
www.springer.com/series/4205

Antonio Robles-Kelly · Cong Phuoc Huynh

Imaging Spectroscopy for Scene Analysis

Springer

Antonio Robles-Kelly
National ICT Australia
Canberra, ACT
Australia

Cong Phuoc Huynh
National ICT Australia
Canberra, ACT
Australia

Series Editors
Prof. Sameer Singh
Research School of Informatics
Loughborough University
Loughborough
UK

Dr. Sing Bing Kang
Microsoft Research
Microsoft Corporation
Redmond, WA
USA

ISSN 2191-6586 ISSN 2191-6594 (electronic)
Advances in Computer Vision and Pattern Recognition
ISBN 978-1-4471-4651-3 ISBN 978-1-4471-4652-0 (eBook)
DOI 10.1007/978-1-4471-4652-0
Springer London Heidelberg New York Dordrecht

Library of Congress Control Number: 2012951401

© Springer-Verlag London 2013
This work is subject to copyright. All rights are reserved by the Publisher, whether the whole or part of the material is concerned, specifically the rights of translation, reprinting, reuse of illustrations, recitation, broadcasting, reproduction on microfilms or in any other physical way, and transmission or information storage and retrieval, electronic adaptation, computer software, or by similar or dissimilar methodology now known or hereafter developed. Exempted from this legal reservation are brief excerpts in connection with reviews or scholarly analysis or material supplied specifically for the purpose of being entered and executed on a computer system, for exclusive use by the purchaser of the work. Duplication of this publication or parts thereof is permitted only under the provisions of the Copyright Law of the Publisher's location, in its current version, and permission for use must always be obtained from Springer. Permissions for use may be obtained through RightsLink at the Copyright Clearance Center. Violations are liable to prosecution under the respective Copyright Law.
The use of general descriptive names, registered names, trademarks, service marks, etc. in this publication does not imply, even in the absence of a specific statement, that such names are exempt from the relevant protective laws and regulations and therefore free for general use.
While the advice and information in this book are believed to be true and accurate at the date of publication, neither the authors nor the editors nor the publisher can accept any legal responsibility for any errors or omissions that may be made. The publisher makes no warranty, express or implied, with respect to the material contained herein.

Printed on acid-free paper

Springer is part of Springer Science+Business Media (www.springer.com)

To my wife, for her love and help over the years. To my children, for the joy they brought into my life. And to my parents and sisters, for their continuous and unconditional support.
Antonio Robles-Kelly

To my parents and brother, for their endless love and continuing support.
Cong Phuoc Huynh

Foreword

Current Computer Vision systems, developed to solve many and varied problems in scene understanding and object recognition, typically use the three colour channels designed to mimic the three-colour vision system of humans. However, human vision is complex and involves significant perceptual learning, knowledge acquisition and, in general, post retinal neural processing to arrive at the type of interpretations of objects, scenes and events that predominate our perception. Indeed, one of the most striking aspects of human visual perception is the ability to apparently "directly" perceive the world around as materials, objects and physical events "out there" and not, purely, as images. The challenge for Computer Vision is to create elegant and efficient ways to achieve these goals and, as this book shows, such achievements are possible with the use of new camera systems that go beyond the fixed three colour channels. Indeed the evolution of Imaging Spectroscopy (in remote sensing called Hyperspectral Imaging), capable of imaging spectral information beyond the visible region and at spectral resolutions far greater than the fixed three colour channels, offers great potential and new ways of extracting the physical properties of materials, objects, scenes and illuminants, much more powerfully than the standard three colour systems can possibly do.

This book is a landmark and timely contribution in this direction as it offers, for the first time, detailed descriptions and analysis of these new cameras and how they can be used to extract physical properties of what is being sensed in elegant and efficient ways relevant to material identification, object recognition and scene understanding.

Imaging Spectroscopy also enables the creation of vision systems "on demand" that best suit specific environments and applications well beyond the systems available in current fixed three-colour system cameras. It is also the ideal camera system for advancing the area of Computational Photography and moving Computer Vision away from the image, per se, to what is being imaged.

In this monograph Antonio and Cong have brought together their work in Imaging Spectroscopy over the past decade to result in a book that will become a standard for the area. Well done.

Victoria Research Laboratory National ICT Australia (NICTA)

Terry Caelli

Preface

The vast majority of ground-based image capture (e.g. for general photography, surveillance and industrial machine vision) is currently performed using RGB filters on imaging sensors. This is particularly true for digital photography which, since its inception in the 1980s, has almost exclusively employed trichromatic cameras based on CMOS or CCD sensors using Bayer arrays, three-CCD cameras or a set of stacked photodiodes with a layer for each colour channel. Despite their differences, all of these sensor types involve capturing light at the wavelengths of the three additive primary colours (red, green and blue). These bands are used because they are a close approximation to the bands to which the human eye is sensitive. Therefore, they are well suited to capturing photos and videos for immediate human perception.

Nowadays, the image data is not only used for immediate human perception as originally intended, but also for a wide range of processing, contributing to camera utility, quality and performance. Over the past five years, many digital cameras have featured integrated circuits and firmware for sophisticated image processing (such as Canon's DIGIC and Nikon's Expeed chips). These chips initially performed functions such as automatic focus, exposure and white balance. More recently, such chips have performed a wider range of higher-level scene analysis features such as face detection, smile detection and object tracking. Such scene analysis features provide a high value to professional and consumer camera users and are typically significant selling points in particular models. Many industrial cameras now also include high-level scene analysis functions such as people counting for surveillance and object recognition for industrial machine vision.

In practise, every scene comprises a rich tapestry of light sources, material reflectance, lighting and other photometric effects due to object curvature and shadows. Despite being reasonably effective for scene analysis, trichromatic (i.e. RGB) technology does have limits in its scene analysis capabilities. For example, a camera with an RGB sensor cannot determine the constituent material of an object in the scene. Similarly, cameras with RGB sensors cannot, in general, deliver photometric invariants characteristic to a material and independent of the lighting condition, for robust tracking, identification and recognition tasks.

In this book, we explore the opportunities, application areas and challenges concerning the use of imaging spectroscopy as a means for scene understanding. This is important, since scene analysis in the scope of imaging spectroscopy involves the ability to robustly encode material properties, object composition and concentrations of primordial components. The combination of spatial and spectral information promises a vast number of application possibilities. For instance, spectroscopic scene analysis can enable advanced capabilities for surveillance by permitting objects to be tracked based on their composition. In computational photography, image colours may be enhanced taking into account each specific material type in the scene. For food security, health and precision agriculture the analysis of spectroscopic images can be the basis for the development of non-intrusive diagnostic, monitoring and surveying tools.

The ability to combine spatial and compositional information of the scene requires solving several difficult problems. With these problems solved, spectroscopic scene analysis offers the possibility of performing shape analysis from a single view for non-diffuse surfaces (Huynh and Robles-Kelly 2009), recovering photometric invariants and material-specific signatures (Fu and Robles-Kelly 2011a), recovering the illuminant power spectrum (Huynh and Robles-Kelly 2010a) and visualising digital media (Kim et al. 2010).

With the availability of imaging spectroscopy in ground-based cameras, it will no longer be necessary to limit the camera data captured to three RGB colour channels. Hyperspectral cameras offer an alternative number and range of bands that provide the best trade-off between functionality, performance and cost for a particular market segment or application need. Rather than having the same spectra captured as displayed, it will be practical to decouple them, capturing a rich spectral representation, performing processing on this representation and then rendering it in trichromatic form when needed.

The use of the information-rich representation of the scene that spectral imaging provides is, by itself, a new approach to scene analysis which makes use of the spectral signatures and their context so as to provide a better understanding of materials and objects throughout the image. The use of polarisation and reflection to recover object profiles akin to those in phase imaging can deliver novel methods capable of recovering an optimal representation of the scene which captures shape, material, object profiles and photometric parameters such as index of refraction.

Furthermore, the high dimensionality inherent in spectroscopy data implies that these algorithms may not be exclusive to imaging spectroscopy, but could also be applied to the processing of other high-dimensional data. Thus, these methods may be extendible to many other sensing technologies beyond just spectral imagery.

References

Fu, Z., & Robles-Kelly, A. (2011). Discriminant absorption feature learning for material classification. *IEEE Transactions on Geoscience and Remote Sensing, 49*(5), 1536–1556.

Huynh, C. P., & Robles-Kelly, A. (2010). A solution of the dichromatic model for multispectral photometric invariance. *International Journal of Computer Vision, 90*(1), 1–27.

Huynh, C. P., & Robles-Kelly, A. (2009). Simultaneous photometric invariance and shape recovery. In *IEEE international conference on computer vision.*

Kim, S. J., Zhuo, S., Deng, F., Fu, C. W., & Brown, M. (2010). Interactive visualization of hyperspectral images of historical documents. *IEEE Transactions on Visualization and Graphics, 16*(6), 1441–1448.

Acknowledgements

We wish to thank Professor Terry Caelli, Professor Edwin Hancock and Mr. Bill Simpson-Young for their support, impartial assessment and constructive criticism on the material in this book. Professor Terry Caelli was instrumental in our efforts towards writing this document. Without his continuing support, some of the research presented here would have never been accomplished. The insights that Professor Edwin Hancock provided into reflectance modelling and polarisation have greatly influenced this book. Mr. Bill Simpson-Young provided a valuable insight into the commercial use and applications of imaging spectroscopy.

We would like to express our sincere appreciation to our collaborators. In particular, we thank Dr. Zhouyu Fu, Dr. Ran Wei, Mr. Lin Gu and Ms. Sejuti Rahman for their patience and dedication to the teamwork that delivered some of the results presented here. We also extend our appreciation to all our colleagues from the computer vision group at NICTA.

Finally, we would like to acknowledge NICTA,[1] and in particular its CEO, Hugh Durrant-Whyte, for fostering this technology. We also thank the ANU and the UNSW@ADFA for allowing us to supervise their graduate students.

---

[1] NICTA is funded by the Australian Government as represented by the Department of Broadband, Communications and the Digital Economy and the Australian Research Council through the ICT Centre of Excellence program.

Contents

1	**Introduction**		1
	1.1 Spectral Imaging		1
	1.2 Applications in Scene Analysis		2
		1.2.1 Photography	3
		1.2.2 Food Security	4
		1.2.3 Defence Technologies	5
		1.2.4 Earth Sciences	6
		1.2.5 Health	6
	1.3 Notes		6
	References		7
2	**Spectral Image Acquisition**		9
	2.1 Spectral Cameras and Sensors		9
	2.2 Dark Current Calibration		14
	2.3 Notes		15
3	**Spectral Image Formation Process**		17
	3.1 Radiometric Definitions		18
		3.1.1 Foreshortening and Solid Angle	19
		3.1.2 Radiance	20
		3.1.3 Irradiance	21
		3.1.4 The Bidirectional Reflectance Distribution Function (BRDF)	21
	3.2 Image Formation Process		23
		3.2.1 Image Irradiance	23
		3.2.2 Colour Response	24
	3.3 Spectral Imagery in Colour		28
		3.3.1 Visualisation	28
		3.3.2 Contrast in Spectral Imagery	30
		3.3.3 Colour Rendering of Spectral Data	30
	3.4 Notes		33
	References		34

4	**Reflectance Modelling**		37
	4.1 Physics-Based Models		39
		4.1.1 Beckmann–Kirchhoff Model	39
		4.1.2 Variants of the Beckmann–Kirchhoff Model	42
	4.2 Phenomenological Models		43
		4.2.1 Lambertian Model	43
		4.2.2 Dichromatic Reflection Model	43
		4.2.3 Wolff Model	44
		4.2.4 Oren–Nayar Model	45
		4.2.5 Torrance–Sparrow Model	46
		4.2.6 Cook–Torrance Model	47
		4.2.7 Micro-Facet Slope Distribution	48
		4.2.8 Multi-Layered Reflectance Modelling	48
	4.3 Notes		49
	References		50
5	**Illuminant Power Spectrum**		53
	5.1 Standard Illuminants		53
	5.2 Correlated Colour Temperature (CCT) and the White Point		55
	5.3 Illuminant Power Spectrum Estimation		57
	5.4 Notes		60
	References		61
6	**Photometric Invariance**		63
	6.1 Separating the Reflection Components		65
		6.1.1 Dichromaticity and Specularity	65
		6.1.2 Logarithmic Differentiation	67
		6.1.3 Entropy Minimisation	68
	6.2 An Optimisation Approach to Illuminant and Photometric Invariant Recovery		70
		6.2.1 Cost Function	70
		6.2.2 Homogeneous Patch Selection	71
		6.2.3 Optimisation Procedure	72
		6.2.4 Illuminant Power Spectrum Recovery	76
		6.2.5 Shading, Reflectance and Specularity Recovery	76
		6.2.6 Imposing Smoothness Constraints	78
	6.3 Band Subtraction		81
	6.4 Notes		84
	References		85
7	**Spectrum Representation**		89
	7.1 Compact Representations of Spectra		90
		7.1.1 Spectrum as a Mixture	90
		7.1.2 A B-Spline Spectrum Representation	93
		7.1.3 Classification and Recognition	101

Contents

- 7.2 Spectrum Descriptors 104
  - 7.2.1 A Log-Linear Shading-Invariant Feature 106
  - 7.2.2 Affine Invariance 107
- 7.3 Automatic Absorption Detection 117
  - 7.3.1 Automatic Absorption Band Recovery 118
  - 7.3.2 Complexity Analysis 128
  - 7.3.3 Absorption Representation and Classification 130
- 7.4 Notes 136
- References 137

8 Material Discovery 141
- 8.1 Scenes in Terms of Materials 141
  - 8.1.1 Imposing End Member Consistency 145
- 8.2 Material Unmixing 148
  - 8.2.1 Linear Spectral Mixture Model 149
  - 8.2.2 Non-Negative Matrix Factorisation (NMF) 150
  - 8.2.3 Geometric Interpretation 151
- 8.3 Material Classification Using Absorptions 153
  - 8.3.1 Spectral Angle Mapper (SAM) 153
  - 8.3.2 Spectral Feature Fitting (SFF) 154
  - 8.3.3 Absorption Feature Based Material Classification ... 155
- 8.4 Statistical Learning and Band Selection 158
  - 8.4.1 Discriminant Absorption Feature Selection 158
  - 8.4.2 Canonical Correlation Analysis 159
  - 8.4.3 Feature Selection and Entropy 161
  - 8.4.4 Material Identification via the α-Entropy 164
- 8.5 Notes 168
- References 171

9 Reflection Geometry 175
- 9.1 General Reflectance Model 176
- 9.2 Parametric Form of Specific Reflectance Models 179
  - 9.2.1 Dichromatic Reflection Model 179
  - 9.2.2 Wolff Model 180
  - 9.2.3 Beckmann–Kirchhoff Model 180
  - 9.2.4 Vernold–Harvey Model 181
  - 9.2.5 Torrance–Sparrow Model 182
  - 9.2.6 Cook–Torrance Model 182
- 9.3 A Variational Approach to Shape Recovery 183
  - 9.3.1 General Irradiance Equation 184
  - 9.3.2 Cost Functional 185
  - 9.3.3 Variational Optimisation 186
  - 9.3.4 Relation to Shape from Shading 188
  - 9.3.5 Relation to Photometric Stereo 191

9.4	Estimating the Illuminant Direction	193
	9.4.1 Shading-Based Methods	193
	9.4.2 Contour-Based Methods	198
	9.4.3 Optimisation Methods	202
9.5	Notes	204
	References	205

10 Polarisation of Light 209
  10.1 Polarisation of Electromagnetic Waves 210
      10.1.1 Transmitted Radiance Sinusoid (TRS) 213
      10.1.2 Decomposing Polarimetric Images 215
  10.2 Polarisation upon Reflection 216
      10.2.1 Polarisation upon Specular Reflection 217
      10.2.2 Polarisation upon Diffuse Reflection 219
  10.3 Polarimetric Reflection Models 222
      10.3.1 A Polarimetric Reflection Model for Rough Surfaces 223
      10.3.2 Relation to the Dichromatic Reflection Model 226
      10.3.3 Reflection Component Separation 227
  10.4 Notes 237
  References 237

11 Shape and Refractive Index from Polarisation 241
  11.1 Azimuth Angle and Phase of Diffuse Polarisation 242
      11.1.1 Disambiguation of the Azimuth Angle 244
      11.1.2 Estimation of the Azimuth Angle 245
  11.2 Zenith Angle and Refractive Index 246
      11.2.1 Integrability Constraint 248
      11.2.2 Material Dispersion 249
      11.2.3 Objective Function 250
  11.3 Optimal Zenith Angle and Refractive Index 251
      11.3.1 Zenith Angle Computation 251
      11.3.2 Refractive Index Computation 255
  11.4 Notes 261
  References 262

Index 265

Chapter 1
Introduction

Imaging spectroscopy relies on associating each pixel in the image with a spectrum representing the intensity at each wavelength. Performing scene analysis on an image with such pixels confronts us with several difficult research problems. These spectra are not only dependent on the materials in the scene, but also depend on the shape of the objects, the illuminant "colour" and the light position. The photometric parameters of the scene influence not only the appearance of an object to an observer, but also the polarisation properties of the emitted radiation. This implies that, to achieve reliable scene understanding we are not only required to focus on higher-level tasks such as recognition or classification, but we also have to recover the object shape, the illuminant power spectrum and the position of the light with respect to the camera.

Thus, the research on spectroscopic scene analysis has a twofold aim. Firstly, it should address the photometric invariance problem so as to recover features devoid of illumination variations, specularities and shadowing. This, in turn, may comprise complex analysis of the spectroscopy data, with goals such as

- Illuminant recovery for invariant imaging spectroscopy
- Shape analysis and reflectance modelling
- Polarimetric scene analysis.

Secondly, such research must explore the use of the imaging spectra for purposes of scene classification and object material recognition aimed at

- Extraction of scale and affine invariant features and spectral signatures
- Spatio-spectral unmixing
- Spatio-spectral feature combination and selection.

1.1 Spectral Imaging

In contrast with trichromatic (RGB) or monochromatic (grey-scale) imagery, hyperspectral and multispectral imaging can comprise a much greater number of wavelength indexed channels or bands. As a result, spectral images become a "cube"

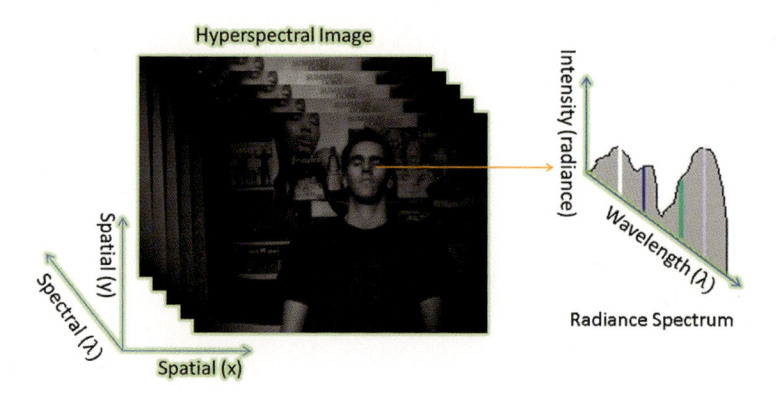

Fig. 1.1 A sample hyperspectral image with its spatial and wavelength dimensions labelled. Note that each pixel on the image cube on the *left-hand side* is, in fact, a wavelength indexed spectrum as depicted in the *right-hand panel*

where two of the spanning dimensions correspond to the coordinates of the pixel on the image lattice whereas the third one corresponds to the wavelength. This can be better appreciated in Fig. 1.1. Note that the image on the left-hand side of the figure is no longer a colour one but rather a higher-dimensional data structure, where each of the bands is a set of pixel intensities at a particular wavelength over a predefined acquisition range.

In general, hyperspectral and multispectral imagery differ in the terms of the number of spectral bands, how wide they are and how densely they cover a given spectral range. In multispectral imaging, these spectral bands are often much wider than those in hyperspectral images, covering a wavelength range which may be discontinuous. In contrast, hyperspectral images can comprise hundreds of narrow wavelength resolved bands covering a continuous wavelength range.

Another example is given in Fig. 1.2. In the figure, we show sample spectral bands for an image depicting four apples of different varieties. The image comprises 66 bands in the range [410 nm, 1050 nm], each sampled in 10 nm intervals. Note that the appearance of each of the apples varies with respect to the wavelength, with bands in the green segment of the spectrum (about 550 nm) being quite different from those in the red (around 700 nm).

1.2 Applications in Scene Analysis

Imaging spectroscopy technology can be used in a wide range of applications. Note that despite the diversity of the following applications, it is feasible to build core imaging spectroscopy technology to handle and process spectral data across all of these application areas, just as some core technologies for processing RGB images are now used in almost all cameras.

1.2 Applications in Scene Analysis

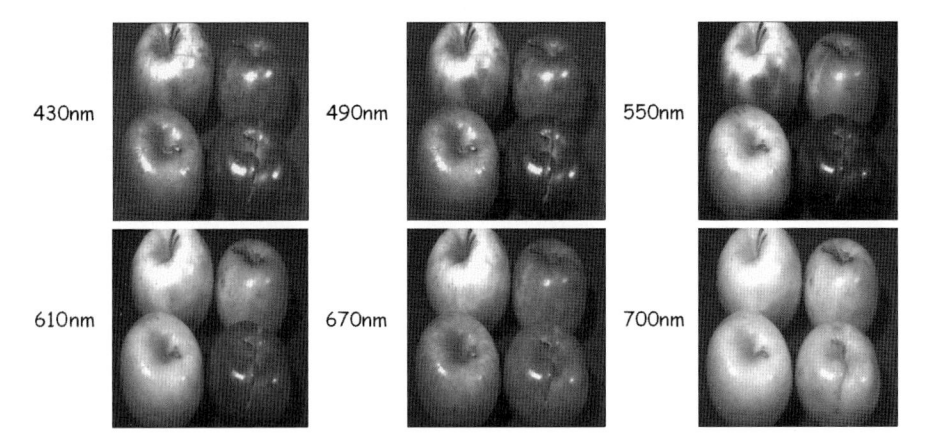

Fig. 1.2 A set of sample bands taken from a hyperspectral image. Note the variation in appearance across the bands shown

1.2.1 Photography

The research on scene analysis using imaging spectroscopy can provide a means for the development of advanced features in digital and computational photography. The illuminant recovery techniques proposed in Huynh and Robles-Kelly (2010a) and the polarimetric analysis methods in Huynh et al. (2010) can be further explored for purposes of scene classification through the integration of spatial and spectral information. This will permit the development of advanced methods for autofocusing, white balance and pre- and postprocessing tasks for digital content generation and media. This is at the core of future developments in computational photography (Raskar et al. 2006). Since resolution is not expected to drive the market in high-end photography in the long term, manufacturers will seek new ways of improving image quality and adding advanced features, such as re-illumination and material substitution.

Imaging spectroscopy can provide advanced functionality and high fidelity colour reproduction while enabling optics to remain unaffected, thus providing a natural extension of current technologies while allowing advanced features. An example of this is shown in Fig. 1.3. In the figure, we show the results obtained by substituting the reflectance of the subject's T-shirt for that of a green cotton cloth. This has been combined with the substitution of the illuminant power spectrum in the scene with that of a neon strip light. The images have been obtained by recovering the illuminant power spectrum using the method in Huynh and Robles-Kelly (2010a). The material recognition has been done using the algorithm in Fu and Robles-Kelly (2011a). Despite being a simple example, postprocessing tasks, such as re-illumination and material substitution, can be employed in digital media generation, whereas colour enhancement and retouching are applicable to digital image processing and post-production.

Fig. 1.3 Example of changing both the material reflectance and the illuminant power spectrum

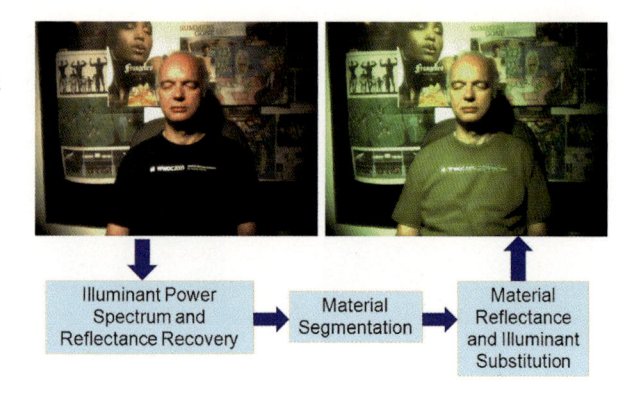

1.2.2 Food Security

Food security is increasingly becoming a national priority for nations around the world. Imaging spectroscopy provides a rich representation of natural environments. Spectral signatures for flora can provide a means for non-destructive biosecurity practises by accurately identifying pests which threaten crops, livestock and habitats.

Imaging spectroscopy technologies may enable early detection of plant pests by allowing diagnosis before symptoms are visible to the naked eye. For instance, in the left-hand column of Fig. 1.4, we show pseudo-colour images of capsicum plants. Some of these have been infected with a virus whose visible symptoms are not yet apparent. Note that, from the colour alone, there is no noticeable difference between the healthy (top row) and infected (bottom row) specimens. Despite this, making use of hyperspectral imagery, the infected instances can be recovered making use of a two-step process, where the plant regions in the image are recovered using chlorophyll recognition and, once the plants have been segmented from the scene, infection can be detected using the method in Fu and Robles-Kelly (2011b).

Further, imaging spectroscopy technologies are not limited to biosecurity but can also be used to solve a set of strategic problems that will provide more automated and informative processes to existing platforms for crop management, plant health and food security. As a result, imaging spectroscopy may find its way into precision agriculture in order to achieve higher yields and improve agribusiness efficiency. For instance, imaging spectroscopy could be used to relate spectral signatures to periodic plant life cycle events and the ways in which they are influenced by seasonal and climatal variations. This will permit the selection and delivery of crops better suited to particular environments.

Another example application is food sorting. In Fig. 1.5, we show results on nut classification using imaging spectroscopy. The figure shows the pseudo-colour image for the hyperspectral images under study along with the example spectra of nuts being classified. In our imagery, there are five nut types that appear in the scene in different quantities. Note that the colour and shape of the nuts do not permit a straightforward classification from RGB data. Nonetheless, by applying the method in Fu and Robles-Kelly (2011a), the nuts can be segmented into different varieties.

1.2 Applications in Scene Analysis

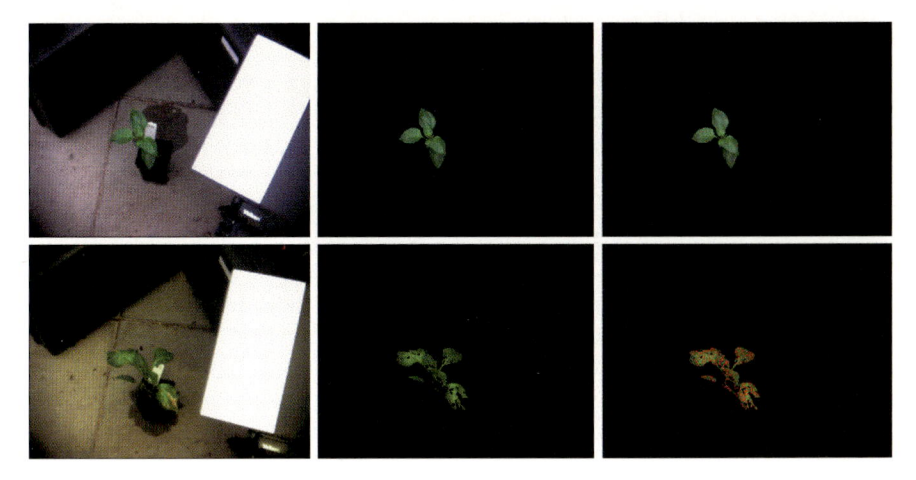

Fig. 1.4 Example results for the detection of viral infection on capsicums from hyperspectral imagery. *Left-hand column*: pseudo-colour images; *Middle column*: foreground regions recovered after chlorophyll recognition; *Right-hand column*: pathogen mapping results after the application of the method in Fu and Robles-Kelly (2011b)

Fig. 1.5 *Left-hand panel*: pseudo-colour input image; *Right-hand panel*: classification map (cashew: *blue*, macadamia: *green*, almond: *red*, peanut: *cyan*, pistachio: *magenta*)

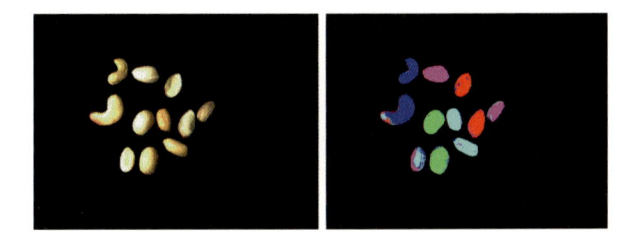

1.2.3 Defence Technologies

Spectroscopic scene analysis also has the potential to transform defence technologies. New spectroscopy methods will provide better judgement on materials and objects in the scene and will deliver advanced tracking techniques robust to illumination changes and confounding factors such as camouflage or makeup. Such capabilities are elusive in current computer vision systems due to their vulnerability to metamerism, i.e. the possibility of having two materials with the same colour but dissimilar composition.

Other functionalities include gender, age and ethnic group recognition from skin biometrics (Huynh and Robles-Kelly 2010b). The extensions of this research to include polarimetric measurements can permit tracking in scattering media to deal with natural phenomena such as mirages. For surveillance, the integration of the spectral and spatial information may open up the possibility of exploiting the interdependency of the spectral data to detect dangerous goods and provide positive access making use of biometric measurements.

1.2.4 Earth Sciences

Imaging spectroscopy technologies can also be applied to earth sciences, for example to measure carbon content in vegetation and biomass. This is related to the natural application of imaging spectroscopy to geosciences and remote sensing, where photometric invariance in hyperspectral imagery for material classification and mapping in aerial imaging is mainly concerned with artifacts induced by atmospheric effects and changing illuminations (Healey and Slater 1999).

For ground-based scene analysis, assumptions often found in remote sensing regarding the illuminant, sensors, scene geometry and imaging circumstances do not apply anymore. This is the light source is no longer a single point at infinity, i.e. the sun, the geometry of the scene is no longer a plane, i.e. the earth's surface, and the spectra of the illuminant and transmission media are not constrained to the atmosphere and sunlight power spectra. Thus close-range ground inspection relevant to areas such as mining and resource exploration can benefit from scene analysis techniques aimed at solving these problems.

1.2.5 Health

Imaging spectroscopy opens up the possibility of tackling problems such as skin cancer detection and non-intrusive saturation measurement for health and high-performance sports. It can, thus, provide an opportunity to develop a non-invasive way of measuring haemoglobin concentration, ratios of oxygenated versus desaturated haemoglobin, carboxyhaemoglobin concentrations or diagnosing melanoma by making use of imaging spectroscopy technology together with 3D shape analysis and models for skin reflectance.

1.3 Notes

In this chapter, we have elaborated on the challenges and opportunities pertaining to the use of imaging spectroscopy for scene analysis. Here, we have focused on spectral imaging rather than the polarisation of light and its implications towards the acquisition, processing and applications of spectro-polarimetric data. In this book, however, we have devoted two chapters to the topic in the recognition that polarisation can be a means for shape analysis.

Shape analysis is a classical problem in computer vision; as a result, recently there has been an increased interest in the use of polarisation as a means to shape analysis. Rahmann and Canterakis (2001) proposed a method based on polarisation imaging for shape recovery of specular surfaces. This approach to depth recovery relies on the correspondences between phase images from multiple views. An important result drawn from their research is that three polarisation views are sufficient for surface reconstruction. Atkinson and Hancock (2005a, 2005b) also made use of the correspondences between the phase and degree of polarisation in two views for shape recovery. As an alternative option to the use of multiple views, polarisation information has been extracted from photometric stereo images for shape recovery. Drbohlav and Šára (2001) have shown how to disambiguate surface orientations from uncalibrated photometric stereo using images corresponding to different polarisation angles of the incident and emitted light.

Thus, topographical and compositional information using polarimetric analysis may be recovered in a number of ways. An option here is to estimate the shape of objects in the scene and index of refraction in tandem through a coordinate search procedure in the space spanned by the reflectance models under consideration. This simultaneous estimation of shape and index of refraction allows for the wavelength-dependency property of the refractive index to be used as a feature for material recognition (see Chaps. 10 and 11 for more details).

References

Atkinson, G., & Hancock, E. R. (2005a). Multi-view surface reconstruction using polarization. In *International conference on computer vision* (pp. 309–316).
Atkinson, G., & Hancock, E. R. (2005b). Recovery of surface height using polarization from two views. In *CAIP* (pp. 162–170).
Drbohlav, O., & Šára, R. (2001). Unambigous determination of shape from photometric stereo with unknown light sources. In *International conference on computer vision* (pp. 581–586).
Fu, Z., & Robles-Kelly, A. (2011a). Discriminant absorption feature learning for material classification. *IEEE Transactions on Geoscience and Remote Sensing, 49*(5), 1536–1556.
Fu, Z., & Robles-Kelly, A. (2011b). MILIS: multiple instance learning with instance selection. *IEEE Transactions on Pattern Analysis and Machine Intelligence, 33*(5), 958–977.
Healey, G., & Slater, D. (1999). Invariant recognition in hyperspectral images. In *IEEE conference on computer vision and pattern recognition* (p. 1438).
Huynh, C. P., & Robles-Kelly, A. (2010a). A solution of the dichromatic model for multispectral photometric invariance. *International Journal of Computer Vision, 90*(1), 1–27.
Huynh, C. P., & Robles-Kelly, A. (2010b). Hyperspectral imaging for skin recognition and biometrics. In *International conference on image processing*.
Huynh, C. P., Robles-Kelly, A., & Hancock, E. R. (2010). Shape and refractive index recovery from single-view polarisation images. In *IEEE conference on computer vision and pattern recognition*.
Rahmann, S., & Canterakis, N. (2001). Reconstruction of specular surfaces using polarization imaging. In *IEEE conference on computer vision and pattern recognition* (Vol. 1, 149–155).
Raskar, R., Tumblin, J., Mohan, A., Agrawal, A., & Li, Y. (2006). Computational photography. In *Proceeding of eurographics: state of the art report STAR*.

Chapter 2
Spectral Image Acquisition

Throughout the book, we make very little, if any, differentiation between the use of spectra in the visible, ultraviolet and near-infrared ranges. This is because applications of imaging spectroscopy for scene understanding using benchtop hyperspectral and multispectral cameras are not necessarily constrained to the range visible to the human eye.

Moreover, we will often refer to the electromagnetic radiation in terms of its wavelength. In particular we will focus on wavelengths spanning from the ultraviolet (UV) to the infrared in its A and B bands (IR-A and IR-B). Here, we use the naming convention of the International Commission on Illumination (CIE) in preference over that proposed by the International Organization for Standardization (ISO). In Table 2.1, we summarise the two main subdivisions of the spectrum, i.e. those corresponding to the CIE and the ISO.

As mentioned earlier, a hyperspectral or multispectral image comprises a set of wavelength indexed bands sampled over a spectral range. In Fig. 2.1, we show the electromagnetic spectrum, divided following the CIE standards, and illustrate how a landscape appears at different bands. As illustrated in Fig. 1.2, the appearance of the scene shows noticeable changes with respect to wavelength.

Since every pixel in the image accounts for a set of wavelength resolved measurements, the acquisition process across the spectral range greatly depends on the camera used to capture the image. Furthermore, the sensor itself may require calibration to account for noise and bias. In this section, we provide an overview of the existing spectral imaging technologies, reduction of the camera sensor noise and rectification of the image spectra with respect to the illumination.

2.1 Spectral Cameras and Sensors

Although imaging spectroscopy has been available as a remote sensing technology since the 1960s, until recently, commercial spectral imaging systems were mainly airborne ones which could not be used for ground-based image acquisition. Furthermore, spectral imaging has often only been available to a limited number of

Table 2.1 Spectral subdivisions according to the CIE and the ISO

Ultraviolet				
ISO[a]	UV-C (100–280 nm)	UV-B (280–315 nm)	UV-A (315–400 nm)	
CIE[b]	UV-C (100–280 nm)	UV-B (280–315 nm)	UV-A1 (315–340 nm)	UV-A2 (340–400 nm)
Visible				
ISO				VIS (400–780 nm)
CIE				VIS (400–700 nm)
Infrared				
ISO[c]	Near-Infrared (780 nm–3 µm)		Mid-Infrared (3–50 µm)	Far-Infrared (50 µm–1 mm)
CIE	IR-A (700 nm–1.400 µm)	IR-B (1.4–3 µm)	IR-C (3 µm–1 mm)	

[a] See ISO-DIS-21348 and, for the infrared range, see ISO 20473:2007
[b] See 134/1 TC 6-26 report: Standardization of the Terms UV-A1, UV-A2 and UV-B
[c] See ISO 20473:2007

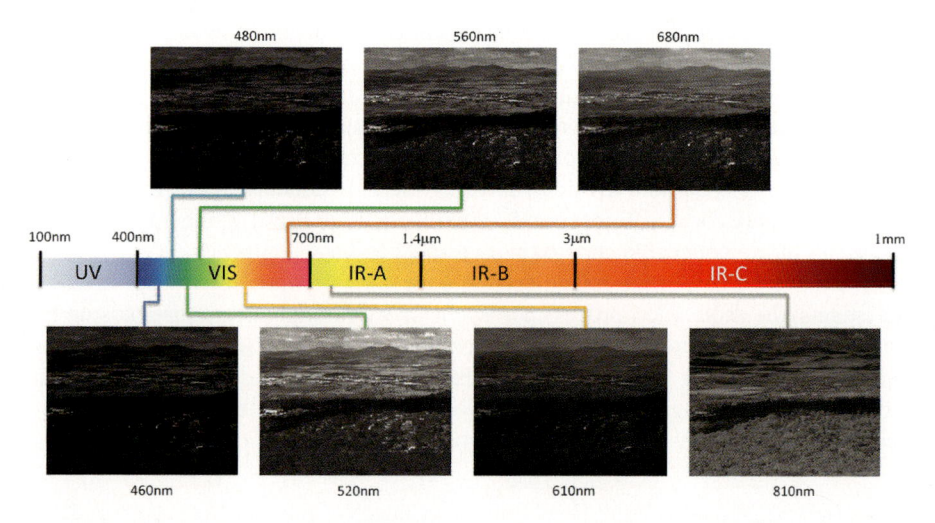

Fig. 2.1 The ultraviolet, visible and infrared spectrum as related to the wavelength resolved bands corresponding to a spectral image of a landscape

researchers and professionals due to the high cost of spectral cameras and the complexity of processing spectral data corresponding to large numbers of bands.

2.1 Spectral Cameras and Sensors

Fig. 2.2 An example benchtop camera system with two sets of filters, optics and detectors (one for the near-infrared and the other one for the visible range). The cameras are connected in daisy chain

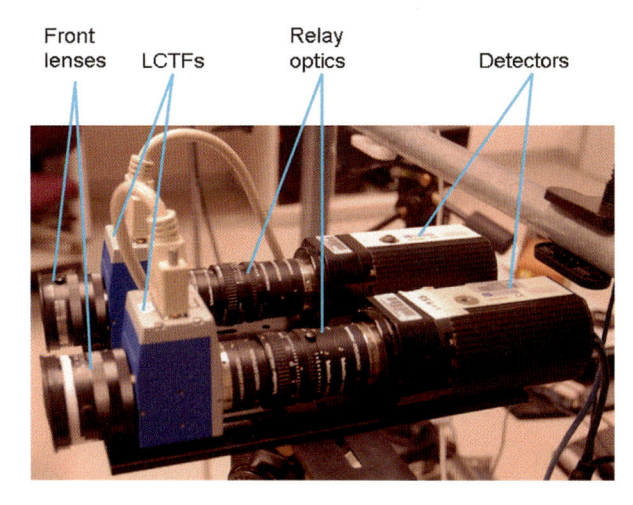

At present, remote sensing imaging technologies coexist with ground-based commercial systems based on liquid crystal tunable filters (LCTFs), acousto-optic tunable filters (AOTFs), Fabry–Perot imagers and multiple-CCD cameras. These technologies comprise the staring arrays, pushbrooms and multiple-CCD spectral cameras used today.

Pushbroom imagers are so named since they gather all of the spectrum at a time for a line on the image. This sequential line acquisition has been used in both remote sensing and benchtop cameras. In a pushbroom imager, the incoming light is gathered by a collimating slit and projected onto a diffraction grating. The diffraction grating disperses the spectrum across the detectors. The main advantage of pushbroom imagers is the fact that all of the spectrum is acquired at the same time. The drawback is that, since one line of the image is acquired at every scan, either the camera or the object should move accordingly. This is, in itself, not a major issue for applications such as remote sensing or industrial machine vision, where the camera is mounted on a moving satellite, a processing line or conveyor belt.

In contrast, staring arrays are band-sequential imagers. That is, the scene is acquired at full spatial resolution at one wavelength indexed band at a time. In a staring array device, the light passes through the focusing optics and then it is filtered, so only a narrowband segment of the spectrum impinges on the focal plane of the sensor, which is typically a CCD. In some cases, a set of relay optics is included to avoid undesired distortions induced by the filters. These filters can be fixed ones mounted on a revolving disk or tunable filters. An example of a staring array camera system is shown in Fig. 2.2. Note that the system comprises the focusing optics, the tunable filters, which in this case are LCTFs, relay optics and a detector built upon a CCD. The figure shows two cameras in daisy chain, one of which operates in the near-infrared range, whereas the other is devoted to the visible range.

It is worth noting that tunable filters in staring array devices are predominantly based on acousto-optic or liquid crystal technologies. AOTFs are solid state optical devices consisting of a tellurium dioxide (TeO_2) or quartz crystal attached to a

Fig. 2.3 *Top*: typical transmission as a function of wavelength for an acousto-optic tunable filter; *Bottom*: typical transmission for a liquid crystal tunable filter

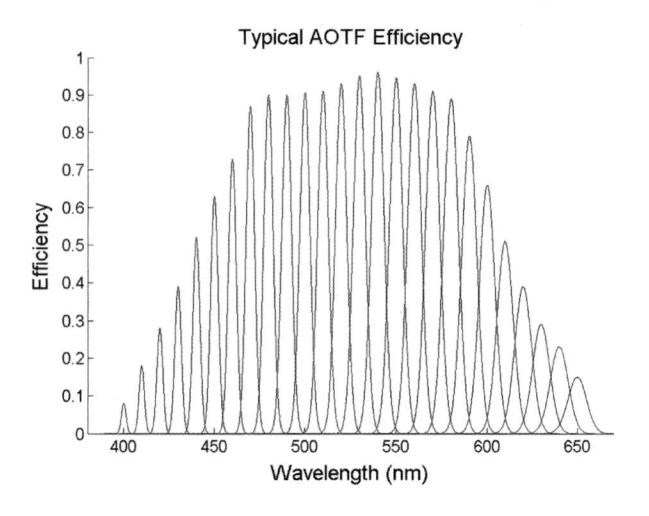

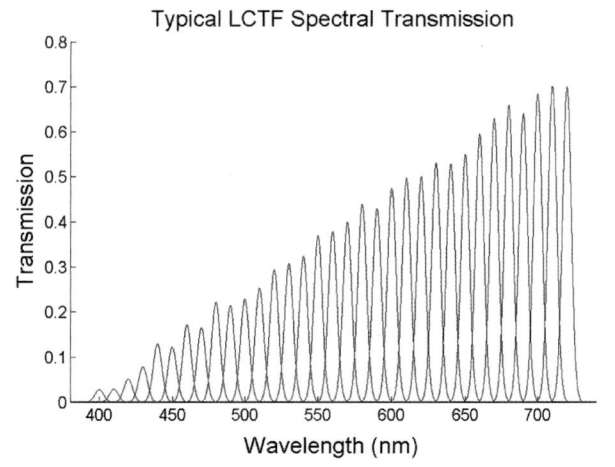

transducer. When a radio frequency signal is applied to the transducer, a high frequency acoustic wave ensues and propagates through the crystal. This ultrasonic acoustic wave induces a change in the refractive index that acts as a transmission diffraction grating. As a result, in practise, the selectivity of the filter is often not fixed across the spectral range. In Fig. 2.3, we show the typical transmission for an AOTF. In the figure, we have tuned the filter to 10 nm intervals. In particular, note that the selectivity of the filter widens as it progresses towards the upper end of the spectrum.

In contrast, LCTFs are based on a set of liquid crystal wave plates (liquid crystal-linear polariser interwoven combinations). In contrast with AOTFs, because they use wave plates rather than being based on changes in the index of refraction, LCTFs provide a linear optical path, which delivers a very low distortion and, hence, high image quality. Thus, intuitively, LCTFs can be viewed as a stack of non-tunable

2.1 Spectral Cameras and Sensors

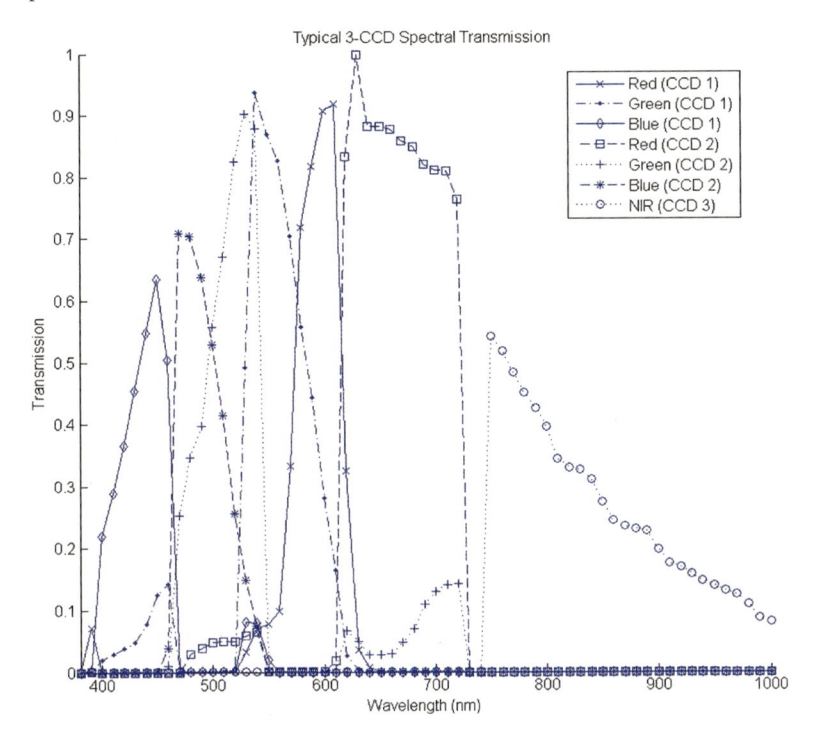

Fig. 2.4 Typical spectral transmission for a seven-band multiple-CCD camera

filters which can be set to on/off states. As a result, they can only be tuned to set intervals. In contrast, AOTFs, in theory, may be tuned to any wavelength across their operating range due to the fact that they effectively operate on an acoustic wave generated by the transducer attached to the quartz or TeO_2 crystal. On the other hand, the use of wave plates with linear polarisers often implies low filter transmissions compared to those of AOTFs. In Fig. 2.4, we show the filter transmission as a function of wavelength. Again, we have tuned the filter at 10 nm intervals. Note that the selectivity of the filter is unchanged with respect to wavelength.

Whereas AOTFs and LCTFs are often employed in hyperspectral cameras, for multispectral imaging, multiple-CCD cameras provide the ability to capture the full set of spectral bands at all pixels in the image at the same time. Multiple-CCD cameras employ a set of beam splitting prisms to separate the light incoming through the lens. The use of multiple CCDs and prisms implies that the complexity of the camera becomes impractical for a large number of bands. Nonetheless, for a few wavelength-resolved bands, multiple-CCD imaging systems provide the capability for "single shot" multispectral acquisition. In Fig. 2.4 we show the typical spectral transmission for a seven-band three-CCD camera. Note that the transmission is not reminiscent of a Gaussian band-pass filter anymore, but rather is a non-linear combination of the beam splitting prism selectivity and the Bayer array on the CCD.

2.2 Dark Current Calibration

Regardless of the method used to separate the incoming light into a set of wavelength indexed bands, hyperspectral and multispectral imaging technologies make use of detectors based on CCDs. The main reason for this is their high quantum efficiency, i.e. the charge per photon impinging on the detector. Nonetheless, CCDs suffer from photon, thermal and read noise. This is particularly important in hyperspectral and multispectral image acquisition since the spectral power per band can be, potentially, quite low due to the narrow nature of the transmission function for the spectral filters.

Recall that, for a CCD, the detector output signal can be expressed as

$$O = (\mathcal{P}Q_e + \mathcal{D})t + \mathcal{R}, \tag{2.1}$$

where $\mathcal{P}$ is the incident photon flux (photons per detector over second), Q_e is the quantum efficiency, $\mathcal{D}$ is the thermal noise, $\mathcal{R}$ is the read noise, i.e. bias, and t is the exposure time.

Since thermal noise accounts for electrons thermally generated within the CCD, it is present even when the shutter is closed and no light is impinging on the detector. As a result, it is also often called the dark current of the sensor. In practise, at acquisition time, this can be removed by taking an image with the shutter closed, i.e. $\mathcal{P} = 0$ over a set integration time t_0. This yields the dark image output given by

$$O_D = \mathcal{D}t_0 + \mathcal{R}. \tag{2.2}$$

Since spectral imaging technologies often employ high-performance CCDs which exhibit a negligible read noise, we can remove $\mathcal{R}$ from further consideration and use the shorthand $\mathcal{D} \approx \frac{O_D}{t_0}$. This permits the subtraction of the dark image output from the detector output signal, which yields

$$\mathcal{P}Q_e t \approx O - \frac{t}{t_0} O_D, \tag{2.3}$$

which implies that, if the quantum efficiency of the sensor is known, the incident photon flux can be approximated using the expression

$$\mathcal{P} \approx \frac{1}{Q_e t}\left(O - \frac{t}{t_0} O_D\right). \tag{2.4}$$

Thus, by effecting the dark image output subtraction we are effectively performing a calibration against noise. Throughout the remainder of this book, we will assume that this dark image calibration has been effected at acquisition time. Further, after calibration, for narrowband filters, we can view the image intensity value $I(u, \lambda)$ at pixel u and wavelength λ as the quantity in Eq. (2.3) weighted by the filter transmission. Thus, we can write

$$I(u, \lambda) = \beta_\lambda \left(O - O_D \frac{t}{t_0}\right) \approx \beta_\lambda Q_e \mathcal{P} t, \tag{2.5}$$

where β_λ is a wavelength-dependent quantity that depends on the geometry of the sensor and the spectral transmission of the optical filter or grating used in the imager.

2.3 Notes

In this chapter, we have focused on the image acquisition process. This is relevant to the exposure time and shutter speed, which are of major importance to capture spectral and trichromatic imagery. Exposure settings usually determine how much, and for how long, light should be allowed to reach the imaging sensors. This is intrinsically linked to the sensitivity of the sensors.

It is also important to understand that natural scenes can vary from very low to very high luminous intensities. The developments presented here have been, so far, devoid of considerations regarding the dynamic range of the sensor, i.e. the ratio between the maximum and minimum intensity value delivered by the imager. In practise, the dynamic range should be taken into consideration and the exposure time should be set carefully, since high contrast may potentially deliver images with reduced dynamic range. Thus, there is a trade-off to be made with respect to the contrast in the image.

Chapter 3
Spectral Image Formation Process

Before proceeding further, it is necessary to provide some background on the image formation process in relation to narrowband spectral images. The concepts involved in the image formation process described in this chapter are fundamental to the formalism in the subsequent chapters. Furthermore, it is important to provide a link between spectral and trichromatic imagery. This relationship further clarifies how the physical aspects of the scene, which can be captured and recovered from spectral imagery, affect the colour perceived by a human observer or that output by a trichromatic sensor. A number of concepts involved in modelling the scene reflectance, such as light scattering and Fresnel reflection theory, are of particular relevance to the physics of polarisation and illumination invariants. The material in this chapter also clearly points out the contributions of separate elements of the physical world, such as geometry, illuminants, and materials, to the subjective colour sensation of a human observer and the colour response of a trichromatic sensor. Moreover, the ability to produce a sensor-dependent colour image from a corresponding spectral image is a highly utilised tool in the subsequent chapters for the purposes of data generation and image display.

Therefore, the focus of this chapter is to establish an understanding of the formation process of spectral imagery and its relationship to colour imaging. To commence, we introduce relevant concepts and formulations from the areas of radiometry, photogrammetry, colourimetry, spectral imaging and reflectance modelling. Subsequently, we show how camera simulation and comparison can be effected with respect to the CIE-1931 colour standard. Here, we use the spectral response of the camera to recover RGB colours corresponding to known illuminant and material reflectance spectra. We have compared these RGB values to those computed making use of the colour matching functions proposed by Stiles and Burch (1955, 1959).

Recall that, in colourimetry, the aim is to capture and reproduce colours to achieve perceptual accordance between the scene image and the observation by the viewer. The simulation and evaluation of this information is important to the understanding of the relation between the scene and the camera image. Moreover, the accurate capture and reproduction of colours as acquired by digital camera sensors is an active area of research which has applications in colour correction (Brainard 1995; Finlayson and Drew 1996; Horn 1984; Wandell 1987), camera simulation (Longere and Brainard 2001) and sensor design (Ejaz et al. 2006).

However, the capture and reproduction of colours that are perceptually consistent with human observations is not a straightforward task. Digital cameras comprise three kinds of spectral broadband colour sensors which provide responses for the three colour channels, i.e. red (R), green (G) and blue (B). In practise, the RGB values recorded by these spectral broadband sensors are not necessarily a linear combination of the CIE colour matching functions of human vision. Rather, colours, as acquired by digital cameras, are device dependent and can be highly non-linear with respect to the CIE-XYZ colour gamut (Wyszecki and Stiles 2000).

While colourimetry focuses on the accuracy of the colours acquired by the camera, spectroscopy studies aspects of the physical world such as the spectrum of light absorbed, transmitted, reflected or emitted by objects and illuminants in the scene. In contrast with trichromatic sensors, multispectral and hyperspectral sensing devices can acquire wavelength indexed reflectance and radiance data in tens of hundreds of bands across a broad spectral range. Making use of photogrammetry and spectroscopy techniques, it is possible to recover the spectral response of the camera under study (Vora et al. 1997). This spectral response, together with material reflectance and illuminant radiance measurements, can be used to perform colourimetric simulations and comparisons.

3.1 Radiometric Definitions

As mentioned earlier, the imagery captured by an imaging sensor is the result of the arrival of energy or photons at the sensor. Before quantifying light energy, we commence with concepts involved in the field of light measurement, also known as radiometry. Radiometry defines terms and units that quantify the energy transfer from light sources to surface areas and from surface areas to image sensors and observers. Much of this process is due to the physical phenomena that occur during and after the impingement of photons on surfaces.

The amount of energy received by a surface depends on both the foreshortened area and the radiant energy of the light source. Here, foreshortening refers to how big a surface area looks from the point of view of the source. It accounts for the fact that surfaces receive different amounts of light from different illumination directions and may radiate different amounts of light in different outgoing directions. Specifically, a surface patch tilting away from the source looks smaller than one that faces perpendicular to the illumination direction. As a result, surfaces tilting away from the source receive a lesser amount of energy than those perpendicular to the light propagation direction due to their smaller areas of light reception.

In the remainder of the chapter the surface reflectance and the illuminant power spectrum are wavelength dependent. We represent them as the functions $S(\lambda)$ and $L(\lambda)$ with respect to the wavelength λ. In addition, let us denote the spectral sensitivity of the red (R), green (G) and blue (B) sensors with respect to the wavelength

3.1 Radiometric Definitions

Notation	Description
$L(\lambda)$	Illuminant spectral power at wavelength λ.
(θ_i, ϕ_i)	Polar and azimuth angles of the incoming light direction.
(θ_s, ϕ_s)	Polar and azimuth angles of the outgoing direction.
$E_i(\theta_i, \phi_i, \lambda)$	Source irradiance.
$E_o(\theta_s, \phi_s, \lambda)$	Outgoing radiance.
$f(\theta_i, \phi_i, \theta_s, \phi_s, \lambda)$	Bidirectional reflectance distribution function (BRDF).
$d\omega_i$	Solid angle of the source viewed from the surface location.
$d\omega_o$	Solid angle of the cone of reflected rays.
dA	Infinitesimal area of a surface patch with refractive index η.
$I_{\text{im}}(\lambda)$	Spectral image irradiance.
$C_c(\lambda)$	Spectral sensitivity of the trichromatic sensor c, where $c \in \{R, G, B\}$.
I_c	Colour response of the trichromatic sensor c.

Fig. 3.1 Notation used in this chapter

Fig. 3.2 The solid angle subtended by a patch of an area dA at a particular point as the projection onto a unit sphere centred at that point

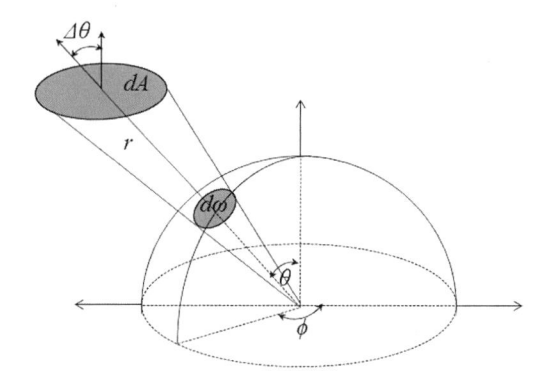

λ as $C_R(\lambda)$, $C_G(\lambda)$ and $C_B(\lambda)$, respectively. For mnemonic purposes, in Fig. 3.1, we provide a list of the notation used throughout the chapter.

3.1.1 Foreshortening and Solid Angle

Foreshortening can be quantified as the area of a surface patch viewed from a light source. We capture this quantity by the notion of the solid angle subtended by the surface patch at the light source position. This notion is based on the intuition that patches that "look the same" on the unit sphere subtend the same solid angle. We illustrate the solid angle using the diagram in Fig. 3.2. Formally, the solid angle subtended by a surface patch at a point is the area of the region obtained by projecting the patch onto the unit sphere centred at that point (Forsyth and Ponce 2002).

Fig. 3.3 The amount of energy transmitted by a patch into an infinitesimal region of solid angle $d\omega_o$ along the direction (θ, ϕ) is proportional to the radiance, the patch area dA, the angle θ between the normal vector $\vec{N}$ and the propagation direction, the solid angle and the transmission time

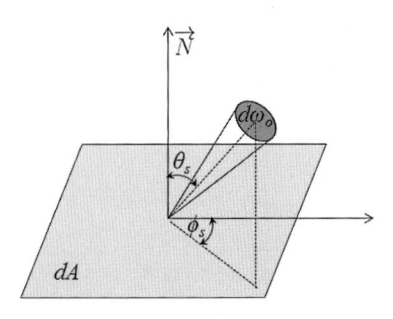

In Fig. 3.2, we consider an infinitesimal patch with an area dA and a distance r from the centre of the unit sphere. In addition, its surface normal forms an angle $\Delta\theta$ with the line of sight. The infinitesimal solid angle $d\omega$ it subtends at the centre of the sphere is quantified in terms of *steradians* (sr) as

$$d\omega = \frac{dA \cos \Delta\theta}{r^2}. \qquad (3.1)$$

Alternatively, the solid angle can be related to the polar coordinates (θ, ϕ) of the line of sight with respect to a unit sphere. Suppose that the projected area onto the unit sphere spans infinitesimal steps $(d\theta, d\phi)$ in the θ and ϕ dimensions. Using trigonometric equalities, the solid angle is computed as

$$d\omega = \sin\theta \, d\theta \, d\phi. \qquad (3.2)$$

3.1.2 Radiance

The imaging process is concerned with the distribution of light in space as a function of position and direction. To this end, we refer to radiance as the measurement of the distribution of light in space. It is defined as the power of light travelling at a point in a specified direction per unit surface area perpendicular to the travel direction, per solid angle unit. Radiance is measured in watts per square meter per steradian ($\text{W m}^{-2} \text{sr}^{-1}$).

Based on this definition, we can quantify the amount of energy transmitted through a region along a particular direction. Figure 3.3 shows a planar surface patch with area dA radiating energy in a direction forming an angle θ_s with the surface normal $\vec{N}$ and a tilt angle ϕ_s with respect to a reference axis on the patch. Assuming that the radiance of the patch in the surface normal direction is L_p, the amount of energy E_o transmitted by the patch along the direction (θ_s, ϕ_s) through an infinitesimal region with a solid angle $d\omega_o$ around the direction of propagation in time dt is

$$E_o = L_p \cos\theta_s \, dA \, d\omega_o \, dt. \qquad (3.3)$$

Fig. 3.4 Incident light source originates from a region of solid angle $d\omega_i$ and arrives at a surface patch with an area dA. The light direction forms an angle θ_i with the surface normal $\vec{N}$ and a tilt angle ϕ_i with a reference axis

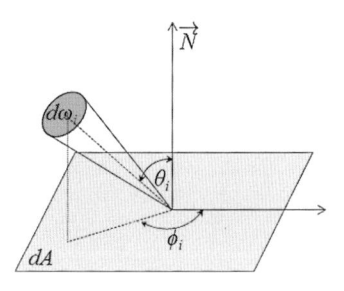

3.1.3 Irradiance

Image sensors are often designed to measure the amount of energy arriving at the image plane as a result of reflection or scattering from different locations and directions in the scene. The amount of incoming energy to the sensor is termed irradiance. By definition, irradiance is the power of incident light per unit surface area not foreshortened, and is measured in watts per square meter (W m^{-2}).

In fact, the amount of incoming energy is proportional to the source radiance. In Fig. 3.4, we consider a source with incident radiance of L coming from a region of solid angle $d\omega_i$ at angles (θ_i, ϕ_i) with respect to the local coordinate system of the patch as illustrated in Fig. 3.3. The irradiance arriving at a surface patch illuminated by this incident flux is, hence, given by Eq. (3.4).

3.1.4 The Bidirectional Reflectance Distribution Function (BRDF)

The most general model of local reflection is the bidirectional reflectance distribution function (BRDF). To formulate the reflectance models herein, let us consider a reflection geometry defined with respect to a right-handed local coordinate system whose origin is the surface location under study and whose z-axis aligns with the local surface normal, as shown in Fig. 3.5. In this coordinate system, the illuminant and viewing directions are described by two pairs of angles (θ_i, ϕ_i) and (θ_s, ϕ_s), where θ_i and θ_s are the zenith angles and ϕ_i and ϕ_s are the azimuth angles of the respective directions. The solid angle of the cone of incident light rays when viewed from the surface is annotated as $d\omega_i$ in Fig. 3.5.

Considering incident light coming from an infinitesimal solid angle $d\omega_i$, we can assume that the source radiance $L(\lambda)$ with wavelength λ is almost constant across all the incoming directions. Thus, the energy flux per unit area perpendicular to the light direction is proportional to the source radiance and the solid angle, i.e. $L(\lambda) d\omega_i$. Due to the foreshortening effect, the irradiance $E_i(\theta_i, \phi_i, \lambda)$ arriving at the surface is, however, measured per unit area perpendicular to the local surface normal,

$$E_i(\theta_i, \phi_i, \lambda) = L(\lambda) \cos\theta_i \, d\omega_i, \qquad (3.4)$$

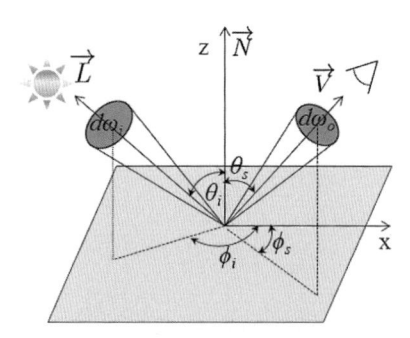

Fig. 3.5 Reflection geometry. The incident light travels from a source with a solid angle $d\omega_i$ from the direction $\vec{L}$, impinging on the surface which reflects it in the viewing direction $\vec{V}$. The right-handed local coordinate system has the origin located at the surface point under study and the surface normal aligned with the z-axis. In this coordinate system, the incident and viewing directions are parameterised by the polar coordinates (θ_i, ϕ_i) and (θ_o, ϕ_o), respectively. θ_i and θ_o are the zenith angles and ϕ_i and ϕ_o are the azimuth angles of these directions

where $\cos\theta_i$ accounts for the ratio of a foreshortened area in the incident light direction (θ_i, ϕ_i) to its projection onto the surface tangent plane at the point under study.

The BRDF $f(\theta_i, \phi_i, \theta_s, \phi_s, \lambda)$ quantifies the fraction of incident light that is reflected from the surface into the air in various directions. Let the light source energy arriving at the surface in the direction (θ_i, ϕ_i) be $E_i(\theta_i, \phi_i, \lambda)$, and the radiance at the same surface point, seen from the direction (θ_s, ϕ_s) be $E_o(\theta_i, \phi_i, \theta_s, \phi_s, \lambda)$. The BRDF of the surface is the ratio of the outgoing radiance to the incoming irradiance,

$$f(\theta_i, \phi_i, \theta_s, \phi_s, \lambda) = \frac{E_o(\theta_i, \phi_i, \theta_s, \phi_s, \lambda)}{E_i(\theta_i, \phi_i, \lambda)}. \quad (3.5)$$

Note that the source irradiance, the surface radiance and the BRDF are dependent on the wavelength λ.

Since the interaction of light with objects may incur complex phenomena, we make the following assumptions to ensure the applicability of a BRDF. Firstly, the radiance leaving a surface point is contributed only by light irradiance arriving at the same point on the surface. This assumption avoids the case of subsurface scattering encountered in translucent or multi-layered surfaces, where light may arrive, penetrate and re-emerge from the surface at different points. Secondly, the surface under study is not fluorescent, i.e. it does not absorb light in the short wavelength range and emits light in a longer wavelength. In other words, the reflected or scattered light at a given wavelength is attributed to incident light at the same wavelength. Thirdly, we assume that surfaces do not themselves emit internal energy and treat the light source energy separately from the surface reflectance.

The BRDF in Eq. (3.5) depends on four angular variables and is therefore cumbersome for directly modelling the characteristics of a surface. For isotropic surfaces, where the BRDF remains constant as the illuminant and viewing directions are rotated about their surface normals by the same tilt (azimuth) angle, the BRDF only depends on three variables, including the polar angles θ_i and θ_s

and the difference of azimuth angles $\phi_s - \phi_i$; i.e. the BRDF can be rewritten as $f(\theta_i, \phi_i, \theta_s, \phi_s, \lambda) = f^*(\theta_i, \theta_s, \phi_s - \phi_i, \lambda)$. On the other hand, the Helmholtz reciprocity principle (Helmholtz 1924) states that the energy travelling from the light source to the surface and being reflected to the observer is equal to that travelling in the reverse direction. Mathematically, this principle is expressed as

$$f(\theta_i, \phi_i, \theta_s, \phi_s, \lambda) = f(\theta_s, \phi_s, \theta_i, \phi_i, \lambda). \tag{3.6}$$

For isotropic surfaces, i.e. those whose reflectance properties are invariant to the surface rotation about an axis of symmetry, we have $f^*(\theta_i, \theta_s, \phi_s - \phi_i, \lambda) = f^*(\theta_i, \theta_s, \phi_i - \phi_s, \lambda)$ or $f(\theta_i, \phi_i, \theta_s, \phi_s, \lambda) = f^*(\theta_i, \theta_s, |\phi_s - \phi_i|, \lambda)$.

A disadvantage of using a non-parametric BRDF model is that irradiance and radiance have to be measured for every pair of illuminant and viewing directions. To reduce the effort for data acquisition, we can resort to parametric reflectance models which provide analytical expressions relating the reflection geometry, the illuminant spectrum and the material properties to the observed radiance.

3.2 Image Formation Process

We now proceed to study the process by which light is transmitted, reflected and scattered upon arrival at a surface before impinging on image sensors. Real-world surfaces exhibit a combination of these effects. A reflectance model describes the fraction of radiance energy reflected from a surface illuminated by a light source. Conventionally, this fraction is dependent on the reflection geometry, i.e. the illuminant and viewing directions and the surface normal, the power spectrum of the illuminant and the material properties such as roughness, albedo, shininess and subsurface structure. In our analysis, we assume that no light energy is absorbed by the surface material, to concentrate our modelling on the transmission, reflection and scattering of light at the surface and subsurface levels.

The photometric process is summarised as follows. A fraction of the light incident on the surface being observed is reflected towards the camera. This fraction depends on the scene geometry and the surface material reflectance. Subsequently, the reflected light passes through the camera lens and is focused onto the image plane of the camera. Finally, the R, G, B values for each pixel in the image are determined by the spectral sensitivity of the R, G, B receptors of the camera to the incoming light.

3.2.1 Image Irradiance

Having modelled the reflection and scattering from the surface under study, we now derive the image irradiance captured by the sensor. We commence by relating the image irradiance to the BRDF and the incident light power spectrum and its direction.

Combining Eqs. (3.4) and (3.5) yields the surface radiance

$$E_o(\theta_i, \phi_i, \theta_s, \phi_s, \lambda) = f(\theta_i, \phi_i, \theta_s, \phi_s, \lambda) L(\lambda) \cos\theta_i \, d\omega_i. \qquad (3.7)$$

We assume that the camera lens has unit transmittance, i.e. the flux radiated from the surface is transmitted through the lens without any loss of energy. The irradiance reaching the image plane is given by (Forsyth and Ponce 2002; Horn 1986)

$$\begin{aligned} I_{\text{im}}(\lambda) &= \frac{\pi}{4} \left(\frac{d}{z}\right)^2 \cos^4\alpha \, E_o(\theta_i, \pi, \theta_s, \phi_s, \lambda) \\ &= m f(\theta_i, \phi_i, \theta_s, \phi_s, \lambda) L(\lambda) \cos\theta_i \cos^4\alpha \, d\omega_i, \qquad (3.8) \end{aligned}$$

where d is the lens diameter, z is the distance between the lens and the image plane, $m \triangleq \frac{\pi}{4}(\frac{d}{z})^2$ and α is the angle between the optical axis of the camera and the line of sight from the surface patch to the centre of the lens.

3.2.2 Colour Response

We now proceed to illustrate the relationship between trichromatic imagery and imaging spectroscopy data. To this end, we examine the RGB responses of trichromatic camera sensors. Here, we focus our attention on the relative spectral distribution of radiance across wavelengths as an alternative to the absolute radiance value. Since the term m in Eq. (3.8) only depends on the camera geometry, it is constant with respect to the spectral and angular variables of the image irradiance equation. Also, note that the derivation of trichromatic responses applies equally to each of the three colour channels. Therefore, we consider a general formulation for a particular colour channel c, where $c \in \{R, G, B\}$.

Let the spectral sensitivity functions, i.e. the colour matching functions, of the red (R), green (G) and blue (B) sensors be denoted $C_R(\lambda)$, $C_G(\lambda)$ and $C_B(\lambda)$, respectively. With this notation, the response for the colour sensor is given by

$$\begin{aligned} I_c &= \kappa_c \int_{\mathcal{W}} C_c(\lambda) I_{\text{im}}(\lambda) \, d\lambda \\ &= m\kappa_c \cos\theta_i \cos^4\alpha \, d\omega_i \times \int_{\mathcal{W}} C_c(\lambda) f(\theta_i, \phi_i, \theta_s, \phi_s, \lambda) L(\lambda) \, d\lambda, \qquad (3.9) \end{aligned}$$

where $\mathcal{W} = [380\text{ nm}, 780\text{ nm}]$ is the human visible spectrum.

In Eq. (3.9), the value of κ_c corresponds to the colour balance factor of the camera against a predetermined reference. By colour balancing, the values of I_B, I_G and I_R are scaled such that a smooth surface, with unit BRDF $f(\theta_i, \phi_i, \theta_s, \phi_s, \lambda) = 1$, placed perpendicular to the camera axis, i.e. $\theta_i = \theta_s = \alpha = 0$, presents a colour perceptually consistent with the reference colour s_c. If the reference colour is white, the tristimulus values I_c, where $c \in \{R, G, B\}$, or the corresponding chromaticity is called the white point under the given illuminant power spectrum $L(\cdot)$.

3.2 Image Formation Process

Thus, the tristimulus value s_c of the colour reference is given by

$$s_c = m\kappa_c \, d\omega_i \int_{\mathcal{W}} C_c(\lambda) L(\lambda) \, d\lambda. \qquad (3.10)$$

By solving the equation above for κ_c and substituting into Eq. (3.9), we get

$$I_c = s_c \cos\theta_i \cos^4\alpha \times \frac{\int_{\mathcal{W}} C_c(\lambda) f(\theta_i, \phi_i, \theta_s, \phi_s, \lambda) L(\lambda) \, d\lambda}{\int_{\mathcal{W}} C_c(\lambda) L(\lambda) \, d\lambda}. \qquad (3.11)$$

To illustrate the differences between sensors, we consider the trichromatic responses of the human eye and three camera models, including a Canon 10D, a Nikon D70 and a Kodak DCS420 sensor. The human eye trichromatic responses were measured as 10-degree colour matching functions in Stiles and Burch (1959). Note that this can also be done using the 2-degree colour matching functions in Stiles and Burch (1955). Also, since the Kodak DCS420 is a discontinued model dating from the mid-1990s, its capacity to reproduce colour with respect to the human eye is expected to be outperformed by the Nikon D70 and the Canon 10D. Nonetheless, we have included the Kodak DCS420 as a matter of comparison with the work on spectral sensitivity function measurement in Vora et al. (1997).

To recover the spectral sensitivity function of the cameras under study, a procedure akin to that in Vora et al. (1997) is used. Narrowband illumination is obtained by passing light from a calibrated tungsten source through a liquid crystal tunable filter (LCTF). This narrowband light is used to image a white reference tile. These white reference tiles are highly diffuse and made of a fluoropolymer, such as Spectralon[1] or a generic type of Halon G-50, i.e. polytetrafluoroethylene (PTFE), which has a reflectance which is approximately unity across the ultraviolet (UV), visible (VIS) and a large segment of the near-infrared (NIR) range, i.e. $f(\theta_i, \phi_i, \theta_s, \phi_s, \lambda) \approx 1$ for $\lambda \in [250 \text{ nm}, 2.5 \text{ μm}]$.

For photometric calibration purposes, the spectrum of the narrowband illuminant is measured on the white reference making use of a spectrometer whose probe is placed at a similar geometric configuration with respect to the white reference as that of the camera used in Vora et al. (1997, 2001). In Fig. 3.6, we show the colour matching functions and the spectral sensitivity functions of the Canon 10D, the Nikon D70 and the Kodak DCS420 cameras. Note that the spectral sensitivity functions for the Kodak DCS420 camera are consistent with those reported in Vora et al. (1997, 2001).

Furthermore, consider the colour temperature characteristic of a Planckian illuminant (see Chap. 5.2), which is given by its chromaticity as compared with a heated black body radiator. The spectral radiant exitance per unit surface area per unit solid angle for a black body radiator is governed by Planck's law. As a function

---

[1] See http://www.labsphere.com/.

(a) Stiles and Burch RGB colour matching functions

(b) Canon 10D

(c) Nikon D70

(d) Kodak DCS420

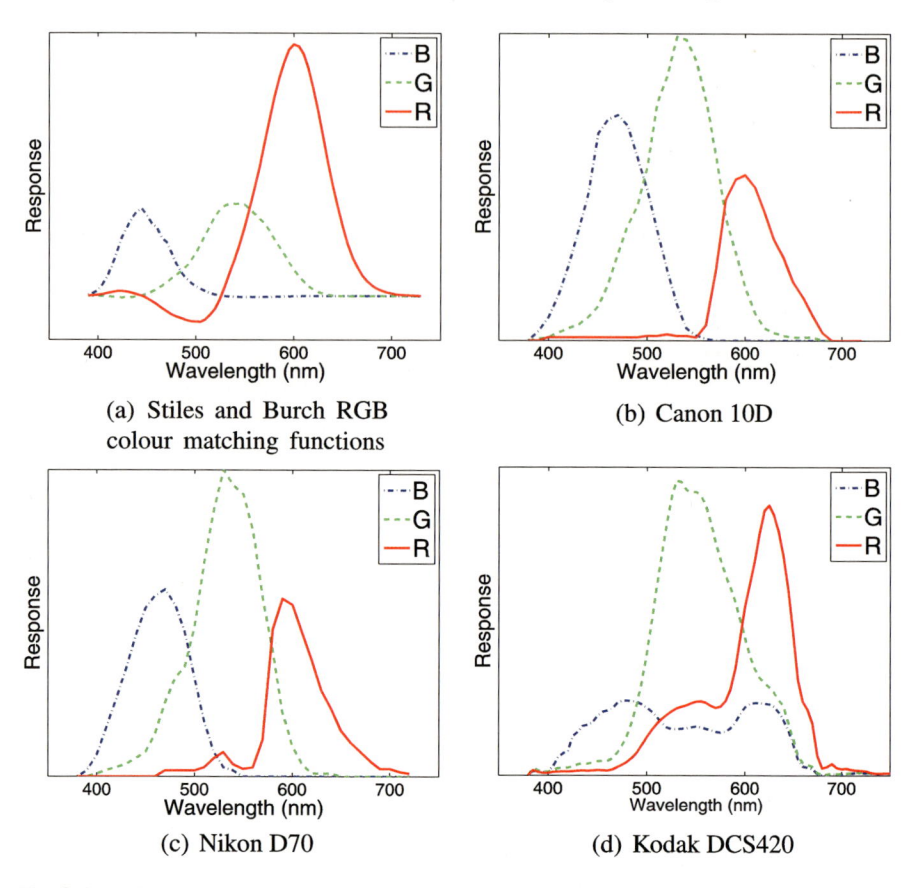

Fig. 3.6 (**a**) Stiles and Burch RGB colour matching functions; (**b**)–(**d**) the RGB sensitivity functions for the Canon 10D, the Nikon D70 and the Kodak DCS420 camera sensors, respectively

of wavelength, let the illuminant power spectrum be given by

$$E_{bb}(\lambda, T) = \frac{c_1}{\lambda^5 (\exp(\frac{c_2}{\lambda T}) - 1)}, \qquad (3.12)$$

where T is the temperature in kelvins (K), λ is the wavelength variable as before and $c_1 = 3.74183 \times 10^{-16}$ W m^2 and $c_2 = 1.4388 \times 10^{-2}$ mK are constants.

Making use of the spectral radiant exitance, the colour temperature captured by the camera can be computed by using $E_{bb}(\lambda, T)$ as an alternative to the image irradiance $I_{\text{im}}(\lambda)$. Note that the colour temperature is determined by the illuminant chromaticity. As a result, we can use Eq. (3.11) to determine the colour temperature by assuming that the RGB spectral sensitivity functions of the camera are normalised to unity, i.e. $s_R = s_G = s_B = 1$, against a white colour reference with unit bidirectional reflectance $f(\theta_i, \phi_i, \theta_s, \phi_s, \lambda) \equiv 1$. Moreover, if the incident light and viewing directions are aligned with the surface normal and the optical axis of

3.2 Image Formation Process

Fig. 3.7 Estimated colour temperature for the range [1000 K, 20000 K] corresponding to the three camera models including Canon 10D, Nikon D70 and Kodak DCS420, and the human eye as measured by Stiles and Burch's colour matching functions

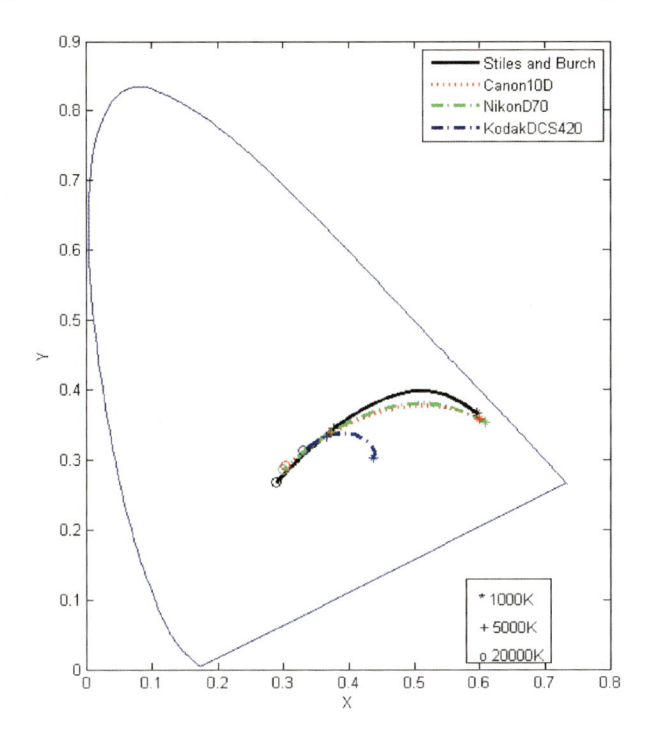

the camera, i.e. $\theta_i = \alpha = 0$, we can simplify Eq. (3.11) so as to obtain the colour of the black body radiator at temperature T by

$$I_B(T) = \frac{\int_W C_B(\lambda) E_{bb}(\lambda, T) d\lambda}{\int_W C_B(\lambda) d\lambda}$$

$$I_G(T) = \frac{\int_W C_G(\lambda) E_{bb}(\lambda, T) d\lambda}{\int_W C_G(\lambda) d\lambda} \qquad (3.13)$$

$$I_R(T) = \frac{\int_W C_R(\lambda) E_{bb}(\lambda, T) d\lambda}{\int_W C_R(\lambda) d\lambda}.$$

In Fig. 3.7 we show a plot on the CIE-XYZ gamut of the colour temperatures for a black body radiator heated at temperatures ranging from 1000 K to 20000 K. This Planckian locus corresponds to the colour temperature yielded by each camera and the colour matching functions corresponding to the human eye's spectral sensitivity responses, where the RGB colour values are converted into the CIE-XYZ colour space using the following linear transformation:

$$\begin{pmatrix} X \\ Y \\ Z \end{pmatrix} = \begin{pmatrix} 0.4124 & 0.3576 & 0.1805 \\ 0.2126 & 0.7152 & 0.0722 \\ 0.0193 & 0.1192 & 0.9505 \end{pmatrix} \begin{pmatrix} R \\ G \\ B \end{pmatrix}. \qquad (3.14)$$

Note that, from the figure, the curves in the temperature range of [5000 K, 6200 K] for the three cameras are in good accordance with that yielded by the colour matching functions. This implies that all the cameras under evaluation are able to perform appropriate white balancing against natural sunlight. Further, the length of each curve across the gamut indicates the dynamic range of the corresponding camera. In particular, the dynamic range of the Kodak DCS420 camera is shorter than those of the others. The second observation is that the Canon 10D and Nikon D70 curves are closer to that for the human eye than the Kodak DCS420 curve. In the cases where the colour temperature is close to its extreme values, the colour produced by Kodak DCS420 is considerably different from those produced by the other two cameras.

3.3 Spectral Imagery in Colour

The visualisation of spectroscopy data is a non-trivial task given that the image does not represent a true colour image, but instead comprises wavelength indexed spectral bands, some of which may be in the non-visible range. In the case where the spectral response of the imaged scene has been decoupled from the environment lighting, it is unclear how best to visualise the spectral image. A trivial solution would be to render the spectral image using synthetic white light (i.e. integrating the visible bands over the RGB basis); however, this approach may not necessarily produce a desirable image with sufficient contrast. In some cases, imaging spectroscopy software such as ENVI,[2] HIAT[3] and Geomatica[4] include a function to render the image as a video where the wavelength index plays the role of the time variable.

3.3.1 Visualisation

Visualising multispectral imagery is a non-trivial task, since the resulting colour tristimulus is dependent not only on the wavelength indexed spectral bands but also on the spectral sensitivity of the colour sensor. To this end, methods such as the one in Chi et al. (2010) allow visualisation by using a predefined set of camera response functions. The method in Chi et al. (2010) employs the radiance of the scene decoupled from the environment lighting rather than the colour and brightness of the image on the visualisation context. Therefore, the method does not always produce an image with sufficient contrast.

Multispectral imagery is often visualised by making use of data-independent linear combinations (Jacobson and Gupta 2005) or subspace projection methods such as independent component analysis (ICA) or principal component analysis (PCA) (Wang and Chang 2006). The main drawback of these methods is that they

---

[2] http://www.ittvis.com/envi/.

[3] http://www.censsis.neu.edu/software/hyperspectral/hyperspectral.html.

[4] http://www.pcigeomatics.com/.

3.3 Spectral Imagery in Colour

Fig. 3.8 *Left-hand panel*: colour image computed with the CIE colour matching functions; *Second panel*: image yielded by the method in Rasche et al. (2005); *Third panel*: pseudo-colour obtained using PCA; *Right-hand panel*: pseudo-colour image obtained by direct assignment of bands to colour channels

are designed for remote sensing, where the aim is to produce a lower dimensional space which can then be used to map data into a colour gamut supported by the visualisation context under consideration, such as a monitor. Moreover, these lower dimensional spaces are not bounded and, even when they are, such as in the case in Cui et al. (2009), they do not reproduce colours that are in perceptual accordance with the imaged scene.

Moreover, tone mapping techniques such as that in Rasche et al. (2005) suffer when the radiance is used directly at input. Other approaches produce what is called a "pseudo-colour" representation of the spectral imagery by selecting three bands to be used as the red, green and blue channels in the output image. This direct assignment of the colour channels can produce very different results depending on the selected bands.

This may cause a discrepancy between the perceptual tristimulus computed from imaging spectroscopy data using colour matching functions and the pseudo-colour delivered by techniques that map spectral radiance to colours in a manner akin to tone mapping. To illustrate these discrepancies, in Fig. 3.8, we show, in the left-hand panel, the perceptual colour of a spectral image rendered with the Stiles and Burch colour matching functions corresponding to the sensitivity of the average human eye.

In the second panel, we show the pseudo-colour computed making use of the tone mapping algorithm in Rasche et al. (2005). Note that tone mapping, when applied directly to the spectra, yields a high contrast image where the luminance is preserved but the colours may be, in general, very different from the pseudo-colour image in the left-hand panel. Furthermore, the third and fourth panels show the pseudo-colour computed by applying PCA and that corresponding to the direct assignment of three bands, i.e. those at 650 nm, 550 nm and 450 nm, to the red, green and blue channels, respectively. In the case of the image yielded by PCA, it shows a strong red hue, which is a result of the first component, i.e. that corresponding to the red colour channel, being dominant. On the other hand, the direct assignment of bands to colour channels results in a pink tint that distorts the final colour.

3.3.2 Contrast in Spectral Imagery

Moreover, a common practise in imaging spectroscopy is to maximise dynamic range at acquisition time. Since high contrast often delivers imagery with reduced dynamic range, there is a trade-off to be made with the contrast in the image. This is actually the case in the left-hand panel of Fig. 3.8, where the colour matching functions deliver a low-contrast pseudo-colour image. Thus, the best imagery is expected to profit from a high dynamic range while having high contrast.

In terms of visual acuity, recall that the contrast sensitivity function is, in general, a band-pass function in cycles per degree (Peli 1990). Thus, the computational treatment of the problem is such that the image contrast at the pixel u is viewed as a convolution on the radiance as follows:

$$C(u) = \sum_{v \in \mathcal{N}_u} \mathcal{I}(v) g(v - u), \qquad (3.15)$$

where $\mathcal{I}(v)$ is the measured radiance value at the pixel v, $\mathcal{N}_u$ is the set of neighbouring pixels of pixel u and g is a matrix of order $|\mathcal{N}| \times |\mathcal{N}|$. Note that the above operation is performed on every spectral band of the input spectral image.

To recover the perceived radiance value $\mathcal{I}_\lambda(u)$ we use the photopic CIE luminosity curve. The photopic luminosity curve $V(\lambda)$ provides the monochromatic stimulus at wavelength λ for a "standard" human observer in the visible spectrum, i.e. $\lambda \in [380 \text{ nm}, 780 \text{ nm}]$ (Wyszecki and Stiles 2000). The luminosity curve acts as a monochromatic matching function, which can be expressed as

$$\mathcal{I}(v) = \sum_\lambda V(\lambda) I_\lambda(v). \qquad (3.16)$$

By substituting the equation above into Eq. 3.15, we can recover the contrast for the multispectral or hyperspectral image as follows:

$$C(u) = \sum_{v \in \mathcal{N}_u} \left(\sum_\lambda V(\lambda) I_\lambda(v) \right) g(v - u). \qquad (3.17)$$

As a result, the recovered contrast can be considered as a filtering operation over the sensitivity function for a standard human observer. To compute the contrast $C(v)$, we perform this filtering operation by a convolution in the spatial domain (Gonzalez and Woods 2001) as follows:

$$C(u) = (\mathcal{I} * g)(u). \qquad (3.18)$$

Note that the Fourier transform of this convolution, which is a product in the frequency domain (Sneddon 1995), can also be employed to allow efficient computation of the contrast.

3.3.3 Colour Rendering of Spectral Data

We now turn our attention to the rendering of 3D meshes with spectroscopy data according to the colour rendering method presented in Sect. 3.2.2. Specifically, we

3.3 Spectral Imagery in Colour

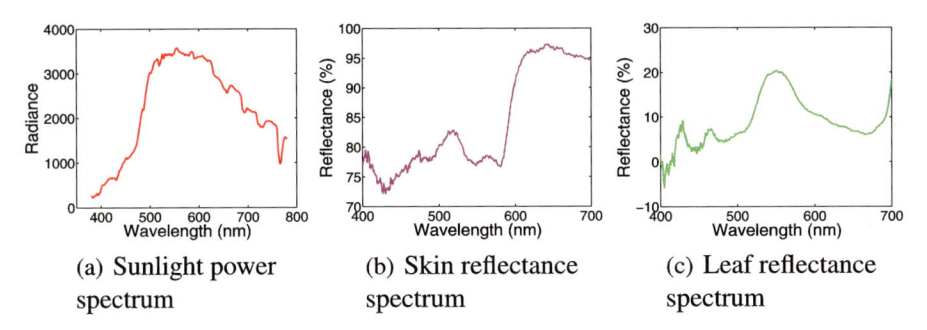

Fig. 3.9 The radiance of the sun (**a**), and the spectral reflectance of human skin (**b**) and a plant leaf (**c**)

(a) Sunlight power spectrum

(b) Skin reflectance spectrum

(c) Leaf reflectance spectrum

illustrate the colour rendering of human skin and plant leaf tissues. In practise, light scattered from natural multi-layered materials is composed of a large subsurface scattering component due to reflection at the bottom surface boundaries and refraction through translucent layers. To this end, a physics-based reflectance model that accounts for refractive attenuation of incident light through several surface layers is suitable for rendering purposes. In particular, we describe a variant of the Beckmann–Kirchoff model proposed by Ragheb and Hancock (2006) in Sect. 4.1.2. This model combines wave scattering theory (Beckmann and Spizzichino 1963) with Fresnel terms introduced in the Wolff reflectance model (Wolff 1994) to explain the physical mechanisms involved in the scattering, transmission and reflection of light. With this model, we substitute the reflectance of polygonal faces of the given mesh for the BRDF $f(.)$ in Eq. (3.11). We note that alternative reflectance models can be employed in place of the above model to provide a means of computing the BRDF term in Eq. (3.11). We elaborate further on reflectance modelling in Chap. 4.

For purposes of illustrating the rendering process, we have used two meshes. These meshes, corresponding to a fern and a human head, are viewed as a collection of planar polygons whose RGB colour is rendered according to Eq. (3.11). This equation involves the illuminant power spectrum, the spectral sensitivity functions of the trichromatic sensor and the surface reflectance. Here, we regard the illuminant as a point light source and instantiate the illuminant power spectrum $L(\lambda)$ as that of natural sunlight. Figure 3.9(a) shows the power spectrum of sunlight. In addition, we employ the reflectance of human skin for the rendering of the head and the reflectance of a leaf for the fern plant. The reflectance spectra of human skin and plant leaves are measured with a spectrometer and are normalised as shown in Fig. 3.9(b) and (c).

For white balancing, we have assumed a light source colour equivalent to the CIE D55 colour temperature, i.e. 5500 K. At 5500 K, the $s_R : s_G : s_B$ proportion for the computation in Eq. (3.11) is $1.1040 : 0.9870 : 0.8233$. Note that the use of these parameters is not overly restrictive, since with a known illuminant power spectrum, this proportion can be computed in a straightforward manner using Eq. (3.10).

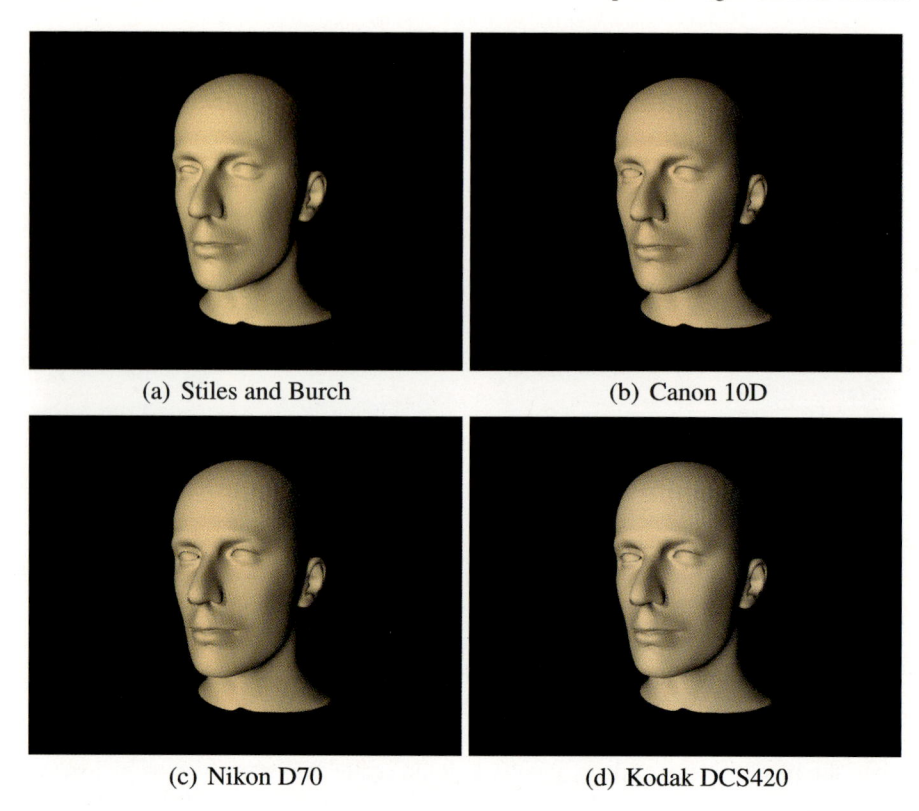

Fig. 3.10 Rendering of a head illuminated by sunlight using (**a**) the Stiles and Burch's colour matching functions, and (**b**) the Canon 10D, (**c**) the Nikon D70 and (**d**) the Kodak DCS420 cameras. All the renderings have been obtained using the skin reflectance spectrum in Fig. 3.9(b)

Figure 3.10 presents the rendering of the head with spectral sensitivity functions of the three camera models above and the Stiles and Burch's colour matching functions. The Canon 10D and the Nikon D70 cameras produce results that closely resemble the one computed using the colour matching functions. The colour achieved by the Kodak DCS420 camera has a higher saturation and tends further to the white, which makes the head appear slightly brighter.

Similarly, in Fig. 3.11, we show the rendering for a bush with the reflectance of a leaf. The surface colour rendered with the Kodak DCS420 response curves is closer to the white reference than that produced for the Canon 10D and Nikon D70 cameras. The latter response curves produce renderings with a stronger green component, which looks more natural than that of the Kodak camera. However, as compared with the results shown in Fig. 3.10, the simulation for these two cameras is perceptually different from that computed with the RGB colour matching functions.

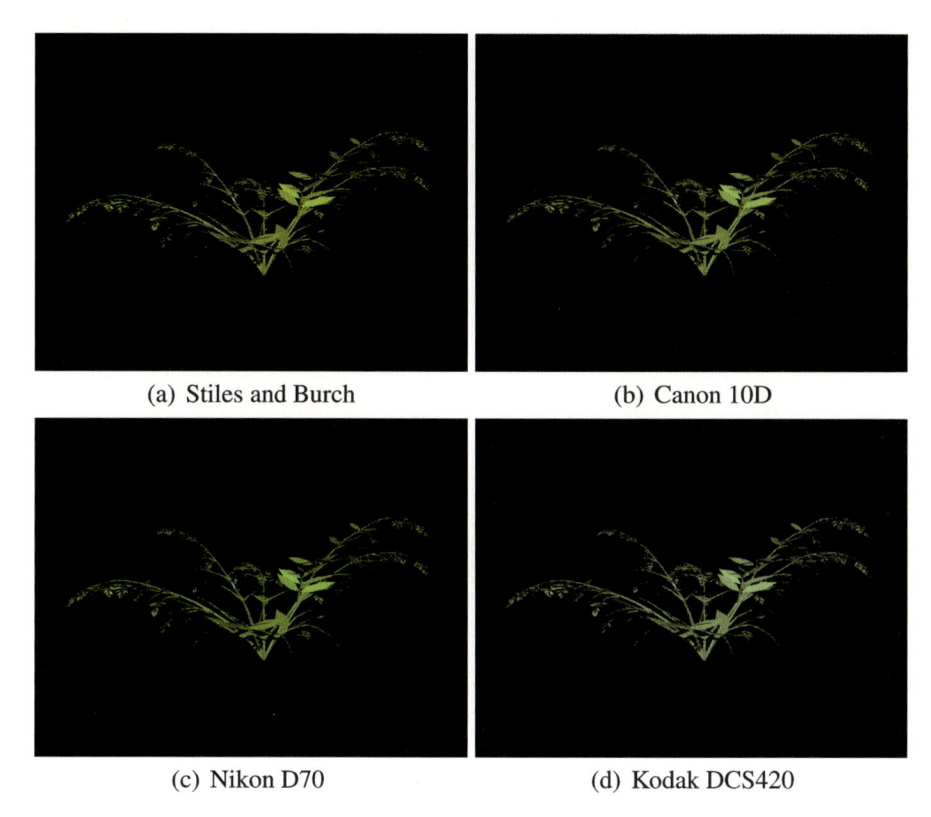

Fig. 3.11 Rendering of a bush illuminated by sunlight using (**a**) the Stiles and Burch's colour matching functions, and (**b**) the Canon 10D, (**c**) the Nikon D70 and (**d**) the Kodak DCS420 cameras. All the renderings have been obtained using the leaf reflectance spectrum in Fig. 3.9(c)

3.4 Notes

The quantification and measurement of the accuracy of the colours acquired by the camera with respect to human perception has evolved in a vast amount of literature on colourimetry and colour science (Wyszecki and Stiles 2000). The acquisition of spectral data, combined with the ability to perform or estimate illuminant power spectrum measurements can be used not only to perform colourimetric simulation (Vora et al. 2001) but also to allow scene colours to be reproduced based upon particular materials and lights.

Colour as acquired with trichromatic cameras is often assumed to be based on a linear combination of colour matching functions. By reproducing colour based on the materials and illuminants in the scene, libraries can be used to achieve imagery whose aesthetics are dependent upon the image spectra and personalised to the user (Gu et al. 2011). Another possibility is the capacity of editing the image with respect to materials in the scene or the modification of the light temperature. Furthermore, traditionally, colour editing (Barret and Cheney 2002), image equalisation (Zuiderveld 1994) and colour transfer (Greenfield and House 2003) approaches based on RGB suffer from the drawbacks introduced by the possibility of having two materials with the same colour but dissimilar reflectance, i.e. metamerism. Since imaging spectroscopy is devoid of metamerism, the use of imaging spectroscopy allows complex photowork (Kirk et al. 2006) in which images are edited, modified, annotated, etc. based on object materials, illuminants and user preferences rather than their RGB values.

References

Barret, A., & Cheney, A. (2002). Object-based image editing. *ACM Transactions on Graphics*, *21*(3), 777–784.
Beckmann, P., & Spizzichino, A. (1963). *The scattering of electromagnetic waves from rough surfaces*. New York: Pergamon.
Brainard, D. H. (1995). *Colorimetry*. New York: McGraw-Hill.
Chi, C., Yoo, H., & Ben-Ezra, M. (2010). Multi-spectral imaging by optimized wide band illumination. *International Journal of Computer Vision*, *86*, 140–151.
Cui, M., Razdan, A., Hu, J., & Wonka, P. (2009). Interactive hyperspectral image visualization using convex optimization. *IEEE Transactions on Geoscience and Remote Sensing*, *47*(7), 1586–1600.
Ejaz, T., Horiuchi, T., Ohashi, G., & Shimodaira, Y. (2006). Development of a camera system for the acquisition of high-fidelity colors. *IEICE Transactions on Electronics*, *E89–C*(10), 1441–1447.
Finlayson, G. D., & Drew, M. S. (1996). The maximum ignorance assumption with positivity. In *Proceedings of the IS&T/SID 4th color imaging conference* (pp. 202–204).
Forsyth, D. A., & Ponce, J. (2002). *Computer vision: a modern approach*. New York: Prentice Hall.
Gonzalez, R. C., & Woods, R. E. (2001). *Digital image processing* (2nd ed.). Boston: Addison-Wesley Longman.
Greenfield, G. R., & House, D. H. (2003). Image recoloring induced by palette color associations. *Journal of WSCG*, *11*, 189–196.
Gu, L., Huynh, C. P., & Robles-Kelly, A. (2011). Material-specific user colour profiles from imaging spectroscopy data. In *IEEE international conference on computer vision*.
Helmholtz, H. V. (1924). *Treatise on physiological optics*.
Horn, B. K. P. (1984). Exact reproduction of colored images. *Computer Vision, Graphics, and Image Processing*, *26*(2), 135–167.
Horn, B. K. P. (1986). *Robot vision*. New York: McGraw-Hill.
Jacobson, N., & Gupta, M. (2005). Design goals and solutions for display of hyperspectral images. *IEEE Transactions on Geoscience and Remote Sensing*, *43*(11), 2684–2692.
Kirk, D. S., Sellen, A. J., Rother, C., & Wood, K. R. (2006). Understanding photowork. In *Proceeding of the SIGCHI conference on human factors in computing systems* (pp. 761–770).
Longere, P., & Brainard, D. H. (2001). Simulation of digital camera images from hyperspectral input. In C. van den Branden Lambrecht (Ed.), *Vision models and applications to image and video processing* (pp. 123–150). Dordrecht: Kluwer.
Peli, E. (1990). Contrast in complex images. *Journal of the Optical Society of America A*, *7*(10), 2032–2040.
Ragheb, H., & Hancock, E. R. (2006). Testing new variants of the Beckmann–Kirchhoff model against radiance data. *Computer Vision and Image Understanding*, *102*(2), 145–168.
Rasche, K., Geist, R., & Westall, J. (2005). Re-coloring images for gamuts of lower dimension. *Computer Graphics Forum*, *24*(3), 423–432.

References

Sneddon, I. N. (1995). *Fourier transforms*. New York: Dover.
Stiles, W. S., & Burch, J. M. (1955). Interim report to the Commission Internationale de l'Éclairage Zurich, 1955, on the National Physical Laboratory's investigation of colour-matching. *Optica Acta*, *2*, 168–181.
Stiles, W. S., & Burch, J. M. (1959). N.P.L. colour-matching investigation: final report 1958. *Optica Acta*, *6*, 1–26.
Vora, P. L., Farrell, J. E., Tietz, J. D., & Brainard, D. H. (1997). Linear models for digital cameras. In *Proceedings of the IS&T's 50th annual conference* (pp. 377–382).
Vora, P., Farrell, J., Tietz, J., & Brainard, D. (2001). Image capture: simulation of sensor responses from hyperspectral images. *IEEE Transactions on Image Processing*, *10*(2), 307–316.
Wandell, B. A. (1987). The synthesis and analysis of color images. *IEEE Transactions on Pattern Analysis and Machine Intelligence*, *9*(1), 2–13.
Wang, J., & Chang, C.-I. (2006). Applications of independent component analysis in endmember extraction and abundance quantification for hyperspectral imagery. *IEEE Transactions on Geoscience and Remote Sensing*, *44*(9), 2601–2616.
Wolff, L. B. (1994). *Diffuse-reflectance model for smooth dielectric surfaces* (No. 11, pp. 2956–2968).
Wyszecki, G., & Stiles, W. (2000). *Color science: concepts and methods, quantitative data and formulae*. New York: Wiley.
Zuiderveld, K. (1994). Graphics gems IV. In *Contrast limited adaptive histogram equalization* (pp. 474–485). New York: Academic Press.

Chapter 4
Reflectance Modelling

In computer vision, the modelling of surface reflectance is a topic of pivotal importance for purposes of shape analysis and pattern recognition. In relation to photometric invariants, a number of physically meaningful parameters of reflectance models are material intrinsic and therefore potentially useful for pattern recognition tasks.

Reflectance models are designed to capture the influence of the illumination condition, the surface geometry and the material properties under study on the observed image irradiance. Reflectance, which is the fraction of incident light reflected from the surface, is dependent on the surface and material properties. The geometric properties include, but are not limited to, the roughness scale and the structure and shape of the constituent micro-facets of the surface. Furthermore, the spectral distribution of surface reflectance is characteristic of the material and could provide useful information for material classification.

Traditionally, reflectance models can be classified into three broad categories, including empirical, physics-based and phenomenological (semi-empirical). Although empirical reflectance models may not be based on a physics theory, they aim to fit the empirical measurements of real-world data. In 1924, Opik (1924) designed an empirical model to estimate the reflection behaviour of the moon. In 1941, Minnaert (1941) modified Opik's model to obtain a reflectance function that was dependent on the polar angles of incidence and reflection, and the surface roughness. This was designed to obey the Helmholtz reciprocity principle (Helmholtz 1924), but did not originate from a theory in physics. Instead, it aimed to predict the light reflection behaviour of realistic non-Lambertian surfaces such as the moon. Phong's model (Phong 1975) achieved rendering realism by an interpolation over the vertex surface normals of polygons constituting the surface being rendered. Ward (1992) introduced a physically plausible, yet computationally simple model for the rendering of surfaces with anisotropic reflectance.

In this chapter, we focus our attention on physics-based and phenomenological models. Physics-based models employ the light scattering theory for modelling surface reflectance. In this category, Kirchhoff's theory of the scattering of electromagnetic waves was first adopted by Beckmann to develop physics-based reflectance

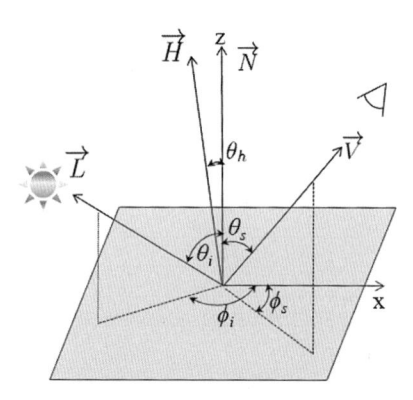

Fig. 4.1 The right-handed local coordinate system defined for each surface point under study. The origin is located at the given point, and the z-axis is aligned with the local surface normal $\vec{N}$. The incident light comes from the direction $\vec{L}$ and is reflected by the surface in the direction $\vec{V}$. In the above coordinate system, the incident direction is parameterised by a zenith angle θ_i formed with the surface normal and an azimuth angle ϕ_i with respect to a reference x-axis on the surface tangent plane at the given point. Similarly, the viewing directions are parameterised by a zenith angle θ_s and an azimuth angle ϕ_s. The half-vector $\vec{H}$ is defined as $\vec{H} = \vec{L} + \vec{V}$ and forms an angle θ_h with the surface normal

models for slightly rough or very rough surfaces (Beckmann and Spizzichino 1963). Semi-empirical or phenomenological models are developed to explain optical phenomena such as off-specular lobe reflection (Torrance and Sparrow 1967) or diffuse reflection (Wolff 1994) from a semi-empirical viewpoint.

Before further formulation, we refer to a local coordinate system commonly used to describe the geometric variables appearing in the reflectance models presented in this chapter. For the purpose of visualisation, in Fig. 4.1, we depict the geometry involved in these reflectance models. Here, the reference local coordinate system is defined with the z-axis aligned to the local surface normal $\vec{N}$ and with the origin as the point of interest on the surface. The incident light direction $\vec{L}$ and viewing direction $\vec{V}$ are defined by the polar angles (θ_i, ϕ_i) and $(\theta_s$ and $\phi_s)$, respectively, where θ_i and θ_s are the zenith angles of the incident and reflected rays and ϕ_i and ϕ_s are the azimuth angles of the respective directions. Without loss of generality, we can assume that $\vec{L}$, $\vec{V}$ and $\vec{N}$ are normalised to unit length and are generally not co-planar. Another vector regularly mentioned in several reflectance models is the bisector $\vec{H}$ between the illuminant direction $\vec{L}$ and the viewing direction $\vec{V}$, which is also known as the halfway vector. The angle between the surface normal and the half-vector is called the halfway angle, denoted as θ_h.

To facilitate quick reference, in Fig. 4.2, we provide a list of notation appearing in the reflectance models presented in this chapter.

4.1 Physics-Based Models

Notation	Description
$\vec{L}$	Illumination direction.
$\vec{V}$	Viewing direction.
$\vec{N}$	Surface normal direction.
$\vec{H}$	The angular bisector of the illumination direction $\vec{L}$ and the viewing direction $\vec{V}$.
(θ_i, ϕ_i)	Polar and azimuth angles of the incoming light direction.
(θ_s, ϕ_s)	Polar and azimuth angles of the outgoing direction.
θ_h	Half-angle vector between the surface normal and the half-vector $\vec{H}$.
$d\omega_i$	Solid angle of the source viewed from the surface location.
u	Pixel location.
λ	Sampled wavelength.
$I(u, \lambda)$	Image radiance at pixel u and wavelength λ.
$L(\lambda)$	Illuminant spectral power at wavelength λ.
$S(u, \lambda)$	Spectral reflectance at pixel u and wavelength λ.
$g(u)$	Shading factor of the dichromatic model at pixel u.
$k(u)$	Specular coefficient of the dichromatic model at pixel u.
σ	Standard deviation of the height of micro-facets.
τ	Surface correlation length.
$r(\theta_i, \theta_s, \lambda, \sigma)$	Surface roughness in the Beckmann–Kirchhoff model.
η	Material refractive index.
$F(\theta_i, \eta)$	Fresnel reflection coefficient of a material with a refractive index η.
G^2_{BK}	The geometric attenuation factor of the original Beckmann–Kirchhoff reflectance model (Beckmann and Spizzichino 1963).
G^2_{VH}	The geometric attenuation factor modified by Vernold and Harvey (1998).
G^2_{FC}	The Fresnel-corrected geometric attenuation factor modified by Ragheb and Hancock (2006).
$\rho(u, \lambda)$	The total diffuse albedo at pixel u and wavelength λ.
ϑ	The angle between the micro-facet normal and the mean surface normal in the Torrance–Sparrow model.

Fig. 4.2 Notation used in Chap. 4

4.1 Physics-Based Models

4.1.1 Beckmann–Kirchhoff Model

Kirchhoff's electromagnetic theory allows a class of physics-based reflectance models to be derived. The earliest work on modelling surface reflectance as a wave scattering process was undertaken by Beckmann (Beckmann and Spizzichino 1963). The model involves two physical parameters. The first of these parameters is the standard deviation σ of the height variation with respect to a mean surface level. The second parameter is the surface correlation length τ, which characterises the random nature of surface profiles in terms of the relative horizontal spacing between peaks and valleys of the surface.

The Beckmann–Kirchhoff model (Beckmann and Spizzichino 1963) describes the reflection from a surface as a sum of two terms. The first of these corresponds to the scattering in the specular direction, and the latter represents the diffuse scattering. The term describing the specular spike is given by

$$f_{BK}^{spec}(u, \lambda) = P_0^2 \exp(-r(\theta_i, \theta_s, \lambda, \sigma)) F(\theta_i, \eta(u, \lambda)). \quad (4.1)$$

According to Eq. (4.1), the specular reflectance at pixel u and wavelength λ is a product of three terms. The first of these is the magnitude of the specular reflectance P_0, which depends on the angle θ_h between the surface normal and the bisector of the illumination and viewing directions. This quantity is nearly zero for all scattering directions except a very narrow range about the halfway vector $\vec{H}$.

In the original formulation of the Beckmann–Kirchhoff model, P_0 is defined based on a rectangular surface patch surrounding the point of interest. Let the area of this patch be $A = XY$, where X and Y are the width and length of the patch. With these ingredients, the magnitude P_0 of the specular component is given by

$$P_0 = \text{sinc}(v_x X) \text{sinc}(v_y Y), \quad (4.2)$$

where $\text{sinc}(x) = \sin(x)/x$, $v_x = k(\sin\theta_i - \sin\theta_s \cos\phi_s)$, $v_y = -k\sin\theta_s \sin\phi_s$. Here, k is the propagation rate of the incident light, related to its wavelength λ through the equation $k = \frac{2\pi}{\lambda}$.

For computational efficiency, Nayar et al. (1991) model the specular magnitude P_0 as a Gaussian distribution function of the half-angle θ_h with a standard deviation m, as an alternative to the original formulation:

$$P_0 = \frac{1}{\sqrt{2\pi}m} \exp\left(-\frac{\theta_h^2}{2m^2}\right). \quad (4.3)$$

In the second term, $r(\cdot)$ plays an important role since it quantifies the surface roughness through $\frac{\sigma}{\lambda}$, the incidence angle θ_i and reflected light angle θ_s. The three cases $r \ll 1$, $r \approx 1$ and $r \gg 1$ correspond to smooth surface, moderately rough and rough surface, respectively. The analytical formula of $r(\cdot)$ is expressed as

$$r(\theta_i, \theta_s, \lambda, \sigma) = \left(\frac{2\pi\sigma}{\lambda}(\cos\theta_i + \cos\theta_s)\right)^2. \quad (4.4)$$

The third term in Eq. (4.1) is the wavelength-dependent Fresnel reflection coefficient $F(\cdot)$ as defined by

$$F(\theta_i, \eta) = \frac{1}{2}\left(\frac{a - \cos\theta_i}{a + \cos\theta_i}\right)^2 \times \left(1 + \left(\frac{a - \sin\theta_i \cos\theta_i}{a + \sin\theta_i \cos\theta_i}\right)^2\right), \quad (4.5)$$

where $a = (\eta^2 - \sin^2\theta_i)^{\frac{1}{2}}$ and η is the refractive index of the surface material.

4.1 Physics-Based Models

Note that, even though there is no Fresnel term in the original model proposed by Beckmann and Spizzichino (1963), we have included it without any loss of generality so as to model the wavelength dependency of the specular component. This also reflects the notion that no surface is a perfect reflector. Therefore, the Fresnel term is used to determine the fraction of the incident light reflected from the surface as a function of the incident angle and the index of refraction.

On the other hand, the modelling of the diffuse scattering component involves the surface correlation function. By far, the two most popular approaches to an approximation of the diffuse reflectance are formulated with the Gaussian and exponential surface correlation functions (Beckmann and Spizzichino 1963). When the correlation function is Gaussian, the diffuse reflectance at a given wavelength λ of incident light from a surface patch of area A is approximated by Beckmann and Spizzichino (1963)

$$f_{BK}^{diff}(u, \lambda) = \frac{\pi \tau^2}{A} G_{BK}^2 e^{-r} \sum_{l=1}^{\infty} \frac{r^l}{l! l} e^{-\frac{\tau^2}{4l} v_{xy}^2}, \quad (4.6)$$

where $v_{xy}^2 = v_x^2 + v_y^2$ and G_{BK} is a geometric attenuation term.

In Eq. (4.6), the geometric factor G_{BK} explains the attenuation of emitted light by the surface orientation with respect to illuminant and viewing directions. The geometric factor is defined as

$$G_{BK} = \frac{1 + \cos\theta_i \cos\theta_s - \sin\theta_i \sin\theta_s \cos\phi_s}{\cos\theta_i (\cos\theta_i + \cos\theta_s)}. \quad (4.7)$$

Since the model in Eq. (4.6) relies on the evaluation of an infinite sum, it is computationally infeasible. For the sake of simplicity and computational efficiency, we adopt an approximation of this model proposed by Ragheb and Hancock (2006) for three different types of surface roughness. These types include slightly rough surfaces, i.e. $r \ll 1$, moderately rough surfaces, i.e. $r \simeq 1$, and very rough surfaces, i.e. $r \gg 1$.

When the surface is moderately rough, i.e. $r \simeq 1$, the diffuse scattering component can be approximated with a finite number of terms in the infinite sum in Eq. (4.6). Further, for cases in which the surface is slightly rough, the diffuse component can be approximated by the first term of the summation in Eq. (4.6), i.e.

$$f_{BK}^{diff}(u, \lambda) = \frac{\pi \tau^2 G_{BK}^2 r}{A} \exp\left[-\left(r + \frac{v_{xy}^2 \tau^2}{4}\right)\right]. \quad (4.8)$$

When the surface is rough or very rough, i.e. $r \gg 1$, and the correlation function is Gaussian, an interesting simplification of the diffuse scattering component in Eq. (4.6) is

$$f_{BK}^{diff}(u, \lambda) = \frac{\pi \tau^2 G_{BK}^2}{A \sigma^2 v_z^2} \exp\left(-\frac{\tau^2 v_{xy}^2}{4 \sigma^2 v_z^2}\right), \quad (4.9)$$

where $v_z = -k(\cos\theta_i + \cos\theta_s)$.

On the other hand, when the surface correlation is modelled with an exponential function $C(\tau) = \exp(-|\tau|/T)$, the diffuse component is approximated by

$$f_{\text{BK}}^{\text{diff}}(u, \lambda) = \frac{2\pi T^2}{A\sigma^2} \frac{G_{\text{BK}}^2}{v_z^2(1 + \frac{T^2 v_{xy}^2}{\sigma^2 v_z^2})^{\frac{3}{2}}}. \tag{4.10}$$

4.1.2 Variants of the Beckmann–Kirchhoff Model

It has been noted that the Beckmann–Kirchhoff model commonly breaks down at large incident and scattering angles (Ogilvy 1991; Vernold and Harvey 1998). This happens because the geometric factor G_{BK} tends to infinity near the grazing angle. Moreover, the model relies on the evaluation of the Kirchhoff integral, which is intractable in closed form. Several variants of the Beckmann–Kirchhoff model aim to overcome its failure at grazing angles and simplify its computation using forms of approximation.

Several authors (Vernold and Harvey 1998; Harvey et al. 2007) have attempted to modify the geometric factor in the Beckmann–Kirchhoff model to improve the model prediction with respect to experimental scattering data for rough surfaces with large incident and scattering angles. Ogilvy (1991) added Lambert's cosine of the incident angle to the original geometric factor of the Beckmann–Kirchhoff model. Nieto-Vesperinas and Garcia (1981) suggested a method to avoid the prediction of the Beckmann–Kirchhoff model from becoming infinite at grazing viewing angles. He (1991) related directional and uniform diffuse reflection to a Fresnel term that depends on the wavelength, the incident angle, the surface roughness and the refractive index of the material under study. Recently, Ragheb and Hancock (2006) replaced the geometric term in the Beckmann–Kirchhoff model with a Fresnel correction term to provide accurate predictions for moderately rough surfaces.

Vernold and Harvey (1998) have proposed a variant of the Beckmann–Kirchhoff model which can cope well with a wide range of angles. In their work, they presented the following modified geometric factor

$$G_{\text{VH}}^2 = \cos\theta_i, \tag{4.11}$$

as an alternative to G_{BK}^2.

Later, to resolve the break-down of the original Beckmann–Kirchhoff model at large incident and scattering angles, Ragheb and Hancock (2006) proposed an alternative model of the geometric attenuation factor. Specifically, they replaced the original geometric attenuation factor in the Beckmann–Kirchhoff model with the one derived by Wolff (1994). In the modified model, the geometric attenuation term becomes

$$G_{\text{FC}}^2 = \big(1 - F(\theta_i, \eta)\big)\left(1 - F\left(\arcsin\left(\frac{\sin\theta_s}{\eta}\right), \frac{1}{\eta}\right)\right)\cos\theta_i, \tag{4.12}$$

where $F(\theta_i, \eta)$ is the Fresnel reflection coefficient defined by Eq. (4.5).

4.2 Phenomenological Models

4.2.1 Lambertian Model

One of the simplest and most widely used reflectance models in computer vision and computer graphics is perhaps the Lambertian model (Lambert 1760). Although being simple, this model has successfully predicted the reflectance of a number of smooth surfaces. Lambert's cosine law states that the diffuse reflection from a surface only depends on the incident angle between the light direction and the surface normal, but not the viewing angle. In addition, the reflected radiance is proportional to the illuminant power and the material reflectance. Mathematically, the reflected diffuse radiance $I(u, \lambda)$ at a location u and a wavelength λ is given by

$$I(u, \lambda) = \frac{1}{\pi} L(\lambda) S(u, \lambda) \cos \theta_i \, d\omega_i, \qquad (4.13)$$

where $S(u, \lambda)$ is the wavelength-dependent surface reflectance, which quantifies the fraction of incident light reflected by the surface, $L(\lambda)$ is light source radiance, θ_i is the incident angle and $d\omega_i$ is the solid angle of the light source when viewed from the location u. Note that the material reflectance and the illuminant power are expressed as wavelength-dependent terms so as to model the various colours of materials and light sources.

Let us denote the unit illumination direction and surface normal vectors at the location u as $\vec{L}$ and $\vec{N}$, respectively. The incident angle θ_i is therefore related to these two vectors as

$$\cos \theta_i = \langle \vec{L}, \vec{N} \rangle. \qquad (4.14)$$

So far, the Lambertian model of diffuse reflection has been incorporated into more complex reflectance models. These latter models serve to explain phenomena absent from the Lambertian model such as specularity and shadow (Shafer 1985; Torrance and Sparrow 1967). In practice, other models such as the Oren–Nayar (1995) and the Wolff (1994) models generalise the Lambertian model for rough or non-Lambertian surfaces.

4.2.2 Dichromatic Reflection Model

Perhaps one of the simplest reflection models that account for both diffuse and specular reflection is the dichromatic model introduced by Shafer (1985). This model assumes uniform illumination across the spatial domain of the observed scene. According to this model, the surface radiance is decomposed into a diffuse component and a specular one. While the diffuse component reflects light equally in all directions, the specular component varies with the viewing direction. We consider an object with reflectance $S(u, \lambda)$ at pixel location u and wavelength λ illuminated by the source power $L(\lambda)$. With these ingredients, the dichromatic model then predicts the radiance $I(u, \lambda)$ emitted from the surface as

$$I(u, \lambda) = g(u)L(\lambda)S(u, \lambda) + k(u)L(\lambda). \quad (4.15)$$

In Eq. (4.15), the shading factor $g(u)$ governs the proportion of diffuse light reflected from the object and depends solely on the surface geometry. For a purely diffuse Lambertian surface, the shading factor is related to the incident angle θ_i as

$$g(u) = \frac{d\omega_i}{\pi} \cos\theta_i. \quad (4.16)$$

As before, $d\omega_i$ is the solid angle of the light source. Due to the fact that θ_i is the angle between the surface normal and the light direction, the shading factor is governed by the surface orientation with respect to the light source.

On the other hand, the factor $k(u)$ models the irregularities of the micro-facet structure that cause specularities in the scene. To be precise, this factor is related to the Fresnel reflection coefficients, which in turn depend on the material refractive index. Although the refractive index is strictly wavelength dependent, its variation within the visible and near-infrared regions is generally small for a wide variety of materials. This results in the common assumption that the specular coefficient is constant across the spectrum.

4.2.3 Wolff Model

Wolff (1994) proposed a diffuse reflection model for smooth dielectric surfaces to explain the peak diffuse reflection effect at large incident and viewing angles observed on many dielectric objects. The model proposed was derived from the theory of radiative transfer through boundaries between dielectric layers. To predict the departure in behaviour from the Lambertian model at large angles between the illuminant and the viewing direction, Wolff viewed the subsurface energy flux as the result of refraction, scattering and reflection phenomena inside the dielectric body. In Fig. 4.3, we illustrate the process of light energy transfer, where incident light penetrates the dielectric surface at an angle θ_i and is scattered by particles inside the dielectric body. Subsequently, a portion of the internally scattered light arrives at the material-air boundary and re-emerges from the surface at an emittance angle θ_s through refraction. Formally, the diffuse reflectance formulated by the Wolff model is given by

$$f_W(u, \lambda) = \rho(u, \lambda)\cos\theta_i \bigl(1 - F\bigl(\theta_i, \eta(u, \lambda)\bigr)\bigr)\left(1 - F\left(\theta'_s, \frac{1}{\eta(u, \lambda)}\right)\right), \quad (4.17)$$

where $\rho(u, \lambda)$ is the total diffuse albedo accumulated by multiple diffuse subsurface scattering, $\eta(u, \lambda)$ is the index of refraction of the dielectric medium and θ'_s is the

4.2 Phenomenological Models

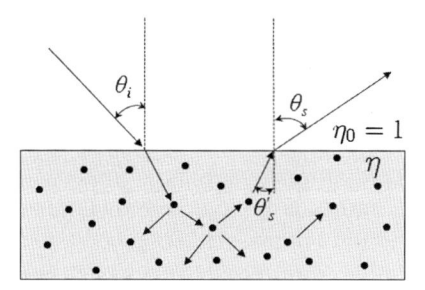

Fig. 4.3 Light propagation process described by the Wolff model. Incident light penetrates the dielectric surface at an angle θ_i and scatters by particles inside the dielectric body. Subsequently, a portion of the internally scattered light arrives at the material-air boundary at an angle θ'_s and re-emerges from the surface at an emittance angle θ_s through refraction

internal angle of incidence on the dielectric-air boundary before the light ray is refracted and re-emerges from the surface.

As seen in Eq. (4.17), the Wolff model consists of Lambert's cosine term multiplied by two Fresnel transmission terms. The $\cos\theta_i$ factor is borrowed from the Lambertian model to account for foreshortening of the incident ray viewed from the point of incidence on the surface. In addition, the Fresnel term $(1 - F(\theta_i, \eta(u, \lambda)))$ accounts for the refraction of the incident light, and $(1 - F(\theta'_s, \frac{1}{\eta(u,\lambda)}))$ accounts for the refraction of the emitted light ray.

The internal angle of incidence is related to the reflection angle through Snell's law by the expression $\theta'_s = \frac{\arcsin(\sin(\theta_s))}{\eta(u,\lambda)}$. Here, we recall that Ragheb and Hancock (2006) have employed the subsurface scattering process described by Wolff to correct the geometric attenuation in the Beckmann–Kirchhoff model. Consequently, the term $\cos\theta_i(1 - F(\theta_i, \eta(u, \lambda)))(1 - F(\theta'_s, \frac{1}{\eta(u,\lambda)}))$ in Eq. (4.17) is treated as the Fresnel correction term of the Beckmann–Kirchhoff model for large incident and scattering angles.

4.2.4 Oren–Nayar Model

Oren and Nayar (1995) observed that the Lambertian model for diffuse reflection did not accurately capture the reflectance from a number of real-world materials such as plaster, clay, sand and cloth. Hence, they extended the Lambertian model to address complex geometric and radiometric phenomena such as masking, shadowing and interreflections between points on the same surface.

This model fundamentally differs from the Lambertian model in that the reflected radiance of illuminated surfaces increases as the viewing direction approaches the source direction. In this model, the rough surface structure is viewed as a collection of long symmetric V-cavities to account for the shadowing, masking and interreflections between particles on the surface. Through experimental validation on a number of rough surface samples, the proposed model showed a significant deviation from the Lambertian model and a strong agreement with reflectance data.

4.2.5 Torrance–Sparrow Model

A well-known model of the off-specular spike phenomenon was formulated by Torrance and Sparrow (1967). The model assumes that the surface is composed of small, randomly distributed, mirror-like micro-facets. With this surface structure, the off-specular peak phenomenon is explained in terms of mutual masking, shadowing, multiple reflection and internal scattering between micro-facets with the V-groove shape. Similar to the Wolff reflectance model, the Torrance and Sparrow model includes a specular reflection component based on the Fresnel reflection theory and a micro-facet distribution function.

The Torrance–Sparrow model (Torrance and Sparrow 1967) describes the reflectance from rough surfaces as a combination of two components: the Lambertian diffuse component and the specular lobe. According to the model, the total reflectance from an infinitesimal area is given by

$$f_{TS}(u,\lambda) = \rho(u,\lambda)\cos\theta_i + F\big(\theta_i, \eta(u,\lambda)\big)\frac{G_{TS}(\theta_{ip}, \theta_{sp})}{\cos\theta_s}D(\vartheta). \quad (4.18)$$

In Eq. (4.18), the first term is the diffuse reflectance component, which obeys Lambert's cosine law. Here, $\rho(u,\lambda)$ is the diffuse albedo of the material and the cosine of the incident angle and $\cos\theta_i$ accounts for the foreshortening of the incident light flux. The latter term in Eq. (4.18) describes the specular lobe exhibited by rough surfaces. The model assumes that specular reflection is the result of direct interface reflection from micro-facets with the normal direction $\vec{N}'$, which is the angular bisector of the incident direction $\vec{L}$ and the viewing direction $\vec{V}$. We illustrate this process in Fig. 4.4. In the figure, we denote the angle between the mean surface normal $\vec{N}$ and the micro-facet normal $\vec{N}'$ as ϑ. Furthermore, θ_{ip} and θ_{sp} are the projections of θ_i and θ_s onto the plane spanned by the facet normal $\vec{N}'$ and the mean surface normal $\vec{N}$. In other words, θ_{ip} and θ_{sp} are, respectively, the angles between $\vec{N}$ and the projections of $\vec{L}$ and $\vec{V}$ onto the above plane.

As before, the Fresnel reflection coefficient $F(\theta_i, \eta(u,\lambda))$ is employed to model the penetration of the incident light into the material body. On the other hand, $G_{TS}(\theta_i, \theta_s, \phi_i, \phi_s)$ is a geometric attenuation factor which represents the fraction of the micro-facet surface that contributes to the reflected flux. The original formulation of this factor is directly related to the normal of each micro-facet. Torrance and Sparrow (1967) derived this factor as a function of the angles θ_{ip} and θ_{sp}. The resulting analytical expression of G_{TS} is quite complex and varies from case to case depending on the range of the angles θ_{ip} and θ_{sp} (Torrance and Sparrow 1967).

Lastly, the term $D(\vartheta)$ represents the distribution of the micro-facet normal with respect to the mean surface normal, in which ϑ is the angle between the micro-facet normal and the mean surface normal. For isotropic surfaces, the micro-facet normal is rotationally symmetric about the mean surface normal. As a result, the distribution $D(\vartheta)$ is often modelled as a Gaussian distribution with zero mean and a standard deviation m, i.e. $D(\vartheta) \sim \mathcal{N}(0, m^2)$.

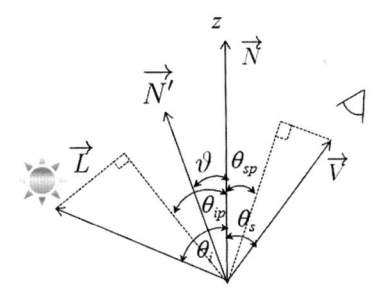

Fig. 4.4 The reflection geometry described by the Torrance–Sparrow model. The incident light arrives in the direction $\vec{L}$ and is reflected in the direction $\vec{V}$ at the respective angles θ_i and θ_s with respect to the mean surface normal $\vec{N}$. Specular reflection is deemed to be mainly contributed by micro-facets with the normal direction $\vec{N}'$, which is the angular bisector of $\vec{L}$ and $\vec{V}$. We denote the angle between the mean surface normal and the micro-facet normal as ϑ. Also, let us consider θ_{ip} and θ_{sp} to be the angles between $\vec{N}$ and the projections of $\vec{L}$ and $\vec{V}$ onto the plane spanned by $\vec{N}$ and $\vec{N}'$, respectively. In other words, θ_{ip} and θ_{sp} are the projections of θ_i and θ_s onto the plane above. With these ingredients, the geometric attenuation can be expressed as a function of the angles θ_{ip} and θ_{sp}

4.2.6 Cook–Torrance Model

Cook and Torrance (1982) introduced a reflectance model for the rendering of synthetic images. This model bears a substantial resemblance to the Torrance–Sparrow model (Torrance and Sparrow 1967) since they are both derived from the same physical foundation. Hence, it consists of an ambient reflectance component, a diffuse Lambertian component and a specular component. Similar to the Torrance–Sparrow model (Torrance and Sparrow 1967), the specular term is a product of a Fresnel reflection coefficient, a geometrical attenuation factor that accounts for shadowing and masking effects and a distribution of micro-facet slopes.

Since the original formulation of the geometrical attenuation factor G_{TS} in the Torrance–Sparrow model is quite complex, it is computationally infeasible for rendering applications. To reduce the computational complexity, Cook and Torrance proposed an alternative formulation of the geometric factor that is only dependent on the mean surface normal $\vec{N}$, the given light source direction $\vec{L}$ and the viewer direction $\vec{V}$. This formulation is based on the intuition that specular reflection is solely dependent on micro-facets with normals aligned with the angular bisector $\vec{H}$ of $\vec{L}$ and $\vec{V}$. With the reflection geometry depicted in Fig. 4.1, the geometric attenuation factor is given by

$$G_{\text{CT}}(\theta_i, \theta_s, \phi_i, \phi_s) = \min\left(1, \frac{2(\vec{N}.\vec{H})(\vec{N} \cdot \vec{V})}{(\vec{V} \cdot \vec{H})}, \frac{2(\vec{N}.\vec{H})(\vec{N} \cdot \vec{L})}{(\vec{V} \cdot \vec{H})}\right). \quad (4.19)$$

As a consequence, the specular component of the Cook–Torrance reflectance model is

$$f_{\text{CT}}^{\text{spec}}(u, \lambda) = \frac{F(\theta_i, \eta(u, \lambda))}{\pi} \frac{D(\theta_h) G_{\text{CT}}(\theta_i, \theta_s, \phi_i, \phi_s)}{\cos\theta_i \cos\theta_s}, \quad (4.20)$$

where θ_h is the angle between the halfway vector $\vec{H}$ and the surface normal $\vec{N}$, and $\eta(u, \lambda)$ is the material refractive index.

4.2.7 Micro-Facet Slope Distribution

The Beckmann–Kirchhoff, Torrance–Sparrow and Cook–Torrance models incorporate a distribution of the micro-facet slope as a factor in their specular component. This can be observed in Eqs. (4.3), (4.18) and (4.20). This distribution is a measure of the proportion of facets that specularly reflect light in the viewing direction, and is characteristic of the micro-facet structure under study. The two typical micro-facet slope distribution models often involved in reflectance modelling are the Gaussian and the Beckmann distribution.

For isotropic surfaces, the micro-facet normal may assume a rotationally symmetric distribution about the mean surface normal. Therefore, the distribution of the half-angle θ_h can be modelled as a Gaussian distribution with zero mean and a standard deviation m,

$$D(\theta_h) = \frac{1}{\sqrt{2\pi}m} \exp\left(-\frac{\theta_h^2}{2m^2}\right). \tag{4.21}$$

Since the standard deviation m indicates the roughness of the surface, it is also called the roughness of the micro-facet structure.

Alternatively, Beckmann (Beckmann and Spizzichino 1963) provided a comprehensive theory of wave scattering for a variety of surfaces from smooth to very rough. For rough surfaces, Beckmann's formulation of the facet slope distribution model is

$$D(\theta_h) = \frac{1}{m^2 \cos^4 \theta_h} \exp\left(-(\tan\theta_h/m)^2\right). \tag{4.22}$$

In Beckmann's formulation in Eq. (4.22), the root means square slope m determines the spread of the specular lobe. The larger the value, the steeper the slope.

4.2.8 Multi-Layered Reflectance Modelling

A number of reflectance models are also designed specifically for multi-layer surfaces such as skin tissues and plant leaves. One of the pioneering works in this area was proposed by Hanrahan and Krueger (1993), who employed a one-dimensional linear transport theory in a subsurface scattering model for layered surfaces. Subsequently, Ng and Li (2001) examined light propagation in a three-layered skin surface model to account for the Fresnel specular reflection, as well as the diffuse reflection caused by subsurface scattering. More recently, Ragheb and Hancock (2006, 2008) modelled the scattering of light through rough surfaces consisting of two layers as a combination of two components. The first component accounts for the light scattering at the upper surface boundary and is modelled using a modified Beckmann model, i.e. the Vernold–Harvey model (Vernold and Harvey 1998). The second component models the incoming and outgoing refraction through the upper surface boundary using Fresnel transmission coefficients and the scattering process occurring at the lower surface boundary. While the refraction involved in the second component was derived in a similar manner to the Wolff reflectance model (Wolff 1994), the scattering at the lower surface boundary was modelled in the same way as the one at the upper surface boundary.

4.3 Notes

The modelling of surface reflectance has attracted considerable attention due to its relevance to scene analysis and image understanding. For instance, Nayar and Bolle (1996) have used photometric invariants derived from the BRDF to recognise objects with different reflectance properties. This work builds on that reported in Nayar and Bolle (1993), where a background-to-foreground reflectance ratio is introduced. In a related development, Dror et al. (2001) have shown how surfaces may be classified from single images through the use of reflectance properties. By contrast, in the graphics community it is the development of computationally efficient tools for the purposes of realistic surface rendering that is of primary interest.

Note that the models presented throughout the chapter and elsewhere in the literature employ the object shape, the index of refraction, reflectance and a measure of surface roughness to embody the surface reflectance. These models (Beckmann and Spizzichino 1963; Torrance and Sparrow 1967; Cook and Torrance 1982; Wolff 1994) also often incorporate a Fresnel term (Born and Wolf 1999). Moreover, the reflectance and index of refraction are, in general, wavelength dependent. The use of imaging spectroscopy opens up the possibility of exploiting this wavelength dependence for purposes of estimating the reflectance parameters.

This is also closely related to the problem of recovering the shape of an object from its shading information. The classic approaches to shape from shading developed by Ikeuchi and Horn (1981), and by Horn and Brooks (1986), hinge on the compliance with the image irradiance equation and local surface smoothness. Zheng and Chellappa (1991) proposed a gradient consistency constraint that penalises differences between the image intensity gradient and the surface gradient for the recovered surface. Moreover, although shape from shading usually relies on the assumption of Lambertian reflectance (Horn and Brooks 1986), photometric correction or specularity subtraction has been applied as a preprocessing step to improve the results obtained. For instance, Brelstaff and Blake (1988) used a simple thresholding strategy to identify specularities on moving curved objects. Other lines of research remove specularities by using additional hardware (Nayar and Bolle 1996), imposing constraints on the input images (Lin and Shum 2001), requiring colour segmentation (Klinker et al. 1990) as postprocessing steps or using reflectance models to account for the distribution of image brightness (Ragheb and Hancock 2003).

References

Beckmann, P., & Spizzichino, A. (1963). *The scattering of electromagnetic waves from rough surfaces*. New York: Pergamon.

Born, M., & Wolf, E. (1999). *Principles of optics: electromagnetic theory of propagation, interference and diffraction of light* (7th ed.). Cambridge: Cambridge University Press.

Brelstaff, G., & Blake, A. (1988). Detecting specular reflection using Lambertian constraints. In *International conference on computer vision* (pp. 297–302).

Cook, R. L., & Torrance, K. E. (1982). A reflectance model for computer graphics. *ACM Transactions on Graphics*, $1(1)$, 7–24.

Dror, R. O., Adelson, E. H., & Willsky, A. S. (2001). Recognition of surface reflectance properties from a single image under unknown real-world illumination. In *Proceedings of the IEEE workshop on identifying objects across variations in lighting*.

Hanrahan, P., & Krueger, W. (1993). Reflection from layered surfaces due to subsurface scattering. In *SIGGRAPH '93: proceedings of the 20th annual conference on computer graphics and interactive techniques* (pp. 165–174). New York: ACM.

Harvey, J. E., Krywonos, A., & Vernold, C. L. (2007). Modified Beckmann–Kirchhoff scattering theory for rough surfaces with large scattering and incident angles. *Optical Engineering*, $46(7)$.

He, X. D. (1991). A comprehensive physical model for light reflection. *Computer Graphics*, $25(4)$, 175–186.

Helmholtz, H. V. (1924). *Treatise on physiological optics*.

Horn, B. K. P., & Brooks, M. J. (1986). The variational approach to shape from shading. *CVGIP*, $33(2)$, 174–208.

Ikeuchi, K., & Horn, B. K. P. (1981). Numerical shape from shading and occluding boundaries. *Artificial Intelligence*, $17(1\text{-}3)$, 141–184.

Klinker, G., Shafer, S., & Kanade, T. (1990). A physical approach to color image understanding. *International Journal of Computer Vision*, $4(1)$, 7–38.

Lambert, J. H. (1760). *Photometria, sive de mensura et gradibus luminus, colorum et umbrae*. Augsburg: Eberhard Klett.

Lin, S., & Shum, H. (2001). Separation of diffuse and specular reflection in color images. In *International conference on computer vision and pattern recognition*.

Minnaert, M. (1941). The reciprocity principle in lunar photometry. *Astrophysical Journal*.

Nayar, S., & Bolle, R. (1996). Reflectance based object recognition. *International Journal of Computer Vision*, $17(3)$, 219–240.

Nayar, S. K., & Bolle, R. M. (1993). Computing reflectance ratios from an image. *Pattern Recognition*, 26, 1529–1542.

Nayar, S. K., Ikeuchi, K., & Kanade, T. (1991). Surface reflection: physical and geometrical perspectives. *IEEE Transactions on Pattern Analysis and Machine Intelligence*, $13(7)$, 611–634.

Ng, C. S. q. L., & Li, L. (2001). A multi-layered reflection model of natural human skin. In *Computer graphics international conference*.

Nieto-Vesperinas, M., & Garcia, N. (1981). A detailed study of the scattering of scalar waves from random rough surfaces. *Optica Acta*, $28(12)$, 1651–1672.

Ogilvy, J. A. (1991). *Theory of wave scattering from random rough surfaces*. Bristol: Adam Hilger.

Opik, E. (1924). Photometric measures of the moon and the moon the earth-shine. *Publications de L'Observatorie Astronomical de L'Universite de Tartu*, $26(1)$, 1–68.

Oren, M., & Nayar, S. K. (1995). Generalization of the Lambertian model and implications for machine vision. *International Journal of Computer Vision*, $14(3)$, 227–251.

Phong, B. T. (1975). Illumination for computer generated pictures. *Communications of the ACM*, $18(6)$, 311–317.

Ragheb, H., & Hancock, E. R. (2003). A probabilistic framework for specular shape-from-shading. *Pattern Recognition*, $36(2)$, 407–427.

Ragheb, H., & Hancock, E. R. (2006). Testing new variants of the Beckmann–Kirchhoff model against radiance data. *Computer Vision and Image Understanding*, $102(2)$, 145–168.

References

Ragheb, H., & Hancock, E. R. (2008). A light scattering model for layered dielectrics with rough surface boundaries. *International Journal of Computer Vision, 79*(2), 179–207.

Shafer, S. A. (1985). Using color to separate reflection components. *Color Research and Application, 10*(4), 210–218.

Torrance, K., & Sparrow, E. (1967). Theory for off-specular reflection from roughened surfaces. *Journal of the Optical Society of America, 57*(9), 1105–1112.

Vernold, C. L., & Harvey, J. E. (1998). A modified Beckmann–Kirchoff scattering theory for non-paraxial angles. *Proceedings of the SPIE. Scattering and surface roughness* (pp. 51–56).

Ward, G. J. (1992). Measuring and modeling anisotropic reflection. *Computer Graphics, 26*(2), 265–272.

Wolff, L. B. (1994). *Diffuse-reflectance model for smooth dielectric surfaces* (pp. 2956–2968).

Zheng, Q., & Chellappa, R. (1991). Estimation of illuminant direction, albedo, and shape from shading. *IEEE Transactions on Pattern Analysis and Machine Intelligence, 13*(7), 680–702.

Chapter 5
Illuminant Power Spectrum

In order to retrieve the spectra of a material, it is first necessary to remove the effect of the illuminants in the scene. The main challenge here stems from the fact that the output of the imager depends on the amount of light that passes through the lens of the camera, not the object reflectance (the ratio of reflected to incoming light impinging on the object surface). Note that whereas irradiance is determined by the light source and viewer directions and the material properties of the surface under study, reflectance is a characteristic of the material. This is well known in remote sensing, where material identification is, in general, a classification problem (Chang et al. 2003).

In spectral imaging, we are often confronted with the need to recover the reflectance invariant to illuminant, viewer directions and sensor choice. Thus, for scene analysis, the problem should be treated as one arising from complex geometric settings found in real-world scenes, with one or more illuminants of different sorts (neon tubes, incandescent light bulbs, sunlight) and inter-reflections between object surfaces, some of them translucent or transparent. Moreover, for reliable scene analysis, methods should be applicable to highly textured surfaces and scenes with multiple illuminants.

In this regard, the case of multiple illuminants, when their directions are known, is also interesting from a scholarly point of view, since the angular variables for isotropic reflection depend solely on the surface normal (Horn and Brooks 1989). The multi-image case, such as that pertaining to stereo vision, where the light source directions are not known, may be a worthy vehicle for application of large-scale optimisation methods to process all the spatial and spectral domain parameters simultaneously in order to recover the illuminant directions, surface shape and photometric parameters.

5.1 Standard Illuminants

In the visible spectrum, the CIE has defined a number of standard illuminant power spectra corresponding to commonly used light sources (Wyszecki and Stiles 2000).

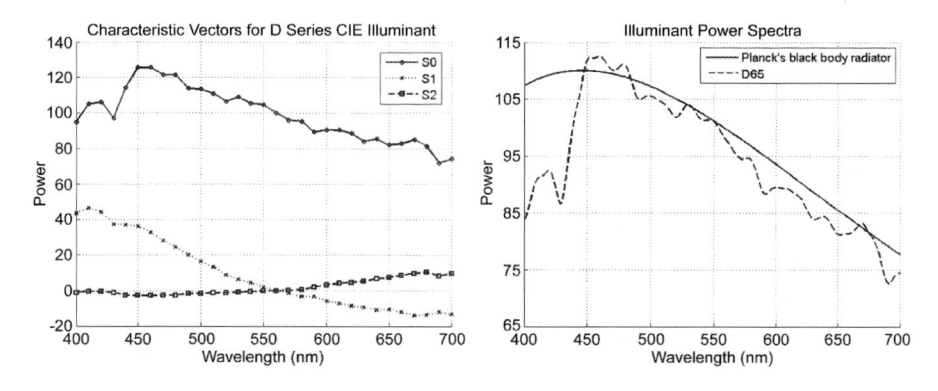

Fig. 5.1 *Left-hand panel*: characteristic vectors of the CIE D illuminant. *Right-hand panel*: power spectra of the D65 illuminant and the Planckian black body radiator

These standard illuminants are denoted by letters; the A, B and C series were defined in 1931 and the D series in 1964. The A series is aimed at representing typical, domestic, tungsten-filament lighting. The A series follows Planck's law for a black body spectral radiance of the form

$$L_A(\lambda) = \frac{c_1 \lambda^{-5}}{\exp(\frac{c_2}{\lambda T}) - 1} \quad (5.1)$$

where $T = 2856$ K, $c_1 = 3.74183 \times 10^{-16}$ W m^2 and $c_2 = 1.4388 \times 10^{-2}$ mK. Note that c_2 acts as a normalisation factor which has been chosen to achieve a spectral power of 100 at 560 nm.

The B and C series correspond to direct and shaded sunlight, respectively. These are derived from illuminant A by using "liquid filters" with a high absorbance in the red end of the visible spectrum. The liquid filters comprise two solutions of copper sulfate, mannite, pyridine, sulfuric acid, cobalt, and ammonium sulfate in distilled water separated by a clear glass. These series do not give a good approximation of common light sources and have been deprecated in favour of the D series.

Illuminants corresponding to the D series represent typical daylight (Judd et al. 1964) and are now widely used in preference to the B and C series. These are somewhat difficult to realise in practise, but easily modelled mathematically. The spectral power spectrum of the D-series illuminants is governed by the correlated colour temperature (CCT) and is modelled as a linear combination of three basis vectors recovered by the characteristic vector analysis in Simonds (1963) on a set of 622 spectra of daylight samples. These vectors are not orthogonal; rather they correspond to the solution to a linear system of equations. In Fig. 5.1, the left-hand panel shows the characteristic vectors of the D illuminant.

Hence, the D series illuminant power spectrum is defined as follows:

$$L_D(\lambda) = \mathbf{S}_0(\lambda) + M_1 \mathbf{S}_1(\lambda) + M_2 \mathbf{S}_2(\lambda). \quad (5.2)$$

In Eq. (5.2), S_i, $i = \{1, 2, 3\}$ are the basis vectors described above. The first vector (S_0) is the mean of all the daylight spectrum samples above. The second vector (S_1) accounts for changes in the CCT due to the presence or absence of clouds or direct sunlight, which is reflected as the yellow–blue variation (Judd et al. 1964). The third vector (S_2) models the pink–green variation caused by the presence of water vapour and haze (Judd et al. 1964).

The mixing coefficients M_1 and M_2 are given as follows:

$$M_1 = \frac{-1.3515 - 1.7703 x_D + 5.9114 y_D}{0.0241 + 0.2562 x_D - 0.7341 y_D},$$
$$M_2 = \frac{0.0300 - 31.4424 x_D + 30.0717 y_D}{0.0241 + 0.2562 x_D - 0.7341 y_D},$$
(5.3)

with x_D and y_D given by

$$x_D = \begin{cases} -4.607 \frac{10^9}{T^3} + 2.968 \frac{10^6}{T^2} + 0.099 \frac{10^3}{T} + 0.244 & \text{for } 4000 \leq T \leq 7000 \text{ K,} \\ -2.006 \frac{10^9}{T^3} + 1.901 \frac{10^6}{T^2} + 0.247 \frac{10^3}{T} + 0.237 & \text{for } 7000 < T \leq 25000 \text{ K,} \end{cases}$$

$$y_D = -3.000 x_D^2 + 2.870 x_D - 0.275,$$

where T is the CCT.

Finally, the E and F illuminants correspond to unit radiance and fluorescent light sources. The E illuminant is of importance as a theoretical reference which gives equal radiance to all wavelengths in the visible range. The F illuminants correspond to various fluorescent lamps. There are 12 standard types of F illuminants, where the first six types, i.e. F1–F6, account for lamps of antimony and manganese in calcium halophosphate phosphor. The following three types, i.e. F7–F9, are full-spectrum lamps, whereas the last three, i.e. F10–F12, are narrow tri-band illuminants.

5.2 Correlated Colour Temperature (CCT) and the White Point

To provide a better understanding of the CCT and the associated white point, we commence by concentrating on modelling the power spectrum of a light source using Planck's black body model. This model has been widely used for deriving several colour constancy algorithms (Finlayson and Schaefer 2001b; Ratnasingam and Collins 2010). The temperature, in kelvins, at which the chromaticity of the black body radiator is equivalent to that of the illuminant under consideration is the illuminant colour temperature. Colour temperature has applications in photography and is relevant to white balancing of digital cameras. Further, it has been used as a standard benchmark in colourimetry (Wyszecki and Stiles 2000).

For a given illuminant characterised by the temperature T, the radiation emitted from Planck's black body can be written as follows:

$$E(\lambda, T) = \frac{2hc^2}{\lambda^5} \frac{1}{(\exp(\frac{hc}{k_b T \lambda}) - 1)}, \qquad (5.4)$$

where λ is the wavelength in nanometres, $h = 6.626 \times 10^{-34}$ joules-second is the Planck constant, $k_b = 1.3806 \times 10^{-23}$ joules per kelvin is the Boltzmann constant and $c = 2.9979 \times 10^8$ metres per second is the speed of light.

Note that, for small values of the product λT, i.e. $\exp(\frac{hc}{k_b T \lambda}) \gg 1$, Eq. (5.4) can be approximated by Wien's radiation law, which is given by

$$E(\lambda, T) \approx \omega_1 \lambda^{-5} e^{-\frac{\omega_2}{T\lambda}}, \qquad (5.5)$$

where $\omega_1 = 2hc^2$ and $\omega_2 = \frac{hc}{k_b \lambda}$. Wien's radiation law is a good approximation for indoor tungsten illuminants and models the overall variation of the daylight spectrum across the visible wavelength range.

Recall that any illuminant power spectrum can be converted into tristimulus (RGB) values, according to Eq. (3.9). In Chap. 3, this process is called colour balance, where a surface with unit reflectance is rendered with a reference colour under the colour matching functions. If the reference colour is the perceived white under the given illuminant power spectrum $L(\cdot)$, its tristimulus values or the corresponding chromaticity is called the white point. This is related to the rationale that, in the absence of fluorescence, the tristimulus values in an image are not greater than those of the scene illuminant. In other words, no pixel can be brighter than the illuminant.

Moreover, the RGB values for any illuminant can be normalised to its CIE chromaticity coordinates, i.e. converted to the XYZ colour space and then represented using the X–Y colour gamut (Wyszecki and Stiles 2000). This implies that, by taking the CIE chromaticity coordinates of the Planckian black body radiator and the white point of the illuminant, one can assign a correlated colour temperature T to the illuminant. In the right-hand panel of Fig. 5.1 we show the plots for the black body radiator and the D65, i.e. the CIE D illuminant with a CCT $T = 6504$ K, normalised at 560 nm.

The use of the CCT is particularly common in regard to the CIE illuminant D, where the temperature, in hundreds of kelvins, is usually appended to the letter so as to denote the illuminant. In the left-hand panel of Fig. 5.2, we show the CIE D40, D55, D100 and D250 illuminants, i.e. $T = \{4000, 5500, 10000, 25000$ K$\}$, across the visible wavelength range normalised at 560 nm. Also, for the readers' reference, in the right-hand panel of the figure we show a sample of noon daylight. Note the variation of the D illuminant spectrum with respect to the CCT and the similarity of the D55 to daylight.

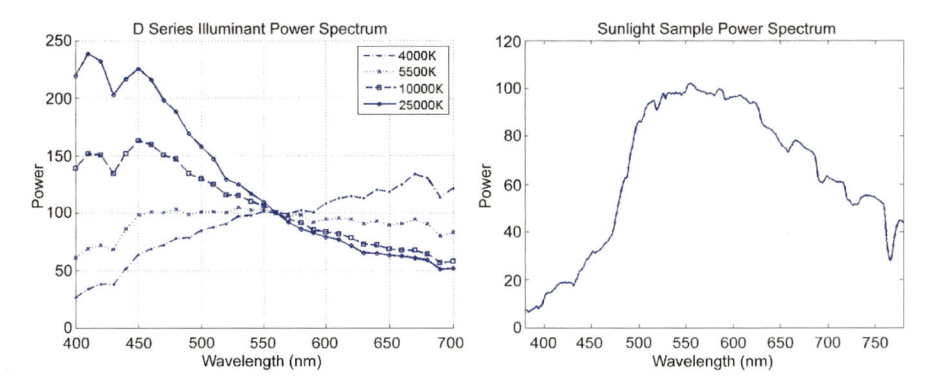

Fig. 5.2 *Left-hand panel*: power spectra of the D illuminants corresponding to a number of CCT values. *Right-hand panel*: sample measurement for the sunlight power spectrum

5.3 Illuminant Power Spectrum Estimation

As mentioned earlier, the CCT can be used to characterise the standard illuminants, in particular the D series. This has been used in colour constancy (Finlayson and Schaefer 2001b) to estimate the colour of the illuminant by intersecting the chromaticity line with the Planckian locus. Here, the Planckian locus refers to the graph of the chromaticity of Planck's black body radiator with respect to the correlated colour temperature T. Although the method in Finlayson and Schaefer (2001b) was originally proposed in trichromatic imagery, this constrained setting can be extended to multispectral and hyperspectral imagery by making use of the dichromatic model (Shafer 1985).

We do this to relate the captured scene radiance to the light spectral power, surface reflectance and the scene geometry. Let us recall from Chap. 4 that the dichromatic model is formulated as

$$I(u,\lambda) = g(u)L(\lambda)S(u,\lambda) + k(u)L(\lambda), \tag{5.6}$$

where $I(u,\lambda)$ is the surface radiance and $S(u,\lambda)$ denotes the surface reflectance at pixel location u and wavelength λ under an illumination spectrum $L(\lambda)$. In this chapter, we assume that the shading factor, specular coefficient and object reflectance are available. In the following chapter, we will elaborate further on how these may be recovered. For now, we focus our attention on the recovery of the illuminant power spectrum.

If the shading factor, specular coefficient and reflectance are known, the problem can then be cast as that of minimising the cost function

$$F(\mathcal{I}) = \sum_{u \in \mathcal{I}} \left[I(u,\lambda) - g(u)L(\lambda)S(u,\lambda) + k(u)L(\lambda) \right]^2$$
$$\text{s.t. } L(\lambda) = E(\lambda, T), \tag{5.7}$$

over all the pixels u in an image $\mathcal{I}$, where $E(\lambda, T)$ is the radiation emitted from the Planckian black body with a given CCT T as provided by Eq. (5.4).

In a related approach, Finlayson and Schaefer (2001a) have estimated the illuminant colours based on the dichromatic model without prior assumptions on the illuminant statistics. Although their experiments were performed on trichromatic imagery, this method can also be adapted to multispectral data in a manner similar to the one above. To do this, we note that their approach relies on the dichromatic plane formalism (Finlayson and Schaefer 2001a; Tominanga and Wandell 1989), which states that spectral radiance vectors at all pixels belonging to the same material lie in a two-dimensional subspace $Q \subset \mathbb{R}^n$, spanned by the light vector $\mathbf{L}$ and the diffuse radiance vector $\mathbf{D}_P \triangleq \mathbf{L} \bullet \mathbf{S}_P$. Here, the diffuse radiance vector is the component-wise multiplication of the illuminant and the surface reflectance spectrum $\mathbf{S}_P = [S_P(\lambda_1), \ldots S_P(\lambda_n)]^T$ common to all the pixels in the surface P.

Having all the pixel radiance vectors $\mathbf{I}(u)$ at hand, one can obtain the subspace Q via singular value decomposition (SVD). Denoting the two basis vectors resulting from this SVD operation as $\mathbf{z}_1$ and $\mathbf{z}_2$, we let the subspace be $Q = span(\mathbf{z}_1, \mathbf{z}_2)$. Since $\mathbf{D}_P^{t-1} \in Q$, we can parameterise $\mathbf{D}_P$ as $\mathbf{D}_P = v\mathbf{z}_1 + \mathbf{z}_2$, where v is a scalar.

Likewise, the light vector $\mathbf{L} \in Q$ can also be decomposed as $\mathbf{L} = w_1 \mathbf{z}_1 + w_2 \mathbf{z}_2$, where the values of w_1 and w_2 are two known scalars. Furthermore, the dichromatic plane hypothesis also implies that, given the light vector $\mathbf{L}$ and the surface diffuse radiance vector $\mathbf{D}_P$, one can decompose any pixel radiance $\mathbf{I}(u)$ into a linear combination of the former two vectors. In other words,

$$\mathbf{I}(u) = g(u)\mathbf{D}_P + k(u)\mathbf{L}$$
$$= \big(g(u)v + k(u)w_1\big)\mathbf{z}_1 + \big(g(u) + k(u)w_2\big)\mathbf{z}_2. \quad (5.8)$$

Utilising this idea, in Finlayson and Schaefer (2001a) the illumination estimation is cast as an optimisation problem so as to maximise the total projection length of the light colour vector onto all the dichromatic planes. Geometrically, this approach predicts the illuminant power spectrum as the least-squares intersection of the dichromatic planes, that is

$$\max_{\mathbf{L}} \sum_{i=1}^{N} \|\mathbf{G}_i \mathbf{L}\|^2, \quad (5.9)$$

where $\mathbf{G}_i = \mathbf{A}_i [\mathbf{A}_i^T \mathbf{A}_i]^{-1} \mathbf{A}_i^T$ is the projection matrix for the ith dichromatic plane recovered from the image, $\mathbf{A}_i = [\mathbf{z}_1, \mathbf{z}_2]$ is a matrix composed of columns corresponding to the basis vectors $\mathbf{z}_1$ and $\mathbf{z}_2$ spanning the dichromatic plane and $\| \cdot \|^2$ denotes the L^2-norm of the argument vectors.

With these ingredients, Finlayson and Schaefer (2001a) show that the light vector that maximises the quantity in Eq. (5.9) is given by the eigenvector corresponding to the largest eigenvalue of the matrix $\sum_{i=1}^{N} \mathbf{G}_i$.

Note that the method in Finlayson and Schaefer (2001a) corresponds to a least-squares solution to the intersection of the dichromatic planes and, therefore, may lead to a numerically unstable solution when the angle between dichromatic planes is small. Huynh and Robles-Kelly (2010) provide a closed-form solution derived

5.3 Illuminant Power Spectrum Estimation

directly from the dichromatic model. To do this, they reformulate the cost function in Eq. (5.7) using the set $\mathcal{P}$ of surface patches with common reflectance in the image as follows:

$$F(\mathcal{I}) = \sum_{P \in \mathcal{P}} \sum_{u \in P} \|\mathbf{I}(u) - g(u)\mathbf{D}_P - k(u)\mathbf{L}\|^2$$

$$= \sum_{P \in \mathcal{P}} \sum_{u \in P} \sum_{\lambda} \big(I(u,\lambda) - g(u)D_P(\lambda) - k(u)L(\lambda)\big)^2, \quad (5.10)$$

where, as before, $\mathbf{D}_P = \mathbf{L} \bullet \mathbf{S}_P$. They note that the expression above is quadratic and, therefore, convex with respect to $\mathbf{L}$ and $\mathbf{D}_P$. Thus, the optimal values of these variables can be obtained by equating the respective partial derivatives of $F(\mathcal{I})$ to zero. These partial derivatives are given by

$$\frac{\partial F(\mathcal{I})}{\partial L(\lambda)} = -2 \sum_{P \in \mathcal{P}} \sum_{u \in P} \big(I(u,\lambda) - g(u)D_P(\lambda) - k(u)L(\lambda)\big)k(u),$$

$$\frac{\partial F(\mathcal{I})}{\partial D_P(\lambda)} = -2 \sum_{u \in P} \big(I(u,\lambda) - g(u)D_P(\lambda) - k(u)L(\lambda)\big)g(u).$$

Equating the above equations to zero, we obtain

$$L(\lambda) = \frac{\sum_{P \in \mathcal{P}} \sum_{u \in P} [k(u)I(u,\lambda) - g(u)k(u)D_P(\lambda)]}{\sum_{P \in \mathcal{P}} \sum_{u \in P} (k(u))^2}, \quad (5.11)$$

$$D_P(\lambda) = \frac{\sum_{u \in P} [g(u)I(u,\lambda) - g(u)k(u)L(\lambda)]}{\sum_{u \in P} (g^t(u))^2}. \quad (5.12)$$

From Eqs. (5.11) and (5.12), the illuminant spectrum can be solved in closed form as

$$L^*(\lambda) = \frac{\sum_{P \in \mathcal{P}} \sum_{u \in P} k(u)I(u,\lambda) - \sum_{P \in \mathcal{P}} [\frac{(\sum_{u \in P} g(u)k(u))(\sum_{u \in P} g(u)I(u,\lambda))}{\sum_{u \in P} (g(u))^2}]}{\sum_{P \in \mathcal{P}} \sum_{u \in P} (k(u))^2 - \sum_{P \in \mathcal{P}} [\frac{(\sum_{u \in P} g(u)k(u))^2}{\sum_{u \in P} (g(u))^2}]}.$$
(5.13)

Note that, so far, we have assumed that all the pixels on a surface P belong to the same dichromatic plane. In practice, these pixels are expected to belong to an image patch, which can be selected by making use of the procedure described in Sect. 6.2.2. With these patches in hand, the methods in Finlayson and Schaefer (2001a) or Huynh and Robles-Kelly (2010) can be applied to the eigensystem of the sum of projection matrices on the dichromatic planes.

This treatment contrasts with the spectrum deconvolution approach commonly employed in remote sensing as proposed by Sunshine et al. (1990) to recover the absorption bands characteristic of the surface material chemistry. This method makes use of the upper bound envelope of a reflectance spectrum, also known as its continuum, which can be regarded as a reflectance spectrum without any absorption

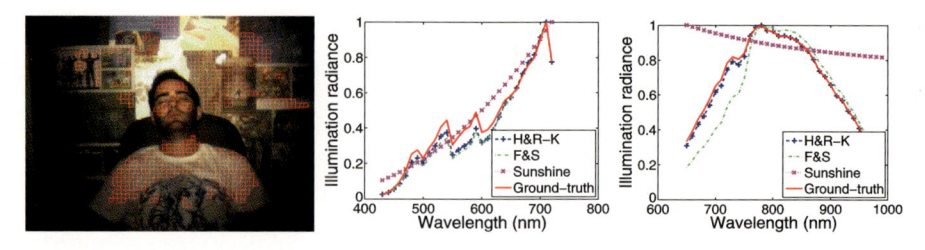

Fig. 5.3 *Left-hand panel*: the pseudo-colour rendering of a spectral image of a human subject captured in the visible range (430–720 nm). The image is overlaid with the rectangular boundaries (coloured in *red*) of all the homogeneous patches detected by the patch selection strategy described in Sect. 6.2.2. *Right-hand panels*: ground truth illuminant spectra and those estimated by the methods in Finlayson and Schaefer (2001a), Huynh and Robles-Kelly (2010) and Sunshine et al. (1990). Here we show the estimated illuminant spectra in both the visible (*middle panel*) and near-infrared (*rightmost panel*)

feature. The work in Sunshine et al. (1990) assumes that the continuum is a linear function of the wave number, i.e. the reciprocal of the wavelength, on the log-reflectance scale. Making use of this assumption, it then fits this parametric form to the continuum of radiance spectra to recover the illuminant.

To illustrate the difference between the methods presented above, we show, in the right-hand panels of Fig. 5.3, the illuminant power spectrum recovered by the methods in Finlayson and Schaefer (2001a) (F & S), Huynh and Robles-Kelly (2010) (H & R-K) and Sunshine et al. (1990) (Sunshine) when applied to the patches selected from the hyperspectral image whose pseudo-colour we show in the left-hand panel. Note that the method of Sunshine fails to capture the complex shape of the illuminant spectrum.

5.4 Notes

The recovery of illuminant and of material reflectance are mutually interdependent. This is not surprising, since the appearance of an object in a scene depends on the reflectance properties of the surface, the power spectrum of the illuminant, the scene geometry and the imaging device. The dependencies on the illuminant power spectrum and the scene geometry cause unwanted variations in the recorded scenes and variations of colour appearance which depend on the environment.

The estimation of the illuminant power spectrum is also closely related to the research on colour constancy. This has traditionally branched into two main trends. One of them relies on the statistics of illuminant and material reflectance (Brainard and Freeman 1997); the other draws upon the physics-based analysis of local shading and specularity of the surface material (Finlayson and Schaefer 2001b; Huynh and Robles-Kelly 2010). These methods often make assumptions regarding shadows (Sato et al. 1999), inter-reflection (Funt et al. 1991), specular reflection (Shafer 1985; Zmura and Lennie 1986) or constraints related to the reflection process (Finlayson and Schaefer 2001b; Tominanga and Wandell 1989).

In contrast with prior literature of colour constancy, the use of imaging spectroscopy integrates the recovery of the illuminant and photometric invariants, i.e. the material reflectance, the shading and the specularity, into a unified setting. This not only extends the colour constancy problem from trichromatic to multispectral and hyperspectral imagery, but also confers several advantages. In such a unified treatment, algorithms are expected to be more generally applicable to surfaces exhibiting both diffuse and specular reflection. In addition, methods of this sort may avoid assumptions on the parametric form or prior knowledge of the illuminant and surface reflectance spectra by casting the problem in an optimisation or a phenomenological setting.

References

Brainard, D. H., & Freeman, W. T. (1997). Bayesian color constancy. *Journal of the Optical Society of America A, 14*(7), 1393–1411.

Chang, J. Y., Lee, K. M., & Lee, S. U. (2003). Shape from shading using graph cuts. In *Proceedings of the international conference on image processing*

Finlayson, G. D., & Schaefer, G. (2001a). Convex and non-convex illuminant constraints for dichromatic colour constancy. In *IEEE conference on computer vision and pattern recognition* (Vol. I, pp. 598–604).

Finlayson, G. D., & Schaefer, G. (2001b). Solving for colour constancy using a constrained dichromatic reflection model. *International Journal of Computer Vision, 42*(3), 127–144.

Funt, B., Drew, M., & Ho, J. (1991). Color constancy from mutual reflection. *International Journal of Computer Vision, 6*(1), 5–24.

Horn, B. K. P., & Brooks, M. J. (1989). *Shape from shading*. Cambridge: MIT Press.

Huynh, C. P., & Robles-Kelly, A. (2010). A solution of the dichromatic model for multispectral photometric invariance. *International Journal of Computer Vision, 90*(1), 1–27.

Judd, D. B., Macadam, D. L., Wyszecki, G., Budde, H. W., Condit, H. R., Henderson, S. T., & Simonds, J. L. (1964). Spectral distribution of typical daylight as a function of correlated color temperature. *Journal of the Optical Society of America, 54*(8), 1031–1036.

Ratnasingam, S., & Collins, S. (2010). Study of the photodetector characteristics of a camera for color constancy in natural scenes. *Journal of the Optical Society of America A, 27*(2), 286–294.

Sato, I., Sato, Y., & Ikeuchi, K. (1999). Illumination distribution from shadows. In *Computer vision and pattern recognition* (pp. 1306–1312).

Shafer, S. A. (1985). Using color to separate reflection components. *Color Research and Application, 10*(4), 210–218.

Simonds, J. L. (1963). Application of characteristic vector analysis to photographic and optical response data. *Journal of the Optical Society of America, 53*(8), 968–971.

Sunshine, J., Pieters, C. M., & Pratt, S. F. (1990). Deconvolution of mineral absorption bands: an improved approach. *Journal of Geophysical Research, 95*(B5), 6955–6966.

Tominaga, S., & Wandell, B. A. (1989). Standard surface-reflectance model and illuminant estimation. *Journal of the Optical Society of America A, 6*, 576–584.

Wyszecki, G., & Stiles, W. (2000). *Color science: concepts and methods, quantitative data and formulae*. New York: Wiley.

Zmura, M. D., & Lennie, P. (1986). Mechanisms of color constancy. *Journal of the Optical Society of America A, 10*(3), 1662–1672.

Chapter 6
Photometric Invariance

The estimation of reflectance parameters is also closely related to the problem of recovering the shape of an object from its shading information. The classic approaches to shape from shading developed by Ikeuchi and Horn (1981), and by Horn and Brooks (1986), hinge on compliance with the image irradiance equation and local surface smoothness. Zheng and Chellappa (1991) proposed a gradient consistency constraint that penalises differences between the image intensity gradient and the surface gradient for the recovered surface. Worthington and Hancock (1999b) imposed the Lambertian radiance constraint in a hard manner. Dupuis and Oliensis (1992) provided a means of recovering probably correct solutions with respect to the image irradiance equation. Prados and Faugeras (2003) presented a shape from shading approach applicable to the case where the light source and viewing directions are no longer coincident. Kimmel and Bruckstein used the apparatus of level set methods to recover solutions to the eikonal equation (Kimmel and Bruckstein 1995).

Specific to imaging spectroscopy, Healey and his colleagues (Healey and Slater 1999; Slater and Healey 1999; Suen and Healey 2001) studied photometric invariance in hyperspectral aerial imagery for material classification and mapping as related to photometric artefacts induced by atmospheric effects and changing solar illuminations. In Stokman and Gevers (1999), a method is presented for hyperspectral edge detection. The method is robust to photometric effects, such as shadows and specularities. In Angelopoulou (2000), a photometrically invariant approach was proposed based on the derivative analysis of the spectra. This local analysis of the spectra was shown to be intrinsic to the surface reflectance. Yet, the analysis in Angelopoulou (2000) was derived from the Lambertian reflection model and, hence, is not applicable to specular reflections. Recently, Huynh and Robles-Kelly (2010) proposed a method based on a dichromatic model (Shafer 1985) aimed at imaging spectroscopy. Although it is effective, the method can be computationally demanding, since it relies on a dual-step gradient descent optimisation scheme governed by a regularised cost function.

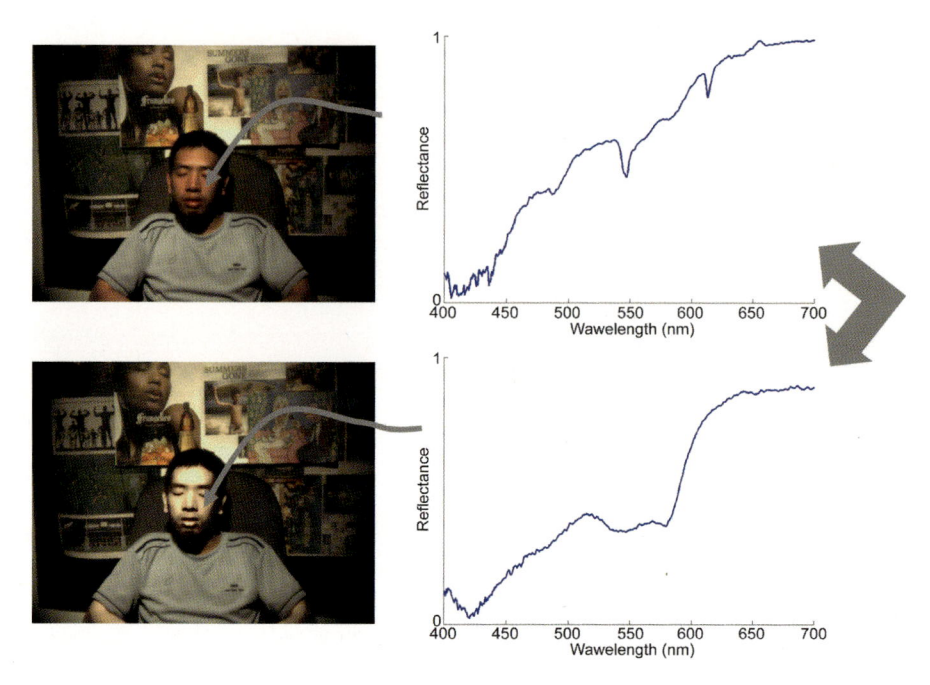

Fig. 6.1 Initial (*top row*) and final (*bottom row*) pseudo-colour images and sample reflectance spectra when the skin reflectance of the subject is substituted by an alternative one

From a practical point of view, the recovery of shape and photometric invariants through imaging spectroscopy can provide a means to advanced editing and postprocessing tasks. By making use of the surface reflectance as a "signature" for identification and recognition purposes, photometric invariants can be used for complex photowork in which images are edited, modified, annotated, etc. based on the object shape, the materials and the illuminants in the scene.

This is illustrated in Fig. 6.1. In the figure, we show the pseudo-colour image of a portrait taken using a hyperspectral camera. By replacing the spectra of the skin on the face of the subject in the left-hand panel, we obtain the image on the right-hand panel of the figure. The skin reflectance of the subject is plotted at the top of the middle panel; the one used as an alternative is shown at the bottom plot. The skin spectra replacement requires the recovery of the reflection parameters. With the reflectance at hand, skin recognition can be effected so as to locate the pixels whose spectra are to be replaced. After the reflectance is replaced, the final step is to recompute the image irradiance using the parameters and the alternative reflectance. Note that a similar sequence of events would apply to the replacement of the illuminant in the scene, where the power spectrum of the light has to be recovered and substituted.

6.1 Separating the Reflection Components

For both imaging spectroscopy and trichromatic imagery, there have been several attempts to remove specularities from images of non-Lambertian objects. Most of the efforts in the literature have been devoted to modelling the effects encountered on shiny or rough surfaces. For shiny surfaces, specular spikes and lobes have to be modelled. In one of the attempts to remove specularities from images of non-Lambertian objects, Brelstaff and Blake (1988) used a simple thresholding strategy to identify specularities on moving curved objects. Other lines of research remove specularities by using additional hardware (Nayar and Bolle 1996), imposing constraints on the input images (Lin and Shum 2001), requiring colour segmentation (Klinker et al. 1990) as postprocessing steps or using reflectance models to account for the distribution of image brightness (Ragheb and Hancock 2003). The main limitation of these methods is that they rely either on predetermined set-ups for the image acquisition or the use of the closed form of the bidirectional reflectance distribution function (BRDF) to characterise the specular spike and lobe.

In a related development, Novak and Shafer (1992) used the colour histogram to relate the shape of the objects in a scene to their roughness in order to estimate the illuminant colour. Sato and Ikeuchi (1994) used the dichromatic reflection model of Shafer (1985) to separate the specular reflection component from a series of colour images. Umeyama and Godin (2004) separated the diffuse from the specular component by maximising the probabilistic independence between these two components via independent component analysis (ICA). Tan and Ikeuchi (2005) used chromaticity for separating the reflection components of textured surfaces using a single image.

6.1.1 Dichromaticity and Specularity

Here we account for specularities in hyperspectral and multispectral imaging by employing the dichromatic model (Shafer 1985) presented in Sect. 4.2.2. We can give an intuitive interpretation to the model as follows. At specular pixels, where the shading factor $g(u)$ is negligible, the surface acts as a mirror, where the radiance spectrum becomes proportional to the illuminant power spectrum $L(\lambda_i)$. On the other hand, if the radiance is diffuse, i.e. $k(u)$ is close to zero, then the surface appears shaded and the reflectance becomes a multiplicative term on the illuminant to determine the radiance. We can easily verify this by taking the ratio of the image irradiance $I(u, \lambda_i)$ to the illuminant power $L(\lambda_i)$ and expressing it as

$$R(u, \lambda_i) \triangleq \frac{I(u, \lambda_i)}{L(\lambda_i)} = g(u)S(u, \lambda_i) + k(u). \qquad (6.1)$$

In Eq. (6.1), the specular component $k(u)$ is constant over the wavelengths. On the other hand, $S(u, \lambda_i)$ is a function of the wavelength. This observation is the key to the following method for extracting the reflectance $S(u, \lambda_i)$. Note that the specular coefficient $k(u)$ can be eliminated by subtracting the mean $\bar{R}(u)$ of $R(u, \cdot)$ across the wavelengths λ_i from Eq. (6.1). Moreover, for an image region of homogeneous reflectance, the variation of radiance across pixels is governed by the shading factor $g(u)$. To commence, we write

$$R(u, \lambda_i) - \bar{R}(u) = g(u)\left(S(u, \lambda_i) - \frac{1}{N}\sum_{i=1}^{N} S(u, \lambda_i)\right)$$
$$= g(u)\left(S(u, \lambda_i) - \bar{S}(u)\right), \qquad (6.2)$$

where N is the number of wavelengths. In the equation above, we have introduced $\bar{S}(u) \triangleq \frac{1}{N}\sum_{i=1}^{N} S(u, \lambda_i)$ for brevity.

In addition, we can express the variances of $S(u, \lambda_i)$ and $R(u, \lambda_i)$ over the wavelength domain as

$$\tau_S^2 = \frac{1}{N}\sum_{i=1}^{N}\left(S(u, \lambda_i) - \bar{S}(u)\right)^2,$$
$$\tau_R^2 = \frac{1}{N}\sum_{i=1}^{N}\left(R(u, \lambda_i) - \bar{R}(u)\right)^2 = g^2(u)\tau_S^2. \qquad (6.3)$$

Let us denote $\hat{S}(u, \lambda_i) \triangleq \frac{S(u,\lambda_i) - \bar{S}(u)}{\tau_S}$. From Eqs. (6.2) and (6.3), we can compute $\hat{S}(u, \lambda_i)$ as

$$\hat{S}(u, \lambda_i) = \frac{S(u, \lambda_i) - \bar{S}(u)}{\tau_S} = \frac{R(u, \lambda_i) - \bar{R}(u)}{\tau_R}. \qquad (6.4)$$

The relation above further suggests that the quantity $\hat{S}(u, \lambda_i)$ can be computed from the image irradiance $I(u, \lambda_i)$ once the illuminant power spectrum $L(\cdot)$ is available. Further, it is independent of the shading factor $g(u)$. With this in mind, we rewrite Eq. (6.1) as

$$R(u, \lambda_i) = g(u)\left(S(u, \lambda_i) - \bar{S}(u)\right) + g(u)\bar{S}(u) + k(u)$$
$$= g(u)\tau_S \hat{S}(u, \lambda_i) + g(u)\bar{S}(u) + k(u)$$
$$= a(u)\hat{S}(u, \lambda_i) + b(u), \qquad (6.5)$$

where in Eq. (6.5) we have employed the following notation:

$$a(u) = g(u)\tau_S, \qquad (6.6)$$
$$b(u) = g(u)\bar{S}(u) + k(u). \qquad (6.7)$$

With $R(u, \lambda_i)$ and $\hat{S}(u, \lambda_i)$ already obtained, Eq. (6.5) permits the use of linear regression over the wavelength domain to recover the variables $a(u)$ and $b(u)$ for

6.1 Separating the Reflection Components

each pixel u. Also, from Eqs. (6.6) and (6.7), we deduce

$$b(u) = \frac{a(u)}{\tau_S} \bar{S}(u) + k(u) = ca(u) + k(u), \qquad (6.8)$$

where $c = \frac{\bar{S}(u)}{\tau_S}$.

Equation (6.8) relates the quantities $a(u)$ and $b(u)$ through a linear equation. We note that for all pixels belonging to the same material, $c = \frac{\bar{S}(u)}{\tau_S}$ is constant since all those pixels have the same material reflectance $S(u, \lambda_i)$, for all λ_i. Making use of Eq. (6.8), we apply linear regression to these pixels to estimate the specular coefficient $k(u)$ and to separate the specular term from Eq. (6.1).

6.1.2 Logarithmic Differentiation

Logarithmic differentiation was used by Tan and Ikeuchi (2005) to separate reflection components. Although their work on specularity removal is based upon chromaticity, their approach can be modified to work with N bands instead of three in a straightforward manner. This is done to account for the fact that we are concerned with imaging spectroscopy data rather than trichromatic imagery.

Tan and Ikeuchi (2005) perform logarithmic differentiation on the diffuse component of the pixel. We can express this operation using Eq. (6.8) as follows:

$$\frac{\partial \log(ca(u))}{\partial u} = \frac{\partial \log(c)}{\partial u} + \frac{\partial \log(a(u))}{\partial u}, \qquad (6.9)$$

where $\frac{\partial \log(c)}{\partial u} \equiv 0$ at regions of continuous reflectance, i.e. for surfaces where there are no material discontinuities.

Let us denote the pseudo-intensity $\hat{I}(u)$ as

$$\hat{I}(u) = \sum_{\lambda} |I(u, \lambda)|. \qquad (6.10)$$

Equation (6.9) implies that the spatial derivative of the pseudo-intensity image $\hat{I}(u)$ should be equal to the logarithmic differentiation in Eq. (6.9) for diffuse pixels. This gives rise to a diffuse verification rule that is used to label pixels for specularity reduction. Note that, for the verification, an initial specularity-free image is required. This image corresponds to the values of $a(u)$ at each pixel u, which can be recovered using linear regression as applied to the line in Eq. (6.8).

The method proceeds by iteratively applying diffuse verification and, once the non-diffuse pixels are identified, specularity reduction ensues. The specularity reduction can be cast, in the hyperspectral or multispectral imaging setting, as a regularisation over the spatial domain where the shading factor for specular pixels is given by the averaging over their neighbours. This is done to bring the shading of the specular pixels "closer" to that of their diffuse neighbours.

Thus, the method proceeds as follows. Firstly, the difference between the logarithmic differentiation of both the pseudo-intensity image and the values of $a(u)$ is computed. For pixels whose difference is greater than a threshold ε, the Euclidean spectral angle between the spectra $I(u,\cdot)$ and $I(v,\cdot)$ at two adjacent pixels u and v, which is computed as $\arccos(\frac{\langle I(u,\cdot), I(v,\cdot)\rangle}{\|I(u,\cdot)\|\cdot\|I(v,\cdot)\|})$, is used to identify material discontinuities. This follows the rationale that, for pixels corresponding to the same material, the Euclidean angles should be approximately null. These discontinuities are removed from further consideration and specular reduction is then applied to the remaining non-diffuse pixels. This process yields an updated pseudo-intensity image. The method iterates until the difference for the logarithmic differentiation at specular pixels is never greater than a threshold ε.

6.1.3 Entropy Minimisation

The formulation in Eq. (6.5) allows the recovery of the variables $a(u)$ and $b(u)$ for each pixel u from $R(u, \lambda_i)$ and $\hat{S}(u, \lambda_i)$. This is often performed as a linear regression over the wavelength domain. Here we show how c and $k(u)$ in Eq. (6.8) can be estimated from $a(u)$ and $b(u)$ by an entropy minimisation method.

Recall that, for a homogeneous reflectance region in the image, the specular coefficient $k(u)$ acts as a bias term, whereas the shading factor $g(u)$ scales up or down the reflectance. This, as mentioned earlier, implies that every pixel in a homogeneous region should lie on the dichromatic line in the $a(u)$–$b(u)$ space. This is illustrated in Fig. 6.2. Note that all the pixels in the region share the same slope c since they have the same material reflectance $S(u, \lambda_i)$, for all λ_i. By projecting the points on the direction of any line in the $a(u)$–$b(u)$ space, we can appreciate that their frequency distribution has a minimal skewness when the intersecting line is perpendicular to $b(u) = ca(u) + k(u)$.

This is illustrated in Fig. 6.2. Note that this implies that the entropy of the frequency distribution of the pixel projections on a line is minimum when the slope of the line is $\frac{1}{c}$. This suggests the use of the Shannon entropy for the recovery of c. Thus, we aim at recovering slope angle θ of the line onto which the projected point distribution have a minimal entropy. Given an arbitrary line going though the origin with a slope $\tan\theta$, the orthogonal distance from the point $(a(u), b(u))$ to the line is

$$d(u, \theta) = a(u)\cos\theta + b(u)\sin\theta. \tag{6.11}$$

To quantify the degree of skewness for the distribution of projected points, we discretise the distribution as follows. To do this, we divide the interval $[\min\{d(u,\theta)\}, \max\{d(u,\theta)\}]$ into K bins such that the ith bin comprises pixels in the range $[(i-1)r, ir)$, where $r = (\max\{d(u,\theta)\} - \min\{d(u,\theta)\})/K$. We then view the cumulative distribution as the probability of a point in the bin indexed i falling in the interval under consideration. This is $p(i|\theta) = \frac{N_i(\theta)}{N_{\text{total}}}$, where N_{total} is the

6.1 Separating the Reflection Components

Fig. 6.2 Distributions for the projection of the $(a(u), b(u))$ points onto the line perpendicular to that with slope c. The *top panel* shows the distribution with optimal Shannon entropy; the *bottom panel* shows a suboptimal projection

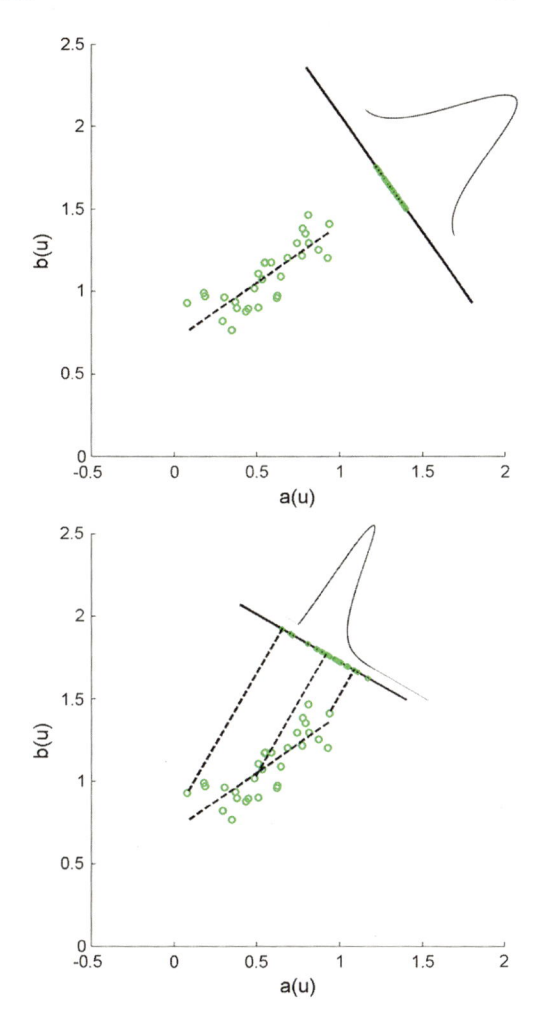

number of pixels in the homogeneous reflectance region under study and $N_i(\theta)$ corresponds to the number of pixels in the interval $[(i-1)r, ir)$. With these ingredients, the Shannon entropy is given by

$$\eta(\theta) = -\sum_i p(i|\theta) \log(p(i|\theta)). \tag{6.12}$$

To implement the algorithm, we first recover the illuminant power spectrum $L(\lambda)$, making use of the method in Huynh and Robles-Kelly (2010), which we will detail in Sect. 6.2. Once the illuminant power spectrum is in hand, we apply k-means to cluster the vectors formed by concatenating $R(u, \lambda i)$ values across the wavelengths. This yields a set of clusters whose pixels can be used to recover the variables $a(u)$ and $b(u)$ by applying a linear regression to Eq. (6.5). With $a(u)$ and $b(u)$ in hand, we perform a search over the range of angle θ to select the one which yields the lowest entropy $\eta(\theta)$. Once the angle θ and the slope c have been determined, we recover the value of $k(u)$ for every pixel in each of the clusters.

6.2 An Optimisation Approach to Illuminant and Photometric Invariant Recovery

Note that, following the model in Eq. (6.1), it would be straightforward to recover the specular coefficient $k(u)$ if the other parameters, i.e. the power spectrum of the light, reflectance and shading factors were in hand. Similarly, by assuming that the illuminant spectrum and the specular coefficient $k(u)$ are available, one could normalise the irradiance and recover the term $R(u, \lambda_i)$. This suggests an iterative process for the recovery of the shading factor, the specular coefficient $k(u)$, the reflectance and the illuminant power spectrum $L(\lambda_i)$.

6.2.1 Cost Function

Thus, with the dichromatic model, we proceed to define our target function for purposes of optimisation. Our algorithm takes as input a multispectral image whose pixel values correspond to the measurements of the spectral irradiance $I(u, \lambda_i)$ indexed to the wavelengths $\lambda \in \{\lambda_1, \ldots, \lambda_n\}$. As mentioned previously, our goal is to fit the observed data to the dichromatic model to recover the parameters $g(u)$, $k(u)$ and $S(u, \lambda_i)$. In general, here we view the dichromatic cost function of a multispectral image $\mathcal{I}$ as the weighted sum of its dichromatic error and a regularisation term $\mathcal{R}(u)$ for each image location. This is

$$F(\mathcal{I}) \triangleq \sum_{u \in \mathcal{I}} \left[\sum_{i=1}^{n} \left[I(u, \lambda_i) - L(\lambda_i)\big(g(u)S(u, \lambda_i) + k(u)\big) \right]^2 + \alpha \mathcal{R}(u) \right]. \quad (6.13)$$

In Eq. (6.13), α is a constant that acts as a balancing factor between the dichromatic error and the regularisation term $\mathcal{R}(u)$ on the right-hand side. The wavelength-independent regularisation term $\mathcal{R}(u)$ is related to the surface shading and will be elaborated on later.

For now, we focus our attention on the solution space of Eq. (6.13). Note that minimising the cost $F(\mathcal{I})$ without further constraints is an under-determined problem. This is due to the fact that, for an image with n spectral bands containing m pixels, we would have to minimise over $2m + n + m \times n$ variables while having only $m \times n$ terms in the summation of Eq. (6.13). However, we notice that this problem can be further constrained if the model is applied to smooth surfaces made of the same material, i.e. if the albedo is uniform across the patch or image region under consideration. This imposes two constraints. Firstly, all locations on the surface share a common diffuse reflectance; i.e. a uniform albedo surface P is assumed to

6.2 An Optimisation Approach to Illuminant and Photometric Invariant

have the same reflectance for each pixel $u \in P$. Note that this constraint significantly reduces the number of unknowns $S(u, \lambda_i)$ from $m \times n$ to $N \times n$, where N is the number of surface albedos in the scene. In addition, the smooth variation of the patch geometry allows us to formulate the regularisation term $\mathcal{R}(u)$ in Eq. (6.13) as a function of the shading factor $g(u)$. In brief, smooth, uniform albedo surface patches naturally provide constraints that reduce the number of unknowns significantly while providing a plausible formulation of the regularisation term $\mathcal{R}(u)$.

Following the rationale above, we proceed to impose constraints on the minimisation problem. For a smooth, uniform-albedo surface patch $P \in \mathcal{I}$, we consider the following cost function:

$$F(P) \triangleq \sum_{u \in P} \left[\sum_{i=1}^{n} [I(u, \lambda_i) - L(\lambda_i)(g(u)S_P(\lambda_i) + k(u))]^2 + \alpha \mathcal{R}(u) \right],$$

where $S(u, \lambda_i) = S_P(\lambda_i)$, for all $u \in P$.

Furthermore, the smoothness constraint on the patch implies that the shading factor $g(u)$ should vary smoothly across the pixels in P. This constraint can be effectively formulated by minimising the variation of gradient magnitude for the shading map. This, effectively, precludes discontinuities in the shading map of P via the regularisation term

$$\mathcal{R}(u) \triangleq \left[\frac{\partial g(u)}{\partial x(u)}\right]^2 + \left[\frac{\partial g(u)}{\partial y(u)}\right]^2, \tag{6.14}$$

where the variables $x(u)$ and $y(u)$ are the column and row coordinates, respectively, for the pixel location u.

Thus, by making use of the set $\mathcal{P}$ of uniform-albedo patches in the image $\mathcal{I}$, we can recover the dichromatic model parameters by minimising the target function

$$F^*(\mathcal{I}) \triangleq \sum_{P \in \mathcal{P}} F(P)$$

$$= \sum_{P \in \mathcal{P}} \sum_{u \in P} \left[\sum_{i=1}^{n} [I(u, \lambda_i) - L(\lambda_i)(g(u)S_P(\lambda_i) + k(u))]^2 + \alpha \mathcal{R}(u) \right], \tag{6.15}$$

as an alternative to $F(\mathcal{I})$.

6.2.2 Homogeneous Patch Selection

In the previous section, we formulated the recovery of the dichromatic model parameters as an optimisation procedure over the surface patch set $\mathcal{P}$. In this section, we describe our method for automatically selecting uniform-albedo surface patches for the minimisation of the cost function in Eq. (6.15).

Our patch selection strategy is performed as follows. We first subdivide the image into patches of equal size in a lattice-like fashion. For each patch, we fit a two-dimensional hyperplane to the radiance vectors of the pixels in the patch. Next, we note that, in perfectly dichromatic patches, the wavelength indexed radiance vector of each pixel lies perfectly in this hyperplane, i.e. the dichromatic plane. To allow for noise effects, we regard dichromatic patches as those containing a percentage of at most t_p pixels whose radiance vectors deviate from their projection given by the singular value decomposition (SVD) in Tominanga and Wandell (1989). We do this by setting a threshold t_a on the angular deviation from the dichromatic plane, where t_p and t_a are global parameters.

However, not all these patches are useful for purposes of illumination spectrum recovery. This is due to the fact that perfectly diffuse surfaces do not provide any information regarding the illuminant spectrum. The reason is that a spectral radiance vector space for this kind of surface is one dimensional, spanned only by the wavelength indexed diffuse radiance vector. On the other hand, the dichromatic model implies that the specularities have the same spectrum as the illuminant, where the specular coefficient can be viewed as a wavelength-independent scaling factor.

Thus, for the recovery of the dichromatic model parameters, we can use highly specular patches by selecting regions with the highest contrast amongst those deemed to have a uniform albedo. We recover the contrast of each patch by computing the variance of the mean radiance over the spectral domain. These highly specular patches provide a means to the recovery of the light spectrum due to the fact that, for highly specular surface patches with uniform albedo, the surface diffuse radiance vector and the illuminant vector span a hyperplane in the radiance vector space. This is a well-known property in colour constancy, where a number of approaches (Healey 1991; Klinker et al. 1988; Lee 1986) have employed subspace projection for purposes of light power spectrum recovery.

6.2.3 Optimisation Procedure

Making use of the notation in Fig. 6.3, we now present an iterative approach to find the variables $\mathbf{L}, \mathbf{S}_P, \mathbf{g}_P$ and $\mathbf{k}_P$ which yield the minimum of the cost function in Eq. (6.15). At each iteration, we minimise the cost function with respect to $\mathbf{L}$ and the triplet $\{\mathbf{g}_P, \mathbf{k}_P, \mathbf{S}_P\}$ in separate steps.

The procedure presented here is, in fact, a coordinate descent approach (Boyd and Vandenberghe 2004) which aims at minimising the cost function. The step sequence of our minimisation strategy is summarised in the pseudo-code of Algorithm 6.1. The coordinate descent approach comprises two interleaved minimisation steps. At each iteration, we index the dichromatic variables to iteration number t and optimise the objective function, in interleaved steps, with respect to the two subsets of variables $\{\mathbf{g}_P, \mathbf{k}_P, \mathbf{S}_P\}, \{\mathbf{L}\}$. Once the former variables are in hand, we can obtain

$\mathbf{I}(u)$: the spectral radiance vector at image pixel u, $\mathbf{I}(u) = [I(u, \lambda_1), \ldots, I(u, \lambda_n)]^T$
$\mathbf{L}$: the spectral power vector of the illuminant, $\mathbf{L} = [L(\lambda_1), \ldots, L(\lambda_n)]^T$
$\mathbf{S}_P$: the common spectral reflectance vector for each patch P,
$\mathbf{S}_P = [S_P(\lambda_1), \ldots, S_P(\lambda_n)]^T$
$\mathbf{g}_P$: the shading map of all pixels in patch P, $\mathbf{g}_P = [g(u_1), \ldots, g(u_l)]^T$
where $u_1, \ldots, u_l$ are all the pixels in the patch P
$\mathbf{g}$: the shading map of all the patches, $\mathbf{g} = [\mathbf{g}_{P_1}^T, \ldots, \mathbf{g}_{P_r}^T]^T$
where $P_1, \ldots, P_r$ are all patches in $\mathcal{P}$
$\mathbf{k}_P$: the specularity map of all pixels in patch P, $\mathbf{k}_P = [k(u_1), \ldots, k(u_l)]^T$
$\mathbf{k}$: the specularity map of all the patches, $\mathbf{k} = [\mathbf{k}_{P_1}^T, \ldots, \mathbf{k}_{P_r}^T]^T$

Fig. 6.3 Notation for Sect. 6.2.3

Algorithm 6.1 Estimate dichromatic variables from a set of homogeneous patches

Require: Image $\mathcal{I}$ with radiance $I(u, \lambda)$ for each band $\lambda \in \{\lambda_1, \ldots, \lambda_n\}$ and location u and the collection of homogeneous patches $\mathcal{P}$
Ensure: $\mathbf{L}, \mathbf{S}_P, \mathbf{g}, \mathbf{k}$, where
  $\mathbf{L}$: the estimated illuminant spectrum.
  $\mathbf{S}_P$: the diffuse reflectance of each surface patch P.
  $\mathbf{g}, \mathbf{k}$: the diffuse and specular reflection coefficients at all locations.
1: $t \leftarrow 1; \mathbf{L}^0 \leftarrow \mathbf{1}^T$
2: **while** true **do**
3:   **for all** $P \in \mathcal{I}$ **do**
4:     $[\mathbf{g}_P^t, \mathbf{k}_P^t, \mathbf{S}_P^t] \leftarrow \operatorname{argmin}_{\mathbf{g}_P, \mathbf{k}_P, \mathbf{S}_P} F(P)|_{\mathbf{L}^{t-1}}$
5:   **end for**
6:   $[\mathbf{L}^t] \leftarrow \operatorname{argmin}_{\mathbf{L}, \mathbf{S}_{P_1}, \ldots, \mathbf{S}_{P_r}} \sum_{P \in \mathcal{P}} F(P)|_{\mathbf{g}^t, \mathbf{k}^t}$
7:   **if** $\angle(\mathbf{L}^t, \mathbf{L}^{t-1}) < t_\mathbf{L}$ **then**
8:     **break**
9:   **else**
10:     $t \leftarrow t + 1$
11:   **end if**
12: **end while**
13: **return** $\mathbf{L}^t, \mathbf{g}^t, \mathbf{k}^t, \mathbf{S}_{P_1}^t, \ldots, \mathbf{S}_{P_r}^t$

---

optimal values for the latter ones. We iterate between these two steps until convergence is reached.

The algorithm commences by initialising the unknown light spectrum $\mathbf{L}$ to an unbiased uniform illumination spectrum, as indicated in Line 1 of Algorithm 6.1. It terminates once the illuminant spectrum does not change, in terms of angle, by an amount beyond a preset global threshold $t_\mathbf{L}$ between two successive iterations. In the following two subsections we show that the two optimisation steps above can be employed to obtain the optimal values of the dichromatic parameters in closed form.

In the first step, we estimate the optimal surface reflectance and shading given the light spectrum $\mathbf{L}^{t-1}$ recovered at iteration $t-1$. This corresponds to Lines 3–5 in Algorithm 6.1. Note that, at iteration t, we can solve for the unknowns $\mathbf{g}_P^t, \mathbf{k}_P^t$ and $\mathbf{S}_P^t$ separately for each surface patch P because, for each patch, these variables appear in a separate term in Eq. (6.15). This step is, therefore, reduced to minimising

$$F(P)|_{\mathbf{L}^{t-1}} = \sum_{u \in P} [\|\mathbf{I}(u) - g(u)\mathbf{D}_P^{t-1} - k(u)\mathbf{L}^{t-1}\|^2 + \alpha \mathcal{R}(u)], \quad (6.16)$$

where the diffuse radiance vector $\mathbf{D}_P^{t-1} \triangleq \mathbf{L}^{t-1} \bullet \mathbf{S}_P$ is the component-wise multiplication of the illuminant and surface reflectance spectra, and $\|\cdot\|$ denotes the L_2-norm of the argument vectors.

Note that the minimisation above involves $2|P| + n$ unknowns, where $|P|$ is the number of pixels in patch P. Hence, it becomes computationally intractable when the surface area is large. In practise, the selected patches need only be large enough to gather useful statistics from the radiance information. Moreover, as mentioned earlier, we can further reduce the degrees of freedom of the unknowns by noting that the spectral radiance vectors at all pixels in the same surface lie in a two-dimensional subspace $Q \subset \mathbb{R}^n$, spanned by the diffuse radiance vector $\mathbf{D}_P^{t-1}$ and the light vector $\mathbf{L}^{t-1}$.

Recall that, having all the pixel radiance vectors $\mathbf{I}(u)$ in hand, one can obtain the subspace Q via SVD. Denote the two basis vectors resulting from this SVD operation $\mathbf{z}_1$ and $\mathbf{z}_2$ and, accordingly, let the subspace be $Q = span(\mathbf{z}_1, \mathbf{z}_2)$. Since $\mathbf{D}_P^{t-1} \in Q$, we can parameterise $\mathbf{D}_P^{t-1}$ up to scale as $\mathbf{D}_P^{t-1} = v\mathbf{z}_1 + \mathbf{z}_2$, where v is a scalar.

Likewise, the light vector $\mathbf{L}^{t-1} \in Q$ can also be decomposed as $\mathbf{L}^{t-1} = w_1\mathbf{z}_1 + w_2\mathbf{z}_2$, where the values of w_1 and w_2 are two known scalars. Furthermore, the dichromatic plane hypothesis also implies that, given the light vector $\mathbf{L}^{t-1}$ and the surface diffuse radiance vector $\mathbf{D}_P^{t-1}$, one can decompose any pixel radiance $\mathbf{I}(u)$ into a linear combination of the former two vectors. In other words,

$$\begin{aligned}\mathbf{I}(u) &= g(u)\mathbf{D}_P^{t-1} + k(u)\mathbf{L}^{t-1} \\ &= \big(g(u)v + k(u)w_1\big)\mathbf{z}_1 + \big(g(u) + k(u)w_2\big)\mathbf{z}_2. \end{aligned} \quad (6.17)$$

Having obtained the basis vectors $\mathbf{z}_1, \mathbf{z}_2$, we can compute the mapping of the pixel radiance $\mathbf{I}(u)$ onto the subspace Q. This is done with respect to this basis by using projection so as to obtain the scalars $\tau_1(u), \tau_2(u)$ such that

$$\mathbf{I}(u) = \tau_1(u)\mathbf{z}_1 + \tau_2(u)\mathbf{z}_2. \quad (6.18)$$

Further, by equating the right-hand sides of Eqs. (6.17) and (6.18), we obtain

$$g(u) = \frac{w_2\tau_1(u) - w_1\tau_2(u)}{w_2 v - w_1}, \quad (6.19)$$

$$k(u) = \frac{\tau_2(u)v - \tau_1(u)}{w_2 v - w_1}. \quad (6.20)$$

6.2 An Optimisation Approach to Illuminant and Photometric Invariant

From Eqs. (6.19) and (6.20), we note that $g(u)$ and $k(u)$ are univariate rational functions of v. Moreover, $\mathbf{D}_P^{t-1}$ is a linear function with respect to v. We also observe that the regularisation term $\mathcal{R}(u)$ is only dependent on $g(u)$. Therefore, the objective function in Eq. (6.16) can be reduced to a univariate rational function of v. Thus, substituting Eqs. (6.19) and (6.20) into the first and second terms on the right-hand side of Eq. (6.16), we have

$$F(P)|_{\mathbf{L}^{t-1}} = \sum_{u \in P} \left\| \mathbf{I}(u) - \frac{w_2\tau_1(u) - w_1\tau_2(u)}{w_2 v - w_1}(v\mathbf{z}_1 + \mathbf{z}_2) - \frac{\tau_2(u)v - \tau_1(u)}{w_2 v - w_1}\mathbf{L}^{t-1} \right\|^2$$

$$+ \sum_{u \in P} \frac{\alpha}{(w_2 v - w_1)^2} \left[\left(\frac{\partial m(u)}{\partial x(u)}\right)^2 + \left(\frac{\partial m(u)}{\partial y(u)}\right)^2 \right]$$

$$= \sum_{u \in P} \frac{1}{(w_2 v - w_1)^2} \left\| \left(\mathbf{I}(u)w_2 - (w_2\tau_1(u) - w_1\tau_2(u))\mathbf{z}_1 - \tau_2(u)\mathbf{L}^{t-1}\right)v \right.$$

$$\left. - \left(\mathbf{I}(u)w_1 - (w_2\tau_1(u) - w_1\tau_2(u))\mathbf{z}_2 - \tau_1(u)\mathbf{L}^{t-1}\right) \right\|^2$$

$$+ \frac{\alpha}{(w_2 v - w_1)^2} \sum_{u \in P} \left[\left(\frac{\partial m(u)}{\partial x(u)}\right)^2 + \left(\frac{\partial m(u)}{\partial y(u)}\right)^2 \right]$$

$$= \sum_{u \in P} \left\| \frac{\mathbf{p}(u)v - \mathbf{q}(u)}{w_2 v - w_1} \right\|^2 + \frac{\alpha N}{(w_2 v - w_1)^2}$$

$$= \sum_{u \in P} \left\| \frac{\mathbf{p}(u)}{w_2} + \frac{\frac{w_1}{w_2}\mathbf{p}(u) - \mathbf{q}(u)}{w_2 v - w_1} \right\|^2 + \frac{\alpha N}{(w_2 v - w_1)^2}$$

$$= \sum_{u \in P} \frac{\|\mathbf{p}(u)\|^2}{w_2^2} + \frac{2}{w_2 v - w_1} \sum_{u \in P} \left\langle \frac{\mathbf{p}(u)}{w_2}, \frac{w_1}{w_2}\mathbf{p}(u) - \mathbf{q}(u) \right\rangle$$

$$+ \frac{1}{(w_2 v - w_1)^2} \left(\sum_{u \in P} \left\| \frac{w_1}{w_2}\mathbf{p}(u) - \mathbf{q}(u) \right\|^2 + \alpha N \right), \quad (6.21)$$

where $\langle .,. \rangle$ denotes the inner product of two vectors, and

$$m(u) = w_2\tau_1(u) - w_1\tau_2(u),$$

$$\mathbf{p}(u) = \mathbf{I}(u)w_2 - (w_2\tau_1(u) - w_1\tau_2(u))\mathbf{z}_1 - \tau_2(u)\mathbf{L}^{t-1},$$

$$\mathbf{q}(u) = \mathbf{I}(u)w_1 - (w_2\tau_1(u) - w_1\tau_2(u))\mathbf{z}_2 - \tau_1(u)\mathbf{L}^{t-1},$$

$$N = \sum_{u \in P} \left[\left(\frac{\partial m(u)}{\partial x(u)}\right)^2 + \left(\frac{\partial m(u)}{\partial y(u)}\right)^2 \right].$$

Note that $\mathbf{p}(u), \mathbf{q}(u), w_1$ and w_2 are known given the vector $\mathbf{L}^{t-1}$. With the change of variable $r = \frac{1}{w_2 v - w_1}$ we can write the right-hand side of Eq. (6.21) as a quadratic function of r whose minimum is attained at

$$r^* = -\frac{\sum_{u \in P} \langle \frac{\mathbf{p}(u)}{w_2}, \frac{w_1}{w_2}\mathbf{p}(u) - \mathbf{q}(u)\rangle}{\sum_{u \in P} \|\frac{w_1}{w_2}\mathbf{p}(u) - \mathbf{q}(u)\|^2 + \alpha N}. \qquad (6.22)$$

This gives the corresponding minimiser $v^* = \frac{1}{w_2}(\frac{1}{r^*} + w_1)$. Hence, given the illuminant spectrum $\mathbf{L}^{t-1}$, one can recover $\mathbf{g}_P, \mathbf{k}_P$ by substituting the optimal value of v into Eqs. (6.19) and (6.20). The diffuse radiance component is computed as $\mathbf{D}_P^t = v^*\mathbf{z}_1 + \mathbf{z}_2$, and the spectral reflectance at wavelength λ is given by $\mathbf{S}_P^t(\lambda) = \frac{\mathbf{D}_P^t(\lambda)}{\mathbf{L}^{t-1}(\lambda)}$.

6.2.4 Illuminant Power Spectrum Recovery

The illuminant spectrum can be solved in closed form as shown in Sect. 5.3. For completeness, here we index the corresponding equation to iteration t so as to obtain

$$L^*(\lambda_i) = \frac{\sum_{P \in \mathcal{P}} \sum_{u \in P} k^t(u) I(u, \lambda_i) - \sum_{P \in \mathcal{P}}[\frac{(\sum_{u \in P} g^t(u) k^t(u))(\sum_{u \in P} g^t(u) I(u, \lambda_i))}{\sum_{u \in P}(g^t(u))^2}]}{\sum_{P \in \mathcal{P}} \sum_{u \in P} (k^t(u))^2 - \sum_{P \in \mathcal{P}}[\frac{(\sum_{u \in P} g^t(u) k^t(u))^2}{\sum_{u \in P}(g^t(u))^2}]}. \qquad (6.23)$$

6.2.5 Shading, Reflectance and Specularity Recovery

In the optimisation scheme above, we recover the reflectance, shading and specularity factors for pixels in each patch $P \in \mathcal{P}$ used for the recovery of the illuminant spectrum. This implies that, although we have only computed the variables $\mathbf{g}(u)$, $\mathbf{k}(u)$ and $\mathbf{S}(u, \cdot)$ for pixel sites $u \in \mathcal{P}$, we have been able to recover the illuminant spectrum $\mathbf{L}$ as a global photometric variable in the scene. With the illuminant in hand, we can recover the remaining dichromatic variables, making use of $\mathbf{L}$ in a straightforward manner.

For this purpose, we assume the input scene is composed of smooth surfaces with slowly varying reflectance. In other words, the neighbourhood of each pixel can be regarded as a locally smooth patch made of the same material, where all the pixels in the neighbourhood share the same spectral reflectance. Given the illuminant spectrum, we can obtain the shading, specularity and reflectance of the neighbourhood at the pixel of interest by applying the procedure corresponding to Line 4 in Algorithm 6.1. This corresponds to the application of the first of the two steps of the optimisation method presented in the previous section.

The pseudo-code of this algorithm is summarised in Algorithm 6.2. Note that the assumption of smooth surfaces with slowly varying reflectance is applicable to a

6.2 An Optimisation Approach to Illuminant and Photometric Invariant

Algorithm 6.2 Estimate the shading, specularity and reflectance of an image knowing the illuminant spectrum

Require: Image $\mathcal{I}$ with radiance $I(u, \lambda)$ for each band $\lambda \in \{\lambda_1, \ldots \lambda_n\}$ and the illuminant spectrum $\mathbf{L}$

Ensure: $\mathbf{g}(u), \mathbf{k}(u), \mathbf{S}(u, \lambda)$ where

    $\mathbf{g}(u), \mathbf{k}(u)$: the shading and specularity at pixel location u.

    $\mathbf{S}(u, \lambda)$: the diffuse reflectance of at pixel u and wavelength λ.

1: **for all** $u \in \mathcal{I}$ **do**
2:     $\mathcal{N} \leftarrow$ Neighbourhood of u
3:     $[\mathbf{g}_\mathcal{N}, \mathbf{k}_\mathcal{N}, \mathbf{S}_\mathcal{N}] \leftarrow \text{argmin}_{\mathbf{g}_\mathcal{N}, \mathbf{k}_\mathcal{N}, \mathbf{S}_\mathcal{N}} F(P)|_\mathbf{L}$
4:     $\mathbf{S}(u) \leftarrow \mathbf{S}_\mathcal{N}$
5: **end for**
6: **return** $\mathbf{g}(u), \mathbf{k}(u), \mathbf{S}(u, \cdot)$

large category of scenes where surfaces have a low degree of texture, edges and occlusion. Following this assumption, the reflectance at each pixel is recovered as the shared reflectance of its surrounding patch. To estimate the shading and specularity, one can apply the closed-form formulas in Eqs. (6.19) and (6.20). These formulas yield exact solutions in the ideal condition, which requires that all the pixel radiance vectors lie in the same dichromatic plane spanned by the illuminant spectrum and the diffuse radiance vector.

However, in practice, it is common for multispectral images to contain noise, which breaks down this assumption and renders the above quotient expressions numerically unstable. Therefore, to enforce a smooth variation of the shading factor across pixels, we recompute the shading and specularity coefficients after obtaining the spectral reflectance. This is due to the observation that the reflectance spectrum is often more stable than the other two variables, i.e. shading and specularity factors. Specifically, one can compute the shading and specular coefficients as those resulting from the projection of pixel radiance onto the subspace spanned by the illuminant spectrum and the diffuse radiance spectrum vectors.

Similar to other photometric methods based on the dichromatic model, this framework breaks down when the dichromatic plane assumption is violated, i.e. the illuminant spectrum is collinear to the diffuse radiance spectrum of the material. This renders the subspace spanned by the radiance spectra of the patch pixels to collapse to a one-dimensional space. As a consequence, an SVD of these radiance spectra does not succeed in finding two basis vectors of the subspace. Since the diffuse component is a product of the illuminant power spectrum and the material reflectance, this failure case only happens when the input scene contains a single material with a uniform reflectance, i.e. one that resembles a shade of grey.

In fact, when the scene contains more than one material, as more uniform-albedo patches are sampled from the scene, there are more opportunities to introduce the non-collinearity between the illuminant spectrum and the surface diffuse radiance spectrum. In short, the method presented in this section guarantees the recovery of dichromatic model parameters on scenes with more than one distinct albedo.

6.2.6 Imposing Smoothness Constraints

In Sect. 6.2.1, we addressed the need of enforcing the smoothness constraint on the shading field $\mathbf{g} = \{\mathbf{g}(u)\}_{u \in \mathcal{I}}$ using the regularisation term $\mathcal{R}(u)$ in Eq. (6.14). In Eq. (6.14), the regulariser encourages the slow spatial variation of the shading field. There are two reasons for using this regulariser in the optimisation framework introduced in the previous sections. Firstly, it yields a closed-form solution for the surface shading and reflectance given the illuminant spectrum. Secondly, it is reminiscent of smoothness constraints imposed upon shape from shading approaches and, hence, it provides a link between other methods in the literature, such as that in Worthington and Hancock (1999a) and the optimisation method in the previous sections. However, we emphasise that the optimisation procedure above by no means implies that the framework is not applicable to alternative regularisers. In fact, the target function is flexible in the sense that other regularisation functions can be formulated dependent on the surface at hand.

In this section, we introduce a number of alternative regularisers on the shading field that are robust to noise and outliers and adaptive to the surface shading variation. To this end, we commence by introducing robust regularisers. We then present extensions based on the surface curvature and the shape index.

To quantify the smoothness of shading, an option is to treat the gradient of the shading field as the smoothness error. In Eq. (6.14), we have introduced a quadratic error function of the smoothness. However, in certain circumstances, enforcing the quadratic regulariser as introduced in Eq. (6.14) causes the undesired effect of over-smoothing the surface. This well-known phenomenon has been experienced in a number of approaches (Brooks and Horn 1985; Ikeuchi and Horn 1981) in the field of shape from shading. We note in passing that ample work exists in the literature addressing the over-smoothing tendency of quadratic regularisers used for enforcing smoothness constraints on gradients (Ferrie and Lagarde 1992; Worthington and Hancock 1999a; Zheng and Chellappa 1991).

As an alternative, we utilise kernel functions stemming from the field of robust statistics. Formally speaking, a robust kernel function $\rho_\sigma(\eta)$ quantifies an energy associated with both the residual η and its influence function, i.e. it measures sensitivity to changes in the shading field. Each residual is, in turn, assigned a weight as defined by an influence function $\Gamma_\sigma(\eta)$. Thus the energy is related to the first moment of the influence function as $\frac{\partial \rho_\sigma(\eta)}{\partial \eta} = \eta \Gamma_\sigma(\eta)$. Table 6.1 shows the formulas for Tukey's biweight (Hoaglin et al. 2000), Li's adaptive potential functions (Li 1995) and Huber's M-estimators (Huber 1981).

6.2 An Optimisation Approach to Illuminant and Photometric Invariant

Table 6.1 Robust kernels and influence functions

Estimator	Robust kernel $\rho_\sigma(\eta)$	Influence function $\Gamma_\sigma(\eta)$
Tukey	$\rho_\sigma(\eta) = \begin{cases} \sigma(1-(1-(\frac{\eta}{\sigma})^2)^3) & \text{if } \|\eta\| < \sigma, \\ \sigma & \text{otherwise} \end{cases}$	$\Gamma_\sigma(\eta) = \begin{cases} (1-(\frac{\eta}{\sigma})^2)^2 & \text{if } \|\eta\| < \sigma, \\ 0 & \text{otherwise} \end{cases}$
Li	$\rho_\sigma(\eta) = \sigma(1-\exp(-\frac{\eta^2}{\sigma}))$	$\Gamma_\sigma(\eta) = \exp(-\frac{\eta^2}{\sigma})$
Huber	$\rho_\sigma(\eta) = \begin{cases} \eta^2 & \text{if } \|\eta\| < \sigma, \\ 2\sigma\|\eta\| - \sigma^2 & \text{otherwise} \end{cases}$	$\Gamma_\sigma(x) = \begin{cases} 1 & \text{if } \|\eta\| < \sigma, \\ \frac{\sigma}{\|\eta\|} & \text{otherwise} \end{cases}$

Robust Shading Smoothness Constraint

Having introduced the above robust estimators, we proceed to employ them as regularisers for the target function. Here, several possibilities exist. One of them is to directly minimise the shading variation by defining robust regularisers with respect to the shading gradient. In this case, the regulariser $\mathcal{R}(u)$ is given by the following formula:

$$\mathcal{R}(u) = \rho_\sigma\left(\left|\frac{\partial g}{\partial x}\right|\right) + \rho_\sigma\left(\left|\frac{\partial g}{\partial y}\right|\right). \tag{6.24}$$

Although being effective, the formula above still employs the gradient of the shading field as a measure of smoothness. In the next section, we explore the use of curvature as a measure of consistency.

Curvature Consistency

Alternatively, one can instead consider the intrinsic characteristics of the surface at hand as given by its curvature. Specifically, Ferrie and Lagarde (1992) have used the global consistency of principal curvatures to refine surface estimates in shape from shading. Moreover, ensuring the consistency of curvature directions does not necessarily imply a large penalty for discontinuities of orientation and depth. Therefore, this measure can avoid over-smoothing, which is a drawback of the quadratic smoothness error.

The curvature consistency can be defined on the shading field by treating it as a manifold. To commence, we define the structure of the shading field using its Hessian matrix,

$$H = \begin{pmatrix} \frac{\partial^2 g}{\partial x^2} & \frac{\partial^2 g}{\partial x \partial y} \\ \frac{\partial^2 g}{\partial y \partial x} & \frac{\partial^2 g}{\partial y^2} \end{pmatrix}.$$

The principal curvatures of the manifold are hence defined as the eigenvalues of the Hessian matrix. Let these eigenvalues be denoted by κ_1 and κ_2, where $\kappa_1 \geq \kappa_2$.

Moreover, we can use the principal curvatures to describe local topology using the shape index (Koenderink and van Doorn 1992) defined as follows:

$$\phi = \frac{2}{\pi} \arctan\left(\frac{\kappa_1 + \kappa_2}{\kappa_1 - \kappa_2}\right).$$

The observation above is important because it permits casting the smoothing process of the shading field as a weighted mean process, where the weight assigned to a pixel is determined by the similarity in local topology, i.e. the shape index, about a local neighbourhood. Effectively, the idea is to favour pixels in the neighbourhood that belong to the same or similar shape class as the pixel of interest. This is an improvement over the quadratic smoothness term defined in Eq. (6.14) because it avoids the indiscriminate averaging of shading factors across discontinuities. Therefore, it is by definition edge preserving.

For each pixel u, we consider a local neighbourhood $\mathcal{N}$ around u and assign a weight to each pixel u^* in the neighbourhood as $w(u^*) = \exp(-\frac{(\phi(u^*)-\mu_\phi(\mathcal{N}))^2}{2\sigma_\phi^2(\mathcal{N})})$, where $\mu_\phi(\mathcal{N})$ and $\sigma_\phi(\mathcal{N})$ are the mean and standard deviation of shape index over the neighbourhood $\mathcal{N}$. Using this weighting process, we obtain an adaptive weighted mean regulariser given by

$$\mathcal{R}(u) = \left(g(u) - \frac{\sum_{u^* \in \mathcal{N}} w(u^*) g(u^*)}{\sum_{u^* \in \mathcal{N}} w(u^*)}\right)^2. \tag{6.25}$$

This approach can be viewed as an extension of the robust regulariser function with a fixed kernel, presented in Eq. (6.24). To regulate the level of smoothing applied to a neighbourhood, we consider the shape index statistics (Koenderink and van Doorn 1992) so as to adaptively change the width of the robust kernel. The rationale behind adaptive kernel widths is that a neighbourhood with a great variation of shape index requires stronger smoothing than one with a smoother variation. The regulariser function is exactly the same as Eq. (6.25), except for the kernel width, which is defined pixel-wise as

$$\sigma_\phi(u) = \exp\left(-\left(\frac{1}{K_\phi |\mathcal{N}|} \sum_{u^* \in \mathcal{N}} (\phi(u^*) - \phi(u))^2\right)^{1/2}\right), \tag{6.26}$$

where $\mathcal{N}$ is a neighbourhood around the pixel u, $|\mathcal{N}|$ is the cardinality of $\mathcal{N}$ and K_ϕ is a normalisation term.

With the above formulation of the kernel width, one can observe that a significant variation of the shape index within the neighbourhood corresponds to a small kernel width, causing the robust regulariser to produce heavy smoothing. In contrast, when the shape index variation is small, a lower level of smoothing occurs due to a wider kernel width.

Note that the use of the robust regularisers introduced earlier in this section as an alternative to the quadratic regulariser does not preclude the applicability of the optimisation framework described previously. In fact, the change of regulariser only affects the formulation of the target function in Eq. (6.21), in which the shading factor $g(u)$ can be expressed as a univariate function as given in Eq. (6.19). Since all the above robust regularisers are only dependent on the shading factor, the resulting target function is still a function of the variable $r \triangleq \frac{1}{w_2 v - w_1}$. Further, by linearisation of the robust regularisers, one can still numerically express the regulariser as a quadratic function of the variable r. Subsequently, the closed-form solution presented earlier stands as originally described.

6.3 Band Subtraction

Note that if the illuminant is known a priori, which is often the case in remote sensing applications or those settings where the environment is a controlled one, we can use the expression

$$R(u, \lambda_i) = \frac{I(u, \lambda_i)}{L(\lambda_i)} = g(u)S(u, \lambda_i) + k(u) \qquad (6.27)$$

to recover the shading, specularity and reflectance by noting that the second term in the equation has been reduced to the specular factor $k(u)$, which does not depend on the wavelength λ_i.

This observation is important since it allows us to eliminate the specular component by taking the difference of reflectance data between two arbitrary bands λ_i and λ_j, i.e.

$$R^*(u, \lambda_i) = R(u, \lambda_i) - R(u, \lambda_j) = g(u)\big(S(u, \lambda_i) - S(u, \lambda_j)\big). \qquad (6.28)$$

Theoretically, an arbitrary reference band λ_j can be used for the difference above. Nonetheless, $R(u, \lambda_i)$ and $R(u, \lambda_j)$ may deviate from the ideal case as modelled by Eq. (6.27) due to measurement noise. Let $\hat{R}(u, \lambda_i)$ be the true spectral value at wavelength λ_i; then the spectral measurement $R(u, \lambda_i)$ is a random variable satisfying the relation $R(u, \lambda_i) = \hat{R}(u, \lambda_i) + e_{\lambda_i}$, where e_{λ_i} is the measurement error at wavelength λ_i. Here we assume that the measurement errors e_{λ_i} are independent and identically distributed random variables governed by Gaussian distributions with zero mean and variance σ^2. Denote $\hat{R}^*(u, \lambda_i) = \hat{R}(u, \lambda_i) - \hat{R}(u, \lambda_j)$ as the true difference value defined in Eq. (6.28). By making use of the expectation operator $\mathbb{E}[\cdot]$, we can obtain the following results regarding $R^*(u, \lambda_i)$ and $\hat{R}^*(u, \lambda_i)$:

$$\begin{aligned} \mathbb{E}\big[R^*(u, \lambda_i)\big] &= \mathbb{E}\big[\hat{R}^*(u, \lambda_i)\big], \\ \mathrm{var}\big(R^*(u, \lambda_i) - \hat{R}^*(u, \lambda_i)\big) &= 2\sigma^2. \end{aligned} \qquad (6.29)$$

Hence, the measurement is unbiased with twice as much variance as that in the measurement of individual bands. Thus, with a slight modification to Eq. (6.28), we can still eliminate the specular component in Eq. (6.27) while greatly reducing the variance. This is achieved by using the average of $R(u, \lambda_i)$ values over the spectrum instead of an arbitrary reference band:

$$R^*(u, \lambda_i) = R(u, \lambda_i) - \frac{1}{N} \sum_{j=1}^{N} R(u, \lambda_j) \quad \forall i, \qquad (6.30)$$

where N is the total number of bands in the spectrum.

The resulting $R^*(u, \lambda_i)$ from the above equation is still unbiased, where the variance for the difference of the measurement $R^*(u, \lambda_i)$ and its true value $\hat{R}^*(u, \lambda_i)$ is given by

$$\operatorname{var}\left(R^*(u, \lambda_i) - \hat{R}^*(u, \lambda_i)\right)$$

$$= \operatorname{var}\left((R(u, \lambda_i) - \hat{R}(u, \lambda_i)) - \frac{1}{N} \sum_{j=1}^{N} (R(u, \lambda_j) - \hat{R}(u, \lambda_j)) \right)$$

$$= \operatorname{var}\left(e_{\lambda_i} - \frac{1}{N} \sum_{j=1}^{N} e_{\lambda_j} \right) = \operatorname{var}\left(\frac{N-1}{N} e_{\lambda_i} - \frac{1}{N} \sum_{j \neq i} e_{\lambda_j} \right)$$

$$= \frac{(N-1)^2}{N^2} \sigma^2 + \frac{N-1}{N^2} \sigma^2 = \frac{N-1}{N} \sigma^2. \qquad (6.31)$$

Hence, we can see that, by using the average value, the variances in the specular-free spectral values are reduced by half, and are even slightly smaller than the variance of the measurement error in the original spectral values $R(u, \lambda_i)$, which equals σ^2.

Next, we aim at removing the shading factor in the specular-free representation $R^*(u, \lambda_i)$ in Eq. (6.30). This is achieved by taking the band ratio values as follows:

$$\hat{F}(u, \lambda_i) = \frac{R^*(u, \lambda_i)}{\frac{1}{N} \sum_{j=1}^{N} |R^*(u, \lambda_j)|} + 1. \qquad (6.32)$$

Here, we have used the average value instead of the value of $R^*(u, \lambda_i)$ at an arbitrary band since it has a lower variance. A bias value of 1 is added simply to make the feature values positive and does not change the nature of the representation.

Consequently, the representation above has three advantages. Firstly, it does not depend on the shading factor $g(u)$. Secondly, it has removed the specular component in Eq. (6.27). Thirdly, it is quite resilient to measurement noise in individual bands. This is due to the subtraction of average band value in the computation of $R^*(u, \lambda_i)$ defined in Eq. (6.30). Moreover, we can use either form of $R^*(u, \lambda_i)$ as presented in Eqs. (6.28) and (6.30) for the computation of $\hat{F}(u, \lambda_i)$ in Eq. (6.32). Since the use of Eq. (6.28) leads to a higher variance than that obtained by using Eq. (6.30), we can expect Eq. (6.30) to produce a lower representation error in Eq. (6.32).

6.3 Band Subtraction

Nonetheless, the corresponding theoretical analysis is intractable due to the normalisation with respect to the sum of random variables; this intuition is confirmed in practice. To this end, we illustrate how the treatment above affects the mean-squared error (MSE) for noise-corrupted spectra making use of a sample spectrum. The original spectrum is shown in the left-hand panel of Fig. 6.4. For this spectrum, we have perturbed the spectral values by adding zero-mean Gaussian noise with increasing variance. Feature vectors were computed for the noise-free and perturbed spectra based on the two different formulations mentioned above. The preceding steps were repeated 20 times for each value of noise variance. The average and standard deviations of the MSEs across different noise levels are shown in the plots on the right-hand panel of the figure. In this panel, the higher error curve corresponds to the feature representation yielded by Eq. (6.32) when Eq. (6.28) is used. Note that, since Eq. (6.28) assumes the use of a single arbitrary band λ_j for reference, in our error computations, and for the sake of comparison, we have randomly selected λ_j for each of the 20 trials for each value of noise variance. The lower error curve corresponds to the feature representation computed by substituting Eq. (6.30) into Eq. (6.32). From the error plots, we can see that the average error values are not only smaller than those using a single, randomly selected band, but more importantly, have a much lower variance as compared with the alternative.

Also, note that the above representation does not explicitly take shadows into account. Shadows in indoor scenes are usually due to the influence of inter-reflection, whereas in outdoor scenes they often correspond to scattered light. In this case, the derivation presented earlier would not exactly remove the shadow component and, moreover, could bias for the removal of shading and specular components. There is no straightforward way to handle this problem. However, the derived representation in Eq. (6.32) is still more robust to the influence of shadow effects than the raw spectral values. The reasons for this are twofold. Firstly, inter-reflections are usually quite small compared to the light source energy $L(\lambda_i)$, making the dichromatic model a good approximation. Secondly, in the case where the pixel location u is foreshadowed by the light source so that the observed radiance is solely contributed by the inter-reflection component, the estimated $R(u, \lambda_i)$ would be much smaller than the true albedo. Taking the ratio in Eq. (6.32) would somehow balance this undesired effect.

The situation is similar for outdoor shadows, which are usually caused by exposure to skylight with a different colour temperature as opposed to sunlight (Marchant and Onyango 2000), where the shadowed areas are also darker. This leads to a lower estimate of $R(u, \lambda_i)$. Taking the ratio would again balance this undesired effect. Moreover, if different light sources can be approximated by black body radiators, then the logarithms of band ratios are shadow-invariant variables whose values are independent of the colour temperature, as shown in Marchant and Onyango (2000). Logarithms can be approximated by linear functions when the ratio values are close to one due to the relation $\log x \approx x - 1$ about $x = 1$.

Fig. 6.4 *Top panel*: sample reflectance spectrum of a leaf; *Bottom panel*: mean-squared error plots as a function of added noise

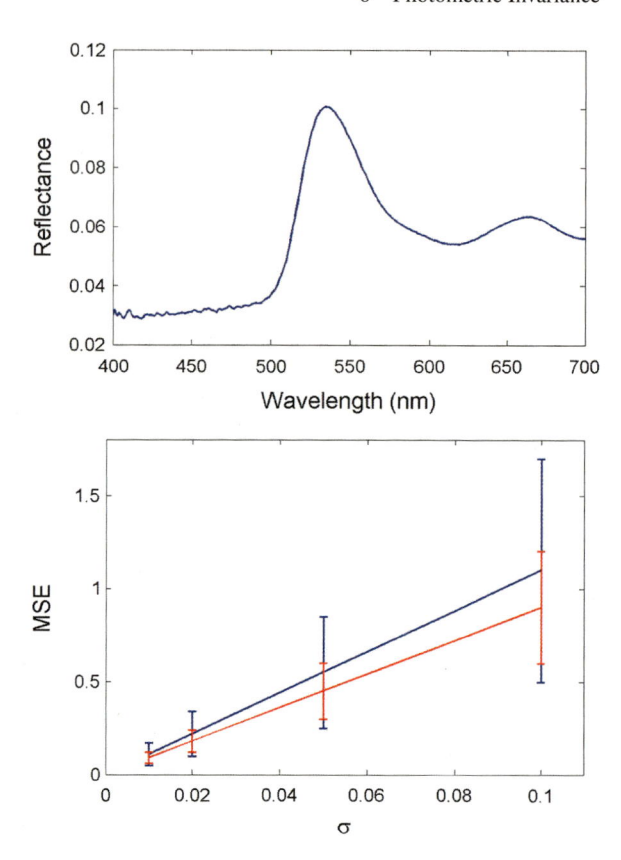

6.4 Notes

So far, we have based our discussion on hyperspectral imaging in the terrestrial setting. We have also assumed that the light source spectrum $L(\lambda_i)$ is known, so the reflectance spectra $R(u, \lambda_i)$ can be obtained by normalising the image radiance value $I(u, \lambda_i)$ with respect to the illuminant $L(\lambda_i)$. Note that the light source spectrum $L(\lambda_i)$ can be measured by making use of a calibration target such as a Spectralon white standard. Hence, we can make use of the estimated reflectance spectra for material identification. However, the case for remote sensing imaging is very different; here atmospheric conditions become the most important factor. There have been many references regarding atmospheric effects and physical assumptions in the literature (Tanre et al. 1979; Hapke 1984, 1993; Schmidt et al. 2007). We do not deal with such effects explicitly in this chapter as we are more interested in establishing an invariant representation to the photometric artifacts in the terrestrial setting.

For remotely sensed imaging data, one can always use existing atmospheric correction methods (Gao et al. 1993; Qu et al. 2003) to recover the scaled reflectance spectra. The invariant feature representation in Eq. (6.32) can be employed to minimise the scaling artefacts. In this case, band subtraction is no longer valid for specularity removal, but still preserves shading invariance. Nevertheless, specularity is much less of a problem for remote sensing than it is for terrestrial imaging. In summary, band subtraction is effective in handling photometric artefacts in the terrestrial imaging setting and does not influence the performance in the remote sensing setting, where the assumptions made with respect to terrestrial imaging models may no longer hold.

With the illuminant in hand, we can compute the surface reflectance using the following equation:

$$S(u, \lambda_i) = \frac{1}{g(u)} \left(\frac{I(u, \lambda_i)}{L(\lambda_i)} - k(u) \right).$$

Furthermore, if the illuminant is known, $R(u, \lambda_i)$ can always be bound between 0 and 1. We can give a physical interpretation to this assumption. The bounded nature of $R(u, \lambda_i)$ reflects the notion that the radiant power of light per unit of surface area, as sensed by the camera, can always be scaled so that its maximum value is unity. This, together with the fact that, from Eq. (6.1), the specular coefficient $k(u)$ is an additive constant to $R(u, \lambda_i)$, implies that we can recover the shading factor by a simple normalisation of the form

$$g(u) = \left(\sum_{i=1}^{N} \left| \frac{I(u, \lambda_i)}{L(\lambda_i)} - k(u) \right|^2 \right)^{1/2}. \tag{6.33}$$

Moreover, the specular coefficient $k(u)$ can be recovered by making use of the methods described earlier in the chapter. We note in passing that the expression above is somewhat reminiscent of that used in Fu and Robles-Kelly (2011). Nonetheless, the work presented in Fu and Robles-Kelly (2011) aims at recognition in a remote sensing setting where the shading factor is accounted for by normalisation.

References

Angelopoulou, E. (2000). Objective colour from multispectral imaging. In *European conference on computer vision* (pp. 359–374).

Boyd, S., & Vandenberghe, L. (2004). *Convex optimization*. Cambridge: Cambridge University Press.

Brelstaff, G., & Blake, A. (1988). Detecting specular reflection using Lambertian constraints. In *International conference on computer vision* (pp. 297–302).

Brooks, M. J., & Horn, B. K. P. (1985). Shape and source from shading. In *International joint conference on artificial intelligence* (pp. 932–936).

Dupuis, P., & Oliensis, J. (1992). Direct method for reconstructing shape from shading. In *Proceedings of the IEEE conference on computer vision and pattern recognition* (pp. 453–458).

Ferrie, F., & Lagarde, J. (1992). Curvature consistency improves local shading analysis. *CVGIP. Image Understanding*, 55(1), 95–105.

Fu, Z., & Robles-Kelly, A. (2011). Discriminant absorption feature learning for material classification. *IEEE Transactions on Geoscience and Remote Sensing*, 49(5), 1536–1556.

Gao, B.-C., Heidebrecht, K. B., & Goetz, A. F. H. (1993). Derivation of scaled surface reflectance from Aviris data. *Remote Sensing of Environment*, 44, 165–178.

Hapke, B. (1984). Bidirectional reflectance spectroscopy, III: correction for macroscopic roughness. *Icarus*, *59*(1), 41–59.
Hapke, B. (1993). *Theory of reflectance and emittance spectroscopy topics in remote sensing*. Cambridge: Cambridge University Press.
Healey, G. (1991). Estimating spectral reflectance using highlights. *Image and Vision Computing*, *9*(5), 333–337.
Healey, G., & Slater, D. (1999). Invariant recognition in hyperspectral images. In *IEEE conference on computer vision and pattern recognition* (p. 1438).
Hoaglin, D. C., Mosteller, F., & Tukey, J. W. (2000). *Understanding robust and exploratory data analysis*. New York: Wiley-Interscience.
Horn, B. K. P., & Brooks, M. J. (1986). The variational approach to shape from shading. *CVGIP*, *33*(2), 174–208.
Huber, P. (1981). *Robust statistics*. New York: Wiley.
Huynh, C. P., & Robles-Kelly, A. (2010). A solution of the dichromatic model for multispectral photometric invariance. *International Journal of Computer Vision*, *90*(1), 1–27.
Ikeuchi, K., & Horn, B. (1981). Numerical shape from shading and occluding boundaries. *Artificial Intelligence*, *17*(1–3), 141–184.
Kimmel, R., & Bruckstein, A. M. (1995). Tracking level sets by level sets: a method for solving the shape from shading problem. *Computer Vision and Image Understanding*, *62*(2), 47–48.
Klinker, G. J., Shafer, S. A., & Kanade, T. (1988). The measurement of highlights in color images. *International Journal of Computer Vision*, *2*(1), 7–32.
Klinker, G., Shafer, S., & Kanade, T. (1990). A physical approach to color image understanding. *International Journal of Computer Vision*, *4*(1), 7–38.
Koenderink, J. J., & van Doorn, A. J. (1992). Surface shape and curvature scales. *Image and Vision Computing*, *10*(8), 557–565.
Lee, H.-C. (1986). Method for computing the scene-illuminant from specular highlights. *Journal of the Optical Society of America A*, *10*(3), 1694–1699.
Li, S. Z. (1995). Discontinuous MRF prior and robust statistics: a comparative study. *Image and Vision Computing*, *13*(3), 227–233.
Lin, S., & Shum, H. (2001). Separation of diffuse and specular reflection in color images. In *International conference on computer vision and pattern recognition*.
Marchant, J. A., & Onyango, C. M. (2000). Shadow-invariant classification for scenes illuminated by daylight. *Journal of the Optical Society of America*, *17*(11), 1952–1961.
Nayar, S., & Bolle, R. (1996). Reflectance based object recognition. *International Journal of Computer Vision*, *17*(3), 219–240.
Novak, C., & Shafer, S. (1992). Anatomy of a color histogram. In *Proceedings of the IEEE conference on computer vision and pattern recognition* (pp. 599–605).
Prados, E., & Faugeras, O. (2003). Perspective shape from shading and viscosity solutions. In *IEEE international conference on computer vision* (Vol. II, pp. 826–831).
Qu, Z., Kindel, B. C., & Goetz, A. F. H. (2003). The high accuracy atmospheric correction for hyperspectral data (HATCH) model. *IEEE Transactions on Geoscience and Remote Sensing*, *41*(9), 1223–1231.
Ragheb, H., & Hancock, E. R. (2003). A probabilistic framework for specular shape-from-shading. *Pattern Recognition*, *36*(2), 407–427.
Sato, Y., & Ikeuchi, K. (1994). Temporal-color space analysis of reflection. *Journal of the Optical Society of America A*, *11*(11), 2990–3002.
Schmidt, F., Doute, S., & Schmitt, B. (2007). Wavanglet: an efficient supervised classifier for hyperspectral images. *IEEE Transactions on Geoscience and Remote Sensing*, *45*(5), 1374–1385.
Shafer, S. A. (1985). Using color to separate reflection components. *Color Research and Application*, *10*(4), 210–218.
Slater, D., & Healey, G. (1999). Material classification for 3D objects in aerial hyperspectral images. In *Proceedings of the IEEE computer vision and pattern recognition* (pp. 268–273).

References

Stokman, H. M. G., & Gevers, T. (1999). Detection and classification of hyper-spectral edges. In *British machine vision conference*.

Suen, P. H., & Healey, G. (2001). Invariant mixture recognition in hyperspectral images. In *International conference on computer vision* (pp. 262–267).

Tan, R. T., & Ikeuchi, K. (2005). Separating reflection components of textured surfaces using a single image. *IEEE Transactions on Pattern Analysis and Machine Intelligence*, *27*(2), 178–193.

Tanre, D., Herman, M., Deschamps, P. Y., & d. Leffe, A. (1979). Atmospheric modeling for space measurements of ground reflectances, including bidirectional properties. *Applied Optics*, *18*, 3587–3594.

Tominanga, S., & Wandell, B. A. (1989). Standard surface-reflectance model and illuminant estimation. *Journal of the Optical Society of America A*, *6*, 576–584.

Umeyama, S., & Godin, G. (2004). Separation of diffuse and specular components of surface reflection by use of polarization and statistical analysis of images. *IEEE Transactions on Pattern Analysis and Machine Intelligence*, *26*(5), 639–647.

Worthington, P. L., & Hancock, E. R. (1999a). Needle map recovery using robust regularizers. *Image and Vision Computing*, *17*, 545–557.

Worthington, P. L., & Hancock, E. R. (1999b). New constraints on data-closeness and needle map consistency for shape-from-shading. *IEEE Transactions on Pattern Analysis and Machine Intelligence*, *21*(12), 1250–1267.

Zheng, Q., & Chellappa, R. (1991). Estimation of illuminant direction, albedo, and shape from shading. *IEEE Transactions on Pattern Analysis and Machine Intelligence*, *13*(7), 680–702.

Chapter 7
Spectrum Representation

Due to the high-dimensional nature of spectral data, many classical algorithms in pattern recognition and machine learning have been naturally borrowed and adapted to perform feature extraction and classification (Landgrebe 2002). Dimensionality reduction techniques such as principle component analysis (PCA) (Jolliffe 2002), linear discriminant analysis (LDA) (Fukunaga 1990), projection pursuit (Jimenez and Landgrebe 1999) and its variants mitigate the curse of dimensionality by treating raw spectra as input vectors in a high-dimensional space, where the dimensionality is given by the number of bands.

However, these methods are not able to interpolate reflectance and radiance values over the full spectrum. This is further exacerbated by the fact that spectra acquired with different sensors are often sampled at disparate spectral resolutions and noise levels. Further, the features extracted from reflectance spectra can be mapped to a low-dimensional space using kernel-based classifiers such as support vector machines (SVMs). This, is often found to be more effective than dimensionality reduction techniques (Fu et al. 2006a; Shah et al. 2003), which is not surprising since dimensionality reduction may potentially result in a loss of discriminant information, whereas the mapping from an input feature space to a kernel space can be viewed as an implicit feature selection.

An alternative to raw reflectance spectra as a means towards classification and recognition is the use of a reflectance descriptor, robust to changes in illumination, noise, geometric and photometric effects. Nayar and Bolle (1996) have proposed a method of object recognition based on the reflectance ratio between object regions. Dror et al. (2001) described a vision system that learnt the relationship between surface reflectance and certain statistics computed from grey-scale images. Slater and Healey (1997) used a set of Gaussian filters to derive moment invariants for recognition. Jacobs et al. (1998) employed image ratios for comparing images under variable illumination. Lin and Lee (1997) used an eigenspace of chromaticity distribution to obtain illumination direction, illumination colour and specularity invariance for three-dimensional object recognition. Lenz et al. (2007) used perspective projections in the canonical space of colour signals to separate intensity from chromaticity so as to recover a three-dimensional colour descriptor. More recently,

Chang et al. (2009) developed a coding scheme for the spectral variation of hyperspectral signatures by means of spectral derivatives.

7.1 Compact Representations of Spectra

The main bulk of work on spectrum representation so far has concentrated on modelling spectra as a linear combination of a predetermined basis. Maloney (1986) validated the use of low-dimensional linear models for representing reflectance spectra through a number of evaluations on empirical measurements of surface reflectance. In Marimont and Wandell (1992), Marimont and Wandell computed a set of linear basis functions which yields the minimal approximating error for surface reflectance spectra and illuminant spectra. Angelopoulou et al. (2001) used spectrophotometric data to model skin colour using several sets of basis functions including Gaussian functions and their first derivatives, wavelets and PCA. More recently, Lansel et al. (2009) proposed a dictionary-based sparse representation of reflectance which can outperform linear models in reflectance spectrum recovery.

7.1.1 Spectrum as a Mixture

An alternative to the treatment above is that of using a combination of basis vectors computed using the mean and covariance for the spectra. In Angelopoulou et al. (2001), a linear combination of M Gaussian basis functions is fitted to each of the image spectra, where the spectrum is a function of the wavelength λ and the kth Gaussian component is associated with a mean μ_k, a standard deviation σ_k and a mixture coefficient α_k. With these ingredients, the spectra is hence represented using the mixture model of the form

$$\mathcal{R}_u = \sum_{k=1}^{K} \alpha_k p(\mathcal{X}|\theta_k), \qquad (7.1)$$

where $p(\mathcal{X}|\theta_k)$ is the probability of the spectral sample $\mathcal{X}$ given the mixture model parameters are $\theta_k = \{\mu_k, \alpha_k\}$. The prior probability function is the bivariate Gaussian distribution given by

$$p(u_i|\theta_k) = \frac{1}{2\pi|\Sigma_k|^{1/2}} e^{-\frac{1}{2}(u_i-\mu_k)^T \Sigma_k^{-1}(u_i-\mu_k)}. \qquad (7.2)$$

This is, in fact, a Gaussian fitting process where the fit is done over the wavelength domain per pixel.

The expression above hints that we can model the spectra in the image as a mixture model which describes the probability of the spectrum $\mathcal{R}_u = \{R(u, \lambda_l)\}$

7.1 Compact Representations of Spectra

for the pixel u in the image $\mathcal{I}$ over the set of N wavelength indexed bands λ_l, $l \in \{1, 2, \ldots, N\}$, where we represent the spectrum as follows:

$$\mathcal{R}_u = \sum_{\mathfrak{N} \in \Omega} \sum_{k=1}^{M} \alpha(u)_{k,\mathfrak{N}} (\mathbf{X}_k + \mu_\mathfrak{N}), \tag{7.3}$$

where the vector $\mathbf{X}_k$ represents the kth basis for the material $\mathfrak{N}$ in the set Ω of materials across the image and the weights $\alpha(u)_{k,\mathfrak{N}}$ correspond to the contribution of each material-basis pair to the spectrum $R(u, \cdot)$.

In Chap. 8, we discuss further the relationship between materials and scenes. For now, we focus on the spectra representation problem above, where $\mu_\mathfrak{N}$ accounts for the statistical mean of the spectra in the material $\mathfrak{N}$. As a result, the model contains M basis vectors per material and the mixing coefficients satisfy $0 \leq \alpha(u)_{k,\mathfrak{N}} \leq 1$ and $\sum_{k=1}^{K} \alpha(u)_{k,\mathfrak{N}} = 1$.

Computing the Basis and the Statistical Mean

This treatment is important, since it permits the use of a probabilistic formulation to capture the distribution of spectra over the image $\mathcal{I}$ in order to arrive at a description that allows high quality colourimetric reconstruction and material recognition. Hence, we express the probability of the image spectrum as

$$P(\mathcal{I}) = \prod_{\mathcal{R}_u \in \mathcal{I}} \sum_{\mathfrak{N} \in \Omega} \sum_{k=1}^{K} \alpha(u)_{k,\mathfrak{N}} P(\mathcal{R}_u | \mathbf{X}_k, \mu_\mathfrak{N}), \tag{7.4}$$

and commence by computing the statistical mean $\mu_\mathfrak{N}$. To this end, we assume the spectra for the material under consideration are normally distributed. Recall that, by employing a *maximum a posteriori* (MAP) approach, it is straightforward to show that, given a current estimate of the mean, i.e. $\mu_\mathfrak{N}^{\text{old}}$, the optimal update is given by

$$\mu_\mathfrak{N}^{\text{new}} = \frac{\sum_{u \in \mathcal{I}} \mathcal{R}_u P(\mathcal{R}_u | \mathbf{X}_k, \mu_\mathfrak{N}^{\text{old}})}{\sum_{u \in \mathcal{I}} P(\mathcal{R}_u | \mathbf{X}_k, \mu_\mathfrak{N}^{\text{old}})}. \tag{7.5}$$

Moreover, by assuming exclusive materials, i.e. a spectrum cannot belong to more than one material $\mathfrak{N}$, we can set

$$P(\mathcal{R}_u | \mathbf{X}_k, \mu_\mathfrak{N}^{\text{old}}) = \begin{cases} 1, & \text{if } \mathcal{R}_u \text{ is the closest neighbour of } \mu_\mathfrak{N}^{\text{old}}, \\ 0, & \text{otherwise,} \end{cases} \tag{7.6}$$

which yields the well-known k-means clustering algorithm.

Once the statistical mean is at hand, we recover the basis vectors $\mathbf{X}_k$ by noting that it is a straightforward task to compute the covariance matrix Σ for the spectra corresponding to the material $\mathfrak{N}$. The covariance matrix is given by

$$\boldsymbol{\Sigma}_{\mathfrak{N}} = \frac{\sum_{u \in \mathcal{I}} P(\mathcal{R}_u | \mathbf{X}_k, \mu_{\mathfrak{N}}^{\text{old}})(\mathcal{R}_u - \mu_{\mathfrak{N}}^{\text{new}})(\mathcal{R}_u - \mu_{\mathfrak{N}}^{\text{new}})^T}{\sum_{u \in \mathcal{I}} P(\mathcal{R}_u | \mathbf{X}_k, \mu_{\mathfrak{N}}^{\text{old}})}, \qquad (7.7)$$

which is consistent with the Gaussian MAP treatment above.

Moreover, we can employ the hard limits of the probabilities $P(\mathcal{R}_u | \mathbf{X}_k, \mu_{\mathfrak{N}}^{\text{old}})$ as given in Eq. (7.6) for the last iteration of the k-means algorithm to arrive at the expression

$$\boldsymbol{\Sigma}_{\mathfrak{N}} = \frac{1}{|\mathfrak{N}|} \sum_{u \in \mathfrak{N}} (\mathcal{R}_u - \mu_{\mathfrak{N}}^{\text{new}})(\mathcal{R}_u - \mu_{\mathfrak{N}}^{\text{new}})^T. \qquad (7.8)$$

We can compute the basis in such a way that the domain spanned by the vectors $\mathbf{X}_k$ is an isometric embedding of our spectra. We do this by applying a factorisation of the covariance matrix in accordance with the Young–Householder theorem (Young and Householder 1938). Let Λ be the diagonal matrix with the eigenvalues of $\boldsymbol{\Sigma}_{\mathfrak{N}}$ ordered by rank and $\Phi = [\phi_1 | \phi_2 | \ldots | \phi_{|V|}]$ be the matrix with the eigenvectors ordered accordingly as columns. As a result we can write $\boldsymbol{\Sigma}_{\mathfrak{N}} = \Phi \Lambda \Phi^T = \mathbf{J}\mathbf{J}^T$, where $\mathbf{J} = \sqrt{\Lambda}\Phi$. The matrix which has the basis vectors $\mathbf{X}_k$ of the spectra as columns is Φ. Hence, $\mathbf{J}\mathbf{J}^T$ is a Gram matrix, i.e. its elements are scalar products of the basis vectors.

The decomposition above also provides a means to understanding the reconstruction of the spectra using PCA. Note that, if PCA is applied, the first principal component can be interpreted as an approximation of the mean, where the spectra can be reconstructed using the least-square projections of the original spectra onto the subspace spanned by the principal components.

Recovering the Mixture Weights

Now we turn our attention to the recovery of the weights $\alpha(u)_{k,\mathfrak{N}}$. To this end, recall that, in the previous sections, we imposed the constraints $\sum_{k=1}^{K} \alpha(u)_{k,\mathfrak{N}} = 1$ and $0 \leq \alpha(u)_{k,\mathfrak{N}} \leq 1$. Moreover, since we have assumed exclusive materials $\mathfrak{N}$, we can write

$$\sum_{k=1}^{M} \alpha(u)_{k,\mathfrak{N}} \mathbf{X}_k = \mathcal{R}_u - \mu_{\mathfrak{N}}, \qquad (7.9)$$

where the equation above follows from Eq. (7.3).

Further, we can write Eq. (7.9) in a compact form as follows:

$$\mathbb{X}\mathbf{a} = \mathbf{y}, \qquad (7.10)$$

where $\mathbf{a}$ is the vector whose kth entry is given by $\alpha(u)_{k,\mathfrak{N}}$, and $\mathbf{y} = \mathcal{R}_u - \mu_{\mathfrak{N}}$ and is a matrix $\mathbb{X} = [\mathbf{X}_1 | \mathbf{X}_2 | \ldots | \mathbf{X}_M]$ whose columns correspond to the basis vectors $\mathbf{X}_k$ for the material $\mathfrak{N}$.

7.1.2 A B-Spline Spectrum Representation

In Huynh and Robles-Kelly (2008), a spectrum is represented as a spline making use of a knot removal scheme with the maximal goodness of fit and minimal representation length, i.e. dimensionality. The representation in Huynh and Robles-Kelly (2008) is designed to cope with very high dimensional data containing millions of pixels and hundreds of spectral bands. Further, long feature vectors have been known to degrade recognition performance since they incur computational cost and learning-theoretic limitations. Thus, it is desirable that such a representation delivers the maximal discriminative power with the lowest possible dimensionality.

B-Splines

A B-spline function is one that has support with respect to degree, smoothness and domain partition. These properties make B-splines a flexible tool for fitting arbitrary shapes and data. The smoothness property makes the interpolating curve robust to noise. The local support property permits the modification of the curve over a local section while keeping the rest of the spline unaffected. The degree, on the other hand, permits the representation of complex shapes by altering the order of the basis functions.

Firstly, we note that reflectance spectra can be treated as functions of wavelength. Therefore we refer to the concepts of B-splines in the two-dimensional (2D) case. A p-degree B-spline curve $\mathcal{C}$ in $\mathbb{R}^2$ composed of n segments is a function in the parameter domain $\mathcal{U}$ of the univariate variable t given by the linear combination $\mathcal{C}(t) = \sum_{i=0}^{n} N_{i,p}(t) P_i$, where $P_i = (x_i, y_i)$ are the 2D control points, and $N_{i,p}(t)$ are the p-degree B-spline basis functions defined on the parameter domain (Piegl and Tiller 1995). The coordinates (x, y) of a point on the curve are expressed in the parametric form

$$x(t) = \sum_{i=0}^{n} N_{i,p}(t) x_i, \qquad y(t) = \sum_{i=0}^{n} N_{i,p}(t) y_i. \qquad (7.11)$$

In Fig. 7.1(a), we illustrate the interpolation of a B-spline curve with degree $p = 3$ to six 2D points $Q_i, i = 0, \ldots, 5$. The control points, denoted $Q_i, i = 0, \ldots, 5$, do not lie exactly on, but are distributed along the curve to govern its overall shape. Here, we note that the control points at both ends coincide with the first and last data points.

A B-spline is characterised not only by the control points but also by a knot vector $U = \{u_0, \ldots, u_m\}$, where $m = n + p + 1$. With these ingredients, we can define the ith B-spline basis function $N_{i,p}(t)$ of degree p as follows:

$$N_{i,0}(t) = \begin{cases} 1, & \text{if } u_i \leq t < u_{i+1}, \\ 0, & \text{otherwise.} \end{cases}$$

$$N_{i,p}(t) = \frac{t - u_i}{u_{i+p} - u_i} N_{i,p-1}(t) + \frac{u_{i+p+1} - t}{u_{i+p+1} - u_{i+1}} N_{i+1,p-1}(t). \qquad (7.12)$$

Fig. 7.1 *Top panel*: a third-degree B-spline curve (in *blue*) is interpolated to the discrete data points $Q_i, i = 0, \ldots, 5$. The control points of the curve are depicted as $P_i, i = 0, \ldots, 5$. Note that the control points at both ends coincide with the first and last data points. *Bottom panel*: the basis functions $N_{i,3}, i = 0, \ldots, 5$, of the B-spline curve in the *top panel* are plotted with respect to the independent parameter t. The *horizontal line* below the graph shows the knot positions, which determine the local support of the basis functions, i.e. where the basis functions assume non-zero values

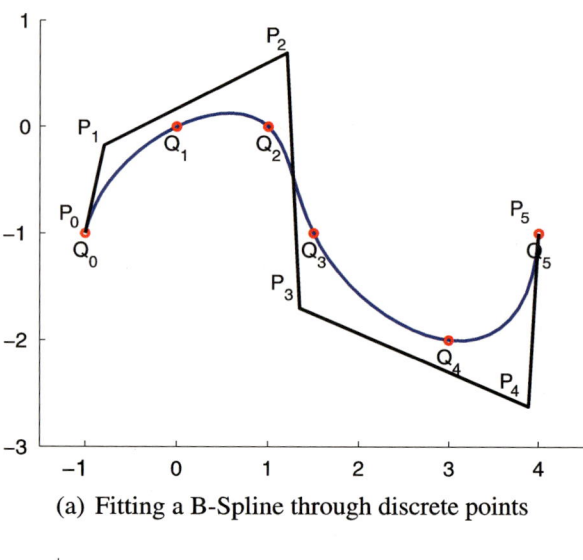

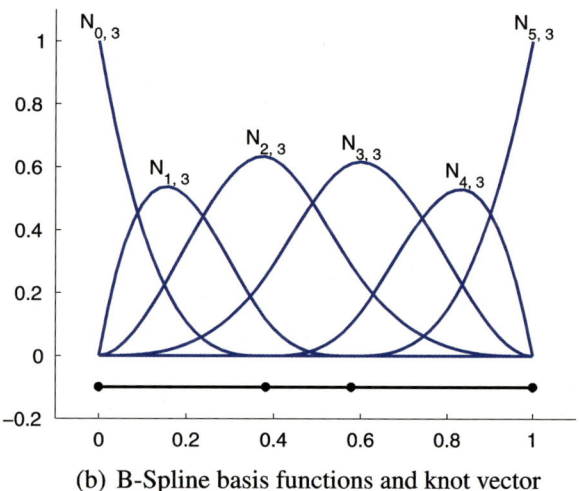

(a) Fitting a B-Spline through discrete points

(b) B-Spline basis functions and knot vector

In Fig. 7.1(b), we show the graph of the B-spline basis functions $N_{i,p}(t), i = 0, \ldots, n$, of the curve in Fig. 7.1(a) with respect to the independent parameter t. We note that each basis function $N_{i,p}(t)$ is a piecewise polynomial assuming non-zero values only in the interval $[u_i, u_{i+p+1})$. Therefore, it only affects the shape of the spline in this local support. The line below the graph shows the distribution of the knot vector in the parameter domain. Note that the local support of each basis function is bounded between a pair of knot locations.

To formulate the B-spline representation, we treat a spectrum as a collection of spectral samples with two coordinates (λ_k, R_k), where R_k is the kth reflectance sample at the wavelength λ_k. Thus, the parametric form of a B-spline curve through the spectra data points is obtained by representing the wavelength and reflectance as two

7.1 Compact Representations of Spectra

functions of an independent parameter t, which we denote $\lambda(t)$ and $\mathcal{R}(t)$, respectively. The descriptor is then governed by a knot vector and a control point set which minimise a cost function defined by the squared differences between the measured reflectance R_k and the reflectance $\mathcal{R}(t)$ reconstructed according to the parametric form of the B-spline. We depart from an initial interpolation of the input reflectance spectrum so as to arrive at a B-spline curve that minimises the cost function through a knot removal algorithm.

Once the control points and knot vector that minimise the cost function are at hand, we proceed to construct the descriptor. We do this by selecting the features from the control point set and the knot vector that are most characteristic of a reflectance spectrum. We observe that each of the knots and the x-coordinates of the control points, i.e. the x_i variables in Eq. (7.11), vary slightly within a local support in the parameter and wavelength domains, respectively, over different reflectance spectra. As a result, these variations may not be as characteristic of the shape of the spectrum as the y-coordinates of the control points, i.e. the y_i variables in Eq. (7.11). Thus, one expects the y-coordinates to provide better discrimination between spectra than the knots and x-coordinates.

Representation of a Single Spectrum

We now formulate an optimality criterion for a B-spline interpolation to the input reflectance spectra. The objective function follows the intuition that a descriptor or a representation of the spectra under study should retain statistical information with a high discriminative power. Hence, given a material with reflectance R_k at each wavelength $\{\lambda_k\}, k = 1, \ldots, l$, we aim to recover an interpolating B-spline curve with control points $P_i = (x_i, y_i)$ and a knot vector U such that

$$\lambda(t_k) = \sum_{i=0}^{n} N_{i,p}(t_k) x_i, \qquad \mathcal{R}(t_k) = \sum_{i=0}^{n} N_{i,p}(t_k) y_i, \qquad (7.13)$$

where the parameter $t_k \in U$ corresponds to the kth wavelength, i.e. $\lambda_k = \lambda(t_k) \forall k$.

In problems such as spectrum reconstruction, interpolation, classification and recognition, the representation length is known beforehand to facilitate comparison and normalisation of data collected from diverse sources. Thus, the optimisation problem herein imposes a predetermined length m on the knot vector U, i.e. $|U| = m$ where $|.|$ denotes the length of the vector argument. In addition, we keep the B-spline degree p fixed. Then the cost of interpolating the points $(\lambda_k, R_k), k = 1, \ldots, l$, using the B-spline curve above is given by

$$K = \sum_{k=1}^{l} \left(\mathcal{R}(t_k) - R_k \right)^2. \qquad (7.14)$$

The right-hand side of Eq. (7.14) is the sum of squared errors in the 2D space of wavelength-reflectance pairs. Thus, the optimal interpolating curve is one that minimises the sum of squared distances $(\mathcal{R}(t_k) - R_k)^2$ subject to a constraint on the given number of knots.

Since the resulting B-spline curve aims to describe the general shape of the input reflectance spectra, the optimisation departs from a global B-spline interpolation that goes through all the sample points of each reflectance spectrum under study. We initialise the B-splines according to the curve interpolation algorithm in Piegl and Tiller (1995) and employ the centripetal method of Lee (1989) to establish an initial knot vector from the input spectral reflectance samples. Subsequently, we apply knot removal so as to reduce the number of knots to a target number, while minimising the interpolation cost introduced in Eq. (7.14).

The knot removal algorithm is summarised in Algorithm 7.1. In Line 1, we perform the curve interpolation algorithm in Piegl and Tiller (1995) to obtain an initial B-spline curve with a zero interpolation error. Once this initial approximation is at hand, we remove knots sequentially using a knot removal method akin to that in Tiller (1992). The algorithm is a two-pass process. In the first pass, removable knots are identified. In the second pass, knots are sequentially removed and new control points are computed. Note that, although it is effective, the algorithm in Tiller (1992) requires the candidate knot for removal to be designated as input. On the other hand, our knot removal procedure evaluates the potential interpolation error contributed by removable knots. Based on these errors, it automatically determines the best knot to be removed in each pass. We then apply Tiller's knot removal algorithm (Tiller 1992) to the selected knots.

The selection of candidate knots for removal can be viewed as a greedy approach which is reminiscent of a gradient descent method. In practise, it is an iterative method in which, at every iteration, we locate the knot whose removal minimises the sum of squared errors. This process is illustrated in Lines 4–13 of Algorithm 7.1, within which the *RemoveKnot*(U, P, u) procedure is an implementation of the knot removal algorithm reported in Tiller (1992). This procedure computes the interpolation error of the removal of the knot u from a B-spline curve with a knot vector U and a control point set P. Subsequently, it will remove the knot u if the interpolation error falls below a tolerance threshold.

Thus, the algorithm selects among the candidate knots the one that yields the minimal interpolation error. Also, note that, following the strategy above, the parameter t_k should be recovered for every wavelength. This is not a straightforward task, since the function $\lambda(t_k)$ is expressed as a linear combination of the basis functions $N_{i,p}$ given in Eq. (7.13). Since this equation cannot be solved analytically, we adopt a numerical approach in order to find an approximate solution with a reduced computational cost. In fact, it is always reasonable to assume that the wavelength $\lambda(t_k)$ is an increasing function in the parameter domain. Therefore, for a given wavelength, we can perform a binary search for t_k such that $\lambda_k \sim \lambda(t_k)$.

As a result of the local support property of B-spline curves, the removal only affects the curve partition in the neighbouring sections of the removed knot. Thus, for efficiency, the upper bound of the interpolation error due to knot removal can be computed by considering only the local neighbourhood of the removal candidate u (Tiller 1992). To do this, we use the spans of the B-spline basis functions (Piegl

7.1 Compact Representations of Spectra

Algorithm 7.1 *KnotRemoval(Q, p, target)*

Require: $Q, p, \alpha, target$
    Q: A spectrum represented as wavelength-reflectance pairs $Q = \{(\lambda_k, R_k)\}, k = 1, \ldots, l$
    p: The degree of basis functions
    α: The balance factor
    target: The target number of knots
    U_0, P_0: the returned knots and control points

1: $(U_0, P_0) \leftarrow Interpolate(Q, p)$
2: **while** $|U_0| > target$ **do**
3:    $SSE_{min} \leftarrow \infty$ //The minimal sum of squared errors resulting from knot removal
4:    **for all** $u \in U_0$ **do**
5:      $(flag, U_1, P_1) \leftarrow RemoveKnot(U_0, P_0, u)$ //*flag* is **true** if knot u is removable
6:      **if** *flag* **then**
7:        $SSE \leftarrow SSE(U_1, P_1)$
8:        **if** $SSE < SSE_{min}$ **then**
9:          $SSE_{min} \leftarrow SSE$
10:          $candidate \leftarrow u$
11:        **end if**
12:      **end if**
13:    **end for**
       //Compute the SSE due to B-spline curve approximation
14:    $(U_2, P_2) \leftarrow Approximate(Q, p, |U_0| - 1)$
15:    $approxSSE \leftarrow error(Q, p, (U_2, P_2))$
       //Compare the lowest SSE with that obtained by B-spline approximation
16:    **if** $SSE_{min} < approxSSE$ **then**
17:      $(flag, U_1, P_1) \leftarrow RemoveKnot(U_0, P_0, candidate)$
18:      $(U_0, P_0) \leftarrow (U_1, P_1)$
19:    **else**
20:      $(U_0, P_0) \leftarrow (U_2, P_2)$
21:    **end if**
22: **end while**
23: **return** U_0, P_0

and Tiller 1995) and employ lists to backtrack their effect across the spline. The knot removal algorithm terminates when removing any knot cannot further reduce the interpolation cost.

It is also worth stressing that the target number of knots resulting from knot removal should be imposed as a hard constraint since the B-spline representation of each spectrum should be of the same length. Note that there may be cases in which no knots can be removed within the interpolation error tolerance. In order to guarantee a target representation length, we employ a global approximation of the original spectra to obtain B-spline curves with the desired number of knots (De Boor 1978). In Algorithm 7.1, we have incorporated an implementation of this algorithm in Line 14. We have followed (De Boor 1978) to preselect a knot vector that is one element shorter than that obtained in the previous iteration. This preselection ensures that the least-square B-spline approximation is well-conditioned. In addition, in each iteration of Algorithm 7.1, we compare the interpolation error produced by the knot removal with the global approximation error and select the method with the lower error (Lines 16–21). At each iteration, the knot removal step proceeds if it yields a lower error than the curve approximation. Otherwise, the method continues from the B-spline curve produced by the global approximation method.

Representation of Multiple Spectra

We turn our attention to a representation of spectral reflectance images or multiple reflectance spectra. Hence, we now extend the developments in Sect. 7.1.2 to the derivation of a B-spline representation for a given collection of reflectance spectra. A common paradigm in pattern recognition and data redundancy reduction is the employment of a common basis across the feature space in order to represent the original data in a compact form. Motivated by this kind of technique, we enforce a common knot vector across all the input spectra as a hard constraint. This implies common B-spline basis functions $N_{i,p}(t), i = 0, \ldots, n$, across the input spectra due to Eq. (7.12). Under this constraint, when a collection of spectral signatures are sampled at the same set of wavelengths, the B-spline representations share the same set of wavelength coordinates $x_i, i = 0, \ldots, n$, as defined in Eq. (7.13).

The purpose of enforcing a common knot vector and a common B-spline basis for representing the spectra is threefold. Firstly, the continuity, differentiability and local support properties of these basis functions enable further functional analysis of spectral signatures. Secondly, this provides a common B-spline basis to represent and recognise the reflectance spectra of different materials in the scene. Thirdly, it opens up the possibility of effecting tasks such as compression, whereby the spectrum at each pixel is represented by a number of control points and a common knot vector which act as a global parameter in each image or a collection of spectra. In this manner, a hyperspectral image, which may comprise tens or hundreds of bands, can be reduced to much fewer control points with a shared knot vector.

Let the input image be treated as a collection of reflectance spectra $\mathbf{R}_v = [R_{v,1}, R_{v,2}, \ldots, R_{v,l}]^T$, where v denotes the pixel index in the spectral image and $R_{v,k}$ is the measured reflectance of the spectrum v at the wavelength $\{\lambda_k\}, k = 1, \ldots, l$. Recall that the feature that renders each spectrum different from one another is the control point coordinates in the reflectance domain. Formally, we aim to recover a B-spline curve $\mathcal{C}_v$ with control points $P_{v,i} = (x_i, y_{v,i}), i = 0, \ldots, n$ for each spectrum index v, and a knot vector U common to all the spectra such that the coordinates of the curve are functions of an independent parameter t as

$$\lambda(t) = \sum_{i=0}^{n} N_{i,p}(t) x_i, \qquad \mathcal{R}_v(t) = \sum_{i=0}^{n} N_{i,p}(t) y_{v,i}. \qquad (7.15)$$

7.1 Compact Representations of Spectra

In Eq. (7.15), the basis functions $N_{i,p}(t)$ and the wavelength interpolation function $\lambda(t)$ are the same for all the given spectra. In contrast, the reflectance interpolation function $\mathcal{R}_v(t)$ and its associated control point coordinates $y_{v,i}$ vary with the spectral signature. Thus, the representation of this collection of spectra effectively consists of the knot vector U, the wavelength coordinates $[x_0, \ldots, x_n]^T$ and the reflectance coordinates $[y_{v,0}, \ldots, y_{v,n}]^T$ of the control points of the interpolating B-splines.

Comparing Eq. (7.15) to Eq. (7.13) suggests the following reformulation of the cost function in Eq. (7.14). Let us consider the set of wavelength indexed reflectance spectra $\{\mathbf{R}_v | \mathbf{R}_v = [R_{v,1}, R_{v,2}, \ldots, R_{v,l}]^T\}$. Let the parameter $t_k \in \mathcal{U}$ correspond to the kth wavelength, i.e. $\lambda_k = \lambda(t_k)$ $\forall k = 1, \ldots, l$. It then becomes straightforward to rewrite Eq. (7.14) as follows:

$$K = \sum_v \sum_{k=1}^{l} \left(\mathcal{R}_v(t_k) - R_{v,k}\right)^2. \qquad (7.16)$$

Equation 7.16 shows a straightforward extension to Eq. (7.14) in the spectrum index or pixel index dimension. Therefore, the cost function can be minimised using the knot removal procedure similar to that presented in Algorithm 7.1, subject to the constraint of a common knot vector across the input spectra.

Enforcing a Common Knot Vector

As mentioned in Sect. 7.1.2, the minimisation of the cost function departs from a global interpolation of B-spline curves to the image reflectance spectra as described by Piegl and Tiller (1995). Note that this procedure is performed before the knot removal algorithm, as shown in Algorithm 7.1. Since this method only applies to a single spectrum, it is not readily applicable to the case of multiple reflectance spectra. To enforce a common knot vector across all the input spectra, we modify the global interpolation method of Piegl and Tiller (1995) as follows.

In Algorithm 7.2, we present a method for the global interpolation of B-splines with a common knot vector to multiple reflectance spectra. At the beginning of Algorithm 7.2, we employ the centripetal method in Lee (1989) (Lines 4–7) to compute parameter values $\bar{t}_{v,k}$ separately for each reflectance spectrum. Note that our choice of the centripetal method is not exclusive; other methods may be used to compute these parameter values. After obtaining the average t_k of these parameter values across all the spectral signatures, a common knot vector across all the input spectra is computed in Lines 9–11. The *FindSpan*(l, p, t_k, U) procedure in Line 14 obtains the knot span index of t_k using a binary search method. Subsequently, the *BasisFunctions*$(span, t_k, p, U)$ procedure in Line 15 evaluates the basis functions at the parametric point t_k. Both procedures have been thoroughly described in Piegl and Tiller (1995). In Lines 17 and 19, the control point coordinates are derived as linear least-squares solutions.

Algorithm 7.2 *Interpolate*($\Lambda, \{\mathbf{R}_v\}, p$)

Require: Λ, R, p
  Λ: The vector of sampled wavelengths, where $\Lambda = [\lambda_1, \ldots, \lambda_l]^T$
  $\{\mathbf{R}_v\}$: A collection of reflectance spectra, where $R_{v,k}$ is the measured reflectance at the kth band of the spectrum with index v, $k = 1, \ldots, l$.
  p: The degree of the basis functions.

1: $l \leftarrow$ the number of bands (wavelengths)
2: $N \leftarrow$ the number of spectra
3: $m \leftarrow l + p + 1$ // l is also the number of control points per interpolating curve
4: **for** reflectance spectrum with index v **do**
5: $\quad d_v \leftarrow \sum_{k=2}^{l} ((R_{v,k} - R_{v,k-1})^2 + (\lambda_k - \lambda_{k-1})^2)^{\frac{1}{4}}$
6: $\quad$ Compute the parameters values $\bar{t}_{v,k}$, where $k = 1, \ldots, l$, such that $\bar{t}_{v,1} = 0$, $\bar{t}_{v,l} = 1$, and
$$\bar{t}_{v,k} = \bar{t}_{v,k-1} + \frac{((R_{v,k}-R_{v,k-1})^2+(\lambda_k-\lambda_{k-1})^2)^{\frac{1}{4}}}{d_v}, \text{ where } k = 2, \ldots, l,$$
7: **end for**
8: $t_k \leftarrow \frac{1}{N} \sum_v \bar{t}_{v,k} \; \forall k = 1 \ldots l$
  //Compute the common knot vector for all spectra
9: $u_k = 0 \; \forall k = 1, \ldots, p+1, u_k = 1 \; \forall k = m-p, \ldots, m$
10: $u_{k+p} = \frac{1}{p} \sum_{i=k}^{k+p-1} t_k \; \forall k = 2, \ldots, l-p$
11: $U \leftarrow [u_1, \ldots, u_m]^T$ //The knot vector
12: Initialise A to a zero-valued array of size $l \times l$
13: **for** $k \leftarrow [1 \ldots l]$ **do**
14: $\quad$ span $\leftarrow$ *FindSpan*(l, p, t_k, U)
15: $\quad A[k][\text{span} - p \ldots \text{span}] \leftarrow$ *BasisFunctions*(span, t_k, p, U)
16: **end for**
  //x: the wavelength coordinates of control points common to all the spectra.
17: $x \leftarrow$ the solution to $Ax = \Lambda$
18: **for** reflectance spectrum with index v **do**
19: $\quad y_v \leftarrow$ the solution to $Ay_v = \mathbf{R}_v$, where $\mathbf{R}_v = [R_{v,1}, R_{v,2}, \ldots, R_{v,l}]^T$ //y_v: the reflectance coordinates of the spectrum with index v.
20: **end for**
21: **return** $U, x, y_v \; \forall v$

In Fig. 7.2, we illustrate the reconstruction error, graded in shades of grey, for two sample images when PCA, B-splines and Gaussian mixtures are used to represent the spectra. The first column shows the pseudo-colour images for the sample subjects. The second and third columns show the absolute and relative reconstruction errors yielded by the B-spline representation. The fourth and fifth columns show the absolute and relative reconstruction errors for the Gaussian mixture representation. The last two columns show the error maps for PCA.

For the sake of visualisation, we have adjusted the grey-scale ranges across the panels in the figure. In the second, fourth and sixth columns, the errors have been averaged across all the bands and displayed in grey-scale intensities, where pure

Fig. 7.2 Absolute and relative reconstruction errors for sample hyperspectral images. *First column*: pseudo-colour images. *Second and third columns*: absolute and relative reconstruction errors for third-degree B-splines with 18 control points. *Fourth and fifth columns*: absolute and relative reconstruction errors for Gaussian mixtures with 6 components. *Sixth and seventh columns*: absolute and relative reconstruction errors resulting from the least-square projection of 18 PCA components

white corresponds to 0.1. In the third, fifth and seventh columns, the relative error maps are depicted as grey-scale images with pure white corresponding to 30 %. In the figure, the absolute error yielded by the spline representation is up to an order of magnitude smaller than that by the Gaussian mixture and slightly larger than that by PCA. The same trend applies to the relative errors. Note that, as expected, the regions with large relative errors are those with low reflectance in the original images.

7.1.3 Classification and Recognition

In this section we illustrate the utility of the representations above for skin recognition and biometrics. These are applications used for illustrative purposes, where we compare the spline representation of the spectra, the Gaussian mixture, raw spectra and PCA and LDA.

We have used an image database of human subjects. The database contains three ethnic groups, with 22 of the subjects identifying themselves as Caucasian, 8 as Indian, 18 as Oriental and 3 as other. To acquire training data, we randomly select a subject representative of each of the main ethnic groups. On the images of these subjects, we randomly sample pixels from skin and non-skin regions, each consisting of 685 skin and 2968 non-skin spectra. We perform the skin recognition with the images that have been acquired under the frontal illumination direction. To obtain the image reflectance, we simply normalised the raw radiance images by the ground truth illumination power, which has been measured on a white calibration target, i.e. a Spectralon target. In practical situations, where a measurement of the illumination power spectrum is unavailable, the image reflectance can still be estimated from the original image as a result of illumination spectrum recovery using the methods in Chap. 5.

For purposes of recognition using the Gaussian mixture, we concatenate the triplet $(\beta_k, \mu_k, \sigma_k)$ for each Gaussian component. We have done this since then in our experiments; these triplets consistently deliver better classification performance than any other tuple combinations involving β_k, μ_k and σ_k. In addition, we

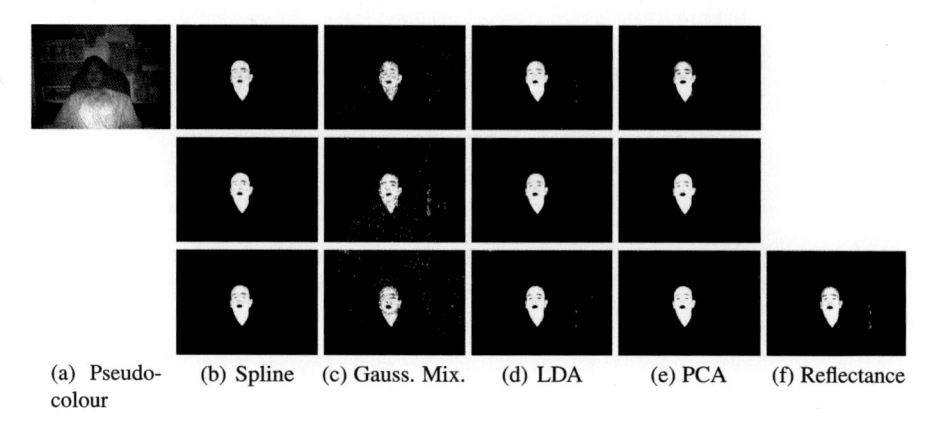

(a) Pseudo-colour (b) Spline (c) Gauss. Mix. (d) LDA (e) PCA (f) Reflectance

Fig. 7.3 Skin probability maps of a sample subject. *First column*: the original input image rendered in pseudo-trichromatic colour. The remaining columns, *from left to right*, correspond to the skin probability maps using the spline-based representation, the Gaussian mixture, LDA, PCA and the input reflectance spectra as discriminant features. The *rows, from top to bottom*, correspond to the description lengths of 12, 24 and 30

compare the performance of the B-splines with that yielded by applying linear discriminant analysis (LDA) and principal component analysis (PCA) to wavelength indexed spectral reflectance vectors. Since the length of the Gaussian mixture descriptor is always a multiple of 3, we perform skin recognition experiments with descriptor lengths of 12, 24 and 30. The spline descriptor is formulated as a vector of the y-coordinates (the reflectance coordinates) of the control points of a third-degree spline curve. All the spline curves corresponding to both the training and test data share the same knot vector and B-spline basis, as described in Sect. 7.1.2.

For classification, we apply a soft-margin support vector machine (SVM) classifier (Boser et al. 1992; Cortes and Vapnik 1995) with a radial basis function (RBF) kernel to the above reflectance descriptors. The parameter values of the SVM classifier are, again, selected by a fourfold cross validation procedure. Subsequently, skin recognition is performed by applying the trained classifier to the spectral reflectance descriptor at each image pixel for all the testing imagery, i.e. that corresponding to the 48 subjects remaining at each trial after excluding the three training images.

In Fig. 7.3, we present the skin probability maps of a sample subject resulting from the employment of the above spectral representations with various lengths. The rows of the figure, from top to bottom, correspond to increasing descriptor lengths of 12, 24 and 30. The first column shows the original input image rendered in pseudo-colour. The remaining columns, from left to right, correspond to skin probability maps obtained by the B-spline descriptor, the Gaussian mixture descriptor, LDA, PCA and the raw reflectance spectra. The Gaussian mixture delivers noisier skin maps than the other features due to over-fitting. On the other hand, the B-spline descriptor produces skin maps similar to those of PCA, LDA and the raw reflectance. Note that the skin maps produced by the B-spline descriptor do not vary significantly with respect to the descriptor length. Therefore, in addition to its re-

7.1 Compact Representations of Spectra

construction capability, the B-spline descriptor offers competitive recognition rates with lower complexity, as compared to the full reflectance spectra.

Finally, we illustrate the application of the representations in ethnic group identification from skin reflectance. To this end, we study the classification of the three ethnic groups, namely Caucasian, Indian and Oriental, based on their skin reflectance spectra. Here, as before, a third-degree B-spline curve with 15 control points and 19 knots is used. Following our skin recognition experiments, we recover a descriptor vector consisting of the y-coordinates (wavelength coordinates) of the control points. For the Gaussian mixture, we employ 5 Gaussian components fitted to the given reflectance spectra (Angelopoulou et al. 2001). For the other alternatives, we use 15 principal LDA and PCA components.

To obtain training data, we randomly select two typical representatives from each ethnic group and collect training data from their skin reflectance. To this end, we use frontally illuminated images of these training subjects and select 5 rectangular skin regions from each image, with an average size of 25×17 pixels. This gives rise to three training classes (one for each ethnic group) which we use to train three SVM classifiers according to a one-versus-all strategy. Each of these uses a first-degree polynomial kernel with its parameters tuned through a fourfold cross validation procedure. During the test phase, the three SVM classifiers assign an ethnic group association probability to each skin pixel in the test image. The overall ethnic identity associated with the test subjects is determined by the majority voting of the ethnic group association across their skin pixels.

In Fig. 7.4, we visualise the ethnic group association probabilities of the test subjects as points in a 3D coordinate system, where the three axes represent the three ethnic groups above. The ground truth ethnic groups of the test subjects are colour coded red, green and blue for Caucasian, Indian and Oriental, respectively. In all the plots except that for the LDA feature, the Indian skin cluster is well separated from the other groups. This is expected, as the Indian skin reflectance spectra from our database have "flatter" slopes and less absorption bands than the other two groups. It is noticeable that the B-spline descriptor provides the highest degree of separation between the Caucasian and the Oriental groups, with the least number of confounded examples, followed by PCA and the raw reflectance spectra, which yield similar distributions. It is also observable that the skin groups obtained with the B-spline descriptor are clustered most tightly, thus allowing more accurate classification of the skin groups. For the B-spline, PCA and the raw reflectance spectra, the Caucasian and Oriental groups are fairly separable from each other with a few examples confounded. This is probably due to image saturation and shadowing. Note also that each ethnic group is correctly located within the proximity of the extreme point corresponding to the pure skin example of the group. The distributions achieved by these three descriptors are superior to those achieved by the Gaussian mixture descriptor and LDA, in which the Caucasian and Oriental groups mingle more closely than the Indian one.

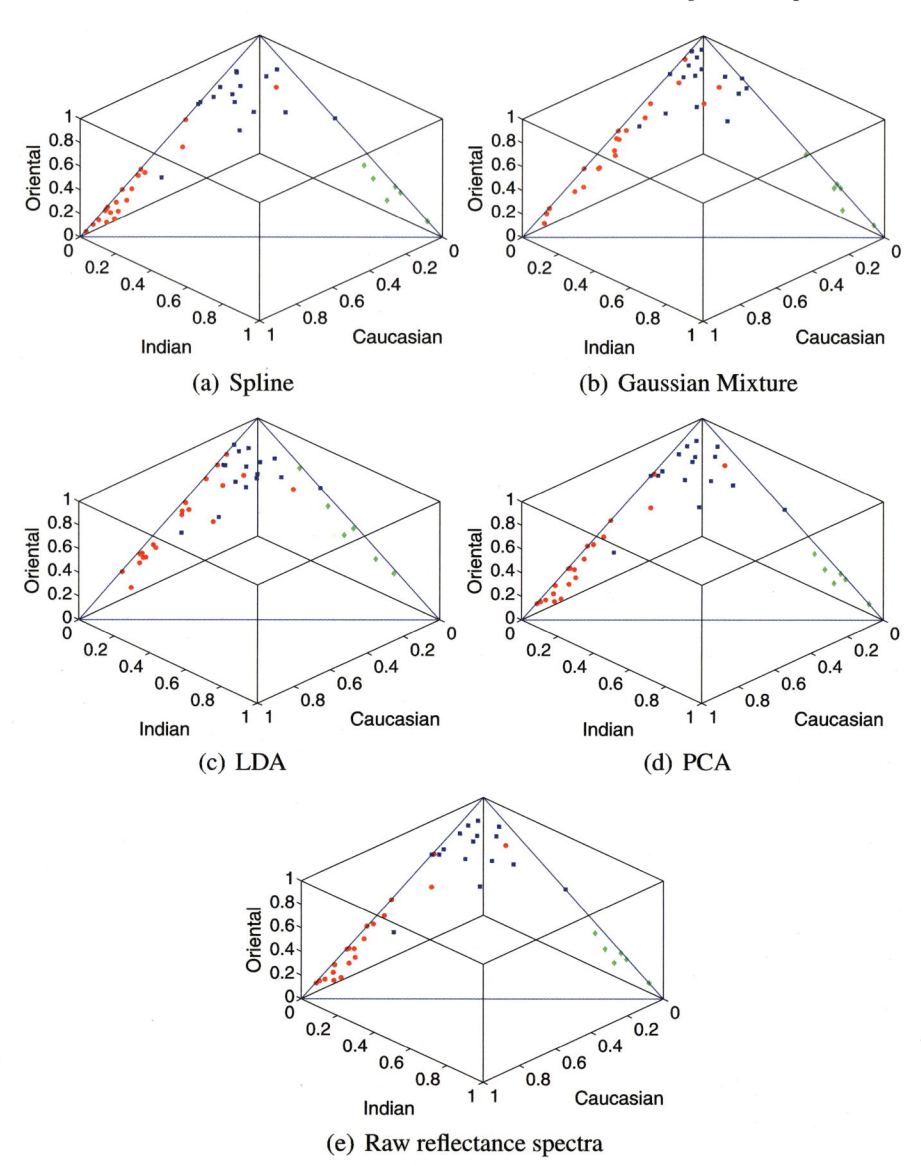

Fig. 7.4 3D visualisation of the profile for three ethnic groups. The *subplots* show the ethnicity association probabilities yielded using (**a**) a third-degree B-spline with 19 knots and 15 control points; (**b**) a Gaussian mixture with 5 components; (**c**) 15 principal LDA components of reflectance spectra; (**d**) 15 principal PCA components of reflectance spectra; (**e**) the raw reflectance spectra

7.2 Spectrum Descriptors

In practise, object classification and image categorisation techniques (Nister and Stewenius 2006; Sivic et al. 2005; Chum et al. 2007) are based on the comparison

7.2 Spectrum Descriptors

of image features between those in a query image and those corresponding to the images in a dataset. This is usually based on a codebook that serves as a means to recover the closest match through classification. Furthermore, descriptors are often combined so as to assign a different importance to each of them in order to maximise performance.

In either case, recovering the optimal descriptors directly (Winder and Brown 2007) or recovering their optimal combination (Nilsback and Zisserman 2006; Varma and Ray 2007), the aim in categorisation is to maximise performance so as to minimise the variance across the dataset with respect to the classifier output. This is done because the classifier output is dependent on the image representation and the similarity measure employed to categorise the images (Vasconcelos 2004).

It is surprising that despite the widespread use of high-level features for recognition and retrieval of monochromatic and trichromatic imagery, image descriptors in imaging spectroscopy are somewhat under-researched, with current methods focusing on the local analysis of individual pixel spectra rather than profiting from the structural information that imagery provides. The development of affine invariant imaging spectroscopy features to capture the structural and spectral information in the imagery for purposes of recognition can greatly enhance the results delivered by material classification methods which operate on pixel spectra alone. Thus, the use of higher-level features in imaging spectroscopy opens up great opportunities in recognition and classification tasks.

For imaging spectroscopy, descriptors and representation of signatures for recognition are mainly limited to edge detection (Stokman and Gevers 1999) or approaches based on the derivative analysis of the spectra (Angelopoulou 2000). Nonetheless this local analysis of the spectra was shown to be intrinsic to the surface albedo, the analysis in Angelopoulou (2000) was derived from the Lambertian reflection model, i.e. where all surfaces reflect light perfectly diffusely. Hence, it is not applicable to complex scenes where inter-reflections and multiple illuminants may occur. Fu et al. (2006a) have proposed the use of band ratios as an alternative to raw spectral bands as features for classification invariant to shading. In Fu et al. (2006b), a subspace projection method for specularity-free spectral representation is presented. Huynh and Robles-Kelly (2008) have presented a method to represent reflectance data making use of a continuous basis by fitting a B-spline to the spectra under study. Recently, Khuwuthyakorn et al. (2009) proposed a texture descriptor for imaging spectroscopy based on Fourier analysis and heavy-tailed probability distributions. This is reminiscent of time-dependent textures, whose probability density functions exhibit first and second-order moments which are space and time-shift invariant (Doretto et al. 2003). Unfortunately, the descriptor presented in Khuwuthyakorn et al. (2009) cannot be used to recover the spectral signatures but rather has been designed for purposes of recognition where precise material matching is not necessary.

7.2.1 A Log-Linear Shading-Invariant Feature

Note that, in Eq. (6.28), the first term on the right-hand side depends on the scene geometry while the second is given by the surface reflectance difference for the two bands λ_i and λ_j. To turn the product on the right-hand side of the equation into a linear combination we can take the logarithm of both sides of Eq. (6.28) to yield a quantity $\Gamma_u(\lambda_i)$ as follows:

$$\begin{aligned}\Gamma_u(\lambda_i) &= \log\bigl(R(u,\lambda_i) - R(u,\lambda_j)\bigr) \\ &= \log\bigl(g(u)\bigr) + \log\bigl(S(u,\lambda_i) - S(u,\lambda_j)\bigr),\end{aligned} \qquad (7.17)$$

where, as before, $R(u,\cdot) \triangleq \frac{I(u,\cdot)}{L(\cdot)}$ is the ratio of the image irradiance $I(u,\cdot)$ to the spectral illuminant power $L(\cdot)$, $S(u,\cdot)$ is the material reflectance, and $g(u)$ corresponds to the shading factor.

To remove the shading factor term from the right-hand side of Eq. (7.17) we can subtract the functions $\Gamma_u(\cdot)$ corresponding to the ith and the kth bands, that is,

$$\begin{aligned}F(u,\lambda_i,\lambda_k) &= \Gamma_u(\lambda_i) - \Gamma_u(\lambda_k) \\ &= \log\bigl(R(u,\lambda_i) - R(u,\lambda_j)\bigr) - \log\bigl(R(u,\lambda_k) - R(u,\lambda_j)\bigr).\end{aligned} \qquad (7.18)$$

Note that, theoretically, an arbitrary reference band λ_j can be used for the differences above. Nonetheless, $\Gamma_u(\cdot)$ may deviate from the ideal case as modelled by Eq. (7.17) due to measurement noise. Assuming that these measurement errors are independently random variables governed by a Gaussian distribution with a zero mean and a band-limited variance, we can use the average value of $R(u,\lambda)$ over the wavelength domain, which is denoted as $R^*(u) \triangleq \frac{1}{N}\sum_{i=1}^{N} R(u,\lambda_i)$, instead of $R(u,\lambda_j)$, so as to reduce the error variance by half. We do this for both the specular-free and the shading-free spectral values. Thus, we rewrite Eq. (7.18) as follows:

$$\hat{F}_u(\lambda_i) = \hat{\Gamma}_u(\lambda_i) - \Gamma_u^* + \kappa_u, \qquad (7.19)$$

where

$$\begin{aligned}\hat{\Gamma}_u(\lambda_i) &= \log\bigl(R(u,\lambda_i) - R^*(u) + \beta_u\bigr), \\ \Gamma_u^* &= \frac{1}{N}\sum_{i=1}^{N} \hat{\Gamma}_u(\lambda_i).\end{aligned} \qquad (7.20)$$

In Eq. (7.19), we have added the per-pixel bias values κ_u and β_u. These additive constants serve to ensure that $\hat{F}_u(\lambda_i)$ is defined, i.e. $\hat{F}_u(\lambda_i) \geq 0$ and $R(u,\lambda_i) - R^*(u) + \beta_u > 0 \ \forall i \in \{1,2,\ldots,N\}$, without changing the nature of the invariant feature $\hat{F}_u(\lambda_i)$.

We note that if the illuminant power spectrum is at hand, Eq. (7.20) can be used to recover a specularity-free pixel-wise spectrum. This can be easily computed by

7.2 Spectrum Descriptors

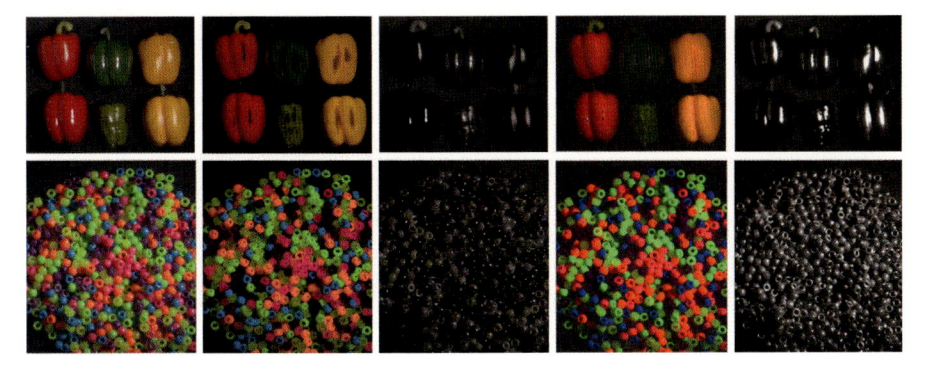

Fig. 7.5 Specularity removal results on two sample images from the CAVE database. *From left to right*: diffuse and specular maps for the method in Tan et al. (2004) and the ones corresponding to the bias β_u

setting the bias β_u to

$$\beta_u = -\min_{\lambda_i}\{R(u,\lambda_i) - R^*(u)\}. \tag{7.21}$$

This is in accordance with the intuition that the specular coefficient at a pixel u in the image does not vary between bands. Therefore, the mean image irradiance normalised by the illuminant spectrum, i.e. $R^*(u)$, can be subtracted from the spectra. Equation (7.21) also follows the notion that the bias β_u should be such that $R(u,\lambda_i) - R^*(u) + \beta_u > 0$ for every wavelength λ_i in the image. As a result, the diffuse component for each pixel u in the image at the band λ_i is given by $R(u,\lambda_i) - R^*(u) + \beta_u$, whereas the specularity can be computed by making use of the expression $R^*(u) - \beta_u$.

In Fig. 7.5, we show sample results for two images in the CAVE Multispectral Image Database (Yasuma et al. 2010). From left to right, the columns show the input image, the diffuse and specular maps delivered by the method in Tan et al. (2004) and those yielded by the bias β_u. Note that the bias β_u delivers better results than those obtained by the method in Tan et al. (2004); in particular, note that the diffuse map for the bias is less "blacked out" by the specularity removal than the imagery delivered by the alternative. Also, the specular component corresponding to the bias appears to capture the highlights better than the alternative. This is particularly visible for the textures on the background, the peppers and the beads.

7.2.2 Affine Invariance

We can also view hyperspectral images as arising from a probability distribution whose observable or occurrences may have long or heavy tails. This implies that the spectra in the image result in values that can be rather high in terms of their deviation from the image spectra mean and variance. As a result, heavy-tailed distributions provide a means to a formulation which can capture high wavelength-dependent variation in the image. This is important, since it allows us to capture information in a descriptor that otherwise may be cast as the product of outliers. Thus, an alternative here is to formulate a descriptor so as to model "rare" stationary wavelength-dependent events on the image plane.

Heavy-Tailed Distributions

Moreover, we can view the pixel values of the hyperspectral image as arising from stochastic processes whose moment generating functions are invariant with respect to shifts in the image coordinates. This means that the mean, covariance, kurtosis, etc. for the corresponding joint probability distribution are required to be invariant with respect to changes of location on the image. Due to the use of heavy-tailed distributions, these densities may have high dispersion and, thus, their probability density functions are, in general, governed by further-order moments. These introduces a number of statistical "skewness" variables that allow modelling high variability spectral behaviour.

This is reminiscent of simulation approaches where importance sampling cannot be effected via an exponential changes in measurement due to the fact that the moments are not exponential in nature. This applies to distributions such as the lognormal, Weibull with increasing skewness and regularly varying distributions such as Pareto, stable and log-gamma distributions (Asmussen et al. 2000). More formally, we formulate the density of the pixel values for the wavelength λ at the pixel u in the image band I_λ of the image as random variables $\mathcal{Y}_u$ whose inherent basis $\mathcal{X}_u = \{x_u(1), x_u(2), \ldots, x_u(|\mathcal{X}_u|)\}$ is such that

$$P(\mathcal{Y}_u) = \sum_{k=1}^{|\mathcal{X}_u|} P(x_u(k)), \tag{7.22}$$

where $x_u(k)$ are identically distributed variables and, as usual for probability distributions of real-valued variables, we have written $P(\mathcal{Y}_u) = Pr[y \leq \mathcal{Y}_u]$ for all $y \in \mathfrak{R}$.

In other words, we view the pixel values for each band in the image under study as arising from a family of heavy-tailed distributions whose variance is not necessarily finite. Note that, for finite variance, the formalism above implies that $P(\mathcal{Y}_u)$ is normally distributed and, as a result, this approach is not exclusive to finite variance distributions. Rather this treatment generalises the stochastic process to a number of independent influences, each of which is captured by the corresponding variable $x_u(k)$.

In practise, the probability density function (PDF) $f(\mathcal{Y}_u)$ is not available in closed form. As a result, we can re-parameterise the PDF by recasting it as a function of the variable ς making use of the characteristic function

$$\psi(\varsigma) = \int_{-\infty}^{\infty} \exp(\mathbf{i}\varsigma \mathcal{Y}_u) f(\mathcal{Y}_u) d\mathcal{Y}_u \tag{7.23}$$

$$= \exp\bigl(\mathbf{i}u\varsigma - \gamma|\varsigma|^\alpha \bigl(1 + \mathbf{i}\beta \operatorname{sign}(\varsigma)\varphi(\varsigma, \alpha)\bigr)\bigr), \tag{7.24}$$

7.2 Spectrum Descriptors

where $i = \sqrt{-1}$, u is, as before, the pixel index on the image plane, $\gamma \in \mathfrak{R}^+$ are function parameters, $\beta \in [-1, 1]$ and $\alpha \in (0, 2]$ are the skewness and characteristic exponents, respectively, and $\varphi(\cdot)$ is defined as follows:

$$\varphi(\varsigma, \alpha) = \begin{cases} \tan(\alpha \frac{\pi}{2}), & \text{if } \alpha \neq 1, \\ -\frac{\pi}{2} \log |\varsigma|, & \text{if } \alpha = 1. \end{cases} \qquad (7.25)$$

For the characteristic function above, some values of α correspond to special cases of the distribution. For instance, $\alpha = 2$ implies a normal distribution, $\beta = 0$ and $\alpha = 1$ correspond to a Cauchy distribution and for the Levy distribution we have $\alpha = \frac{1}{2}$ and $\beta = 1$. Thus, although the formalism above can capture a number of cases in exponential families, it is still quite general in nature; thus it allows the modelling of a large number of distributions that may apply to hyperspectral data and whose characteristic exponents α are not those of distributions whose tails are exponentially bounded.

So far, we have limited ourselves to the image plane for a fixed wavelength λ. That is, we have been concentrating on the distribution of spectral values across every wavelength resolved band in the image. Note that, without loss of generality, we can extend Eq. (7.24) to the wavelength domain, i.e. the spectra of the image across a segment of bands. This is a straightforward task if we note that the equation above can be viewed as the cross-correlation between the function $f(\mathcal{Y}_u)$ and the exponential given by $\exp(i\varsigma \mathcal{Y}_u)$. Hence, we can write the characteristic function for the image parameterised with respect to the wavelength λ as follows:

$$\vartheta(\lambda) = \int_{-\infty}^{\infty} \int_{-\infty}^{\infty} \exp(i\lambda\varsigma) \exp(i\varsigma \mathcal{Y}_u) f(\mathcal{Y}_u) d\mathcal{Y}_u d\varsigma \qquad (7.26)$$

$$= \int_{-\infty}^{\infty} \exp(i\lambda\varsigma) \psi(\varsigma) d\varsigma, \qquad (7.27)$$

where the second line in the equation above corresponds to the substitution of Eq. (7.24) into Eq. (7.26).

Equation (7.27) captures the spectral cross-correlation for the characteristic functions for each band. In this manner, we view the characteristic function for the hyperspectral image as a heavy-tailed distribution of another set of heavy-tailed PDFs, which correspond to each of the bands in the image. This can also be interpreted as a composition of two heavy-tailed distributions, where Eq. (7.24) corresponds to the image band domain ς of the image and Eq. (7.27) is determined by the wavelength-dependent domain λ.

This composition operation suggests a two-step process for the computation of the image descriptor. Firstly, at the band level, the information can be represented in a compact fashion by making use of harmonic analysis and then rendered invariant to geometric distortions on the object surface plane. Secondly, the wavelength-dependent correlation between bands can be computed by making use of the operation in Eq. (7.27).

Harmonic Analysis

In this section, we explore the use of harmonic analysis and the fundamentals of integral transforms (Sneddon 1995) to provide a means of computing the descriptor. We commence by noting that Eqs. (7.23) and (7.26) are characteristic functions obtained via the integral of the product of the function $g(\eta)$, i.e. $f(\mathcal{Y}_u)$ and $\psi(\varsigma)$, multiplied by a kernel, given by $\exp(i\lambda\varsigma)$ and $\exp(i\varsigma\mathcal{Y}_u)$, respectively.

To appreciate this more clearly, consider the function given by

$$F(\omega) = \int_{-\infty}^{\infty} g(\eta) K(\omega, \eta)\, d\eta, \tag{7.28}$$

where $K(\omega, \eta)$ is a harmonic kernel of the form

$$K(\omega, \eta) = \sum_{k=1}^{\infty} a_k \phi_k(\omega) \phi_k(\eta), \tag{7.29}$$

where a_k is the kth real scalar corresponding to the harmonic expansion and $\phi_k(\cdot)$ are orthonormal functions such that $\langle \phi_k(\omega), \phi_n(\eta) \rangle = 0 \;\forall n \neq k$. Moreover, we consider cases in which the functions $\phi_k(\cdot)$ constitute a basis for a Hilbert space (Young 1988). Therefore, the right-hand side of Eq. (7.29) is convergent to $K(\omega, \eta)$ as k tends to infinity.

To see the relation between Eq. (7.28) and the equations in previous sections, we can examine $\psi(\varsigma)$ in more detail and write

$$\log[\psi(\varsigma)] = iu\varsigma - \gamma |\varsigma|^\alpha \bigl(1 + i\beta\, \text{sign}(\varsigma)\varphi(\varsigma, \alpha)\bigr) \tag{7.30}$$

$$= iu\varsigma - |\varsigma|^\alpha \gamma^{*\alpha} \exp\left(-i\beta^* \frac{\pi}{2}\vartheta\,\text{sign}(\varsigma)\right), \tag{7.31}$$

where $\vartheta = 1 - |1 - \alpha|$ and parameters γ^* and β^* are given by

$$\gamma^* = \left(\frac{\gamma \sqrt{\Omega}}{\cos(\alpha \frac{\pi}{2})}\right)^{\frac{1}{\alpha}}, \tag{7.32}$$

$$\beta^* = \frac{2}{\pi \vartheta} \arccos\left(\frac{\cos(\alpha \frac{\pi}{2})}{\sqrt{\Omega}}\right), \tag{7.33}$$

and $\Omega = \cos^2(\alpha \frac{\pi}{2}) + \beta^2 \sin^2(\alpha \frac{\pi}{2})$.

To obtain the kernel for Eq. (7.28), we can use Fourier inversion on the characteristic function and, making use of the shorthand notation defined above, the PDF may be computed via the equation

$$f(\mathcal{Y}_u; u, \beta^*, \gamma^*, \alpha) = \frac{1}{\pi \gamma^*} \int_0^\infty \cos\left(\frac{(u - \mathcal{Y}_u)s}{\gamma^*} + s^\alpha \sin(\phi)\right) \exp(-s^\alpha \sin(\phi))\, ds, \tag{7.34}$$

where $\phi = \frac{\beta^* \pi \eta}{2}$.

7.2 Spectrum Descriptors

This treatment not only opens up the possibility of applying functional analysis on the characteristic function using the techniques in the Fourier domain, but also allows the use of other harmonic kernels for compactness and ease of computation. This is due to the fact that, we can view the kernel $K(\omega, \eta)$ as the exponential $\exp(-s^\alpha \sin(\phi))$, whereas the function $g(\eta)$ is given by the cosine term. Thus, we can use other harmonic kernels to induce a change of basis without any loss of generality. Actually, the expression above can be greatly simplified by making use of the shorthand terms $A = \frac{(u-\mathcal{Y}_u)}{\gamma^*}$, $\eta = s^\alpha$ and $\omega\eta = As + s^\alpha \sin(\phi)$, which yields

$$s^\alpha \sin(\phi) = \omega\eta - A\eta^{\frac{1}{\alpha}}. \tag{7.35}$$

Substituting Eq. (7.34) into Eq. (7.35), the PDF can be expressed as

$$f(\mathcal{Y}_u; u, \beta^*, \gamma^*, \alpha) = \sqrt{\frac{2}{\pi}} \int_0^\infty \frac{\exp(-\omega\eta + A\eta^{\frac{1}{\alpha}})}{\sqrt{2\pi}\gamma^*\alpha\eta^{(\frac{\alpha-1}{\alpha})}} \cos(\omega\eta)\, d\eta, \tag{7.36}$$

where the kernel then becomes

$$K(\omega, \eta) = \cos(\omega\eta). \tag{7.37}$$

This can be related, in a straightforward manner, to the Fourier cosine transform (FCT) of the form

$$F(\omega) = \sqrt{\frac{2}{\pi}} \int_0^\infty \frac{\exp(-\omega\eta + \frac{(u-\mathcal{Y}_u)}{\gamma^*}\eta^{\frac{1}{\alpha}})}{\sqrt{2\pi}\gamma^*\alpha\eta^{(\frac{\alpha-1}{\alpha})}} \cos(\omega\eta)\, d\eta, \tag{7.38}$$

which is analogous to the expression in Eq. (7.34). Nonetheless, the transform above does not have imaginary coefficients. This can be viewed as a representation in the power spectrum rather than in the phase spectrum. Moreover, it has the advantage of compacting the spectral information in the lower order Fourier terms, i.e. those for which ω is close to the origin. This follows the strong "information compaction" property of FCTs introduced in Rao and Yip (1990) and ensures a good trade-off between discriminative power and complexity.

We emphasise that, due to the harmonic analysis treatment given to the problem in this section, other kernels may be used for purposes of computing other integral transforms (Sneddon 1995) spanning Hilbert spaces. These include wavelets and the Mellin ($K(\omega, \eta) = \eta^{\omega-1}$) and Hankel transforms. In fact, other kernels may be obtained by performing an appropriate substitution on the term $\cos(\omega\eta)$.

Invariance to Affine Distortions

Having introduced the notion of the harmonic analysis and shown how the probability density function can be recovered using a Fourier transform, we now focus on the relation between distortions on the object surface plane and the Fourier domain. To this end, we follow (Bracewell et al. 1993) and relate the harmonic kernel above to affine transformations on the object's locally planar shape. As mentioned earlier, the function $f(\mathcal{Y}_u)$ corresponds to the band-dependent component of the image and, as a result, it is prone to affine distortion. This hinges on the notion that a distortion on the object's surface will affect the geometric factor for the scene, but not its photometric properties. In other words, the material index of refraction, roughness, etc. remains unchanged, whereas the geometry of the reflective process does vary with respect to affine distortions on the image plane. The corresponding 2D integral transform of the function $f(\mathcal{Y}_u)$ which, as introduced in the previous sections, corresponds to the pixel value image under study is given by

$$F(\xi) = \int_\Gamma f(\mathcal{Y}_u) K(\xi^T, u) \, du, \tag{7.39}$$

where $u = [x, y]^T$ is the vector of 2D coordinates for the compact domain $\Gamma \in \mathfrak{R}^2$ and, in the case of the FCT, $K(\xi^T, u) = \cos(2\pi(\xi^T u))$.

In practise, the coordinate vectors u will be given by discrete quantities on the image lattice. For purposes of analysis, we consider the continuous case and note that the affine coordinate transformation can be expressed in matrix notation as follows:

$$u' = \begin{pmatrix} x' \\ y' \end{pmatrix} = \begin{pmatrix} a & b \\ d & e \end{pmatrix} \begin{pmatrix} x \\ y \end{pmatrix} + \begin{pmatrix} c \\ f \end{pmatrix}. \tag{7.40}$$

This observation is important because we can relate the kernel for the FCT in Eq. (7.39) to the transformed coordinate $u' = [x', y']^T$. Also, note that, for patches centred at key points in the image, the locally planar object surface patch can be considered devoid of translation. Thus, we can set $c = f = 0$ and write

$$\xi^T u = \xi^T \begin{pmatrix} x \\ y \end{pmatrix} \tag{7.41}$$

$$= \begin{pmatrix} \xi_x & \xi_y \end{pmatrix} \begin{pmatrix} a & b \\ d & e \end{pmatrix}^{-1} \begin{pmatrix} x' \\ y' \end{pmatrix} \tag{7.42}$$

$$= \frac{1}{ae - bd} \begin{pmatrix} (e\xi_x - d\xi_y) & (-b\xi_x + a\xi_y) \end{pmatrix} \begin{pmatrix} x' \\ y' \end{pmatrix}, \tag{7.43}$$

where $\xi = [\xi_x, \xi_y]^T$ is the vector of spectral indices for the 2D integral transform.

Hence, after some algebra, and using the shorthand $\Delta = (ae - bd)$, we can show that for the coordinates u', the integral transform is given by

$$F(\xi) = \frac{1}{|\Delta|} \int_{-\infty}^{\infty} \int_{-\infty}^{\infty} f(\mathcal{Y}_{u'}) K\left(\frac{1}{\Delta}[(e\xi_x - d\xi_y), (b\xi_x - a\xi_y)], [x', y']^T\right) dx' dy'. \tag{7.44}$$

7.2 Spectrum Descriptors

This implies that

$$F(\xi) = \frac{1}{|\Delta|} F(\xi'), \qquad (7.45)$$

where ξ' is the "distorted" analogue of ξ. The distortion matrix $\mathbb{T}$ is such that

$$\xi = \begin{pmatrix} \xi_x \\ \xi_y \end{pmatrix} = \begin{pmatrix} a & d \\ b & e \end{pmatrix} \begin{pmatrix} \xi'_x \\ \xi'_y \end{pmatrix} = \mathbb{T}\xi'. \qquad (7.46)$$

As a result, from Eq. (7.44), we can conclude that the effect of the affine coordinate transformation matrix $\mathbb{T}$ is to produce a distortion equivalent to $(\mathbb{T}^T)^{-1}$ in the ξ domain for the corresponding integral transform. This observation is an important one since it permits achieving invariance to affine transformations on the locally planar object surface patch. This can be done in practise via a ξ-domain distortion correction operation of the form

$$F(\xi) = \left(\mathbb{T}^T\right)^{-1} F(\xi'). \qquad (7.47)$$

Thus, the descriptor computation is a two-step process. Firstly, we compute the affine-invariant 2D integral transform for every band in the hyperspectral image under study. This is equivalent to computing the band-dependent component of the characteristic function $\psi(\varsigma)$. Secondly, we capture the wavelength-dependent behaviour of the hyperspectral image by computing the cross-correlation with respect to the spectral domain for the set of distortion-invariant integral transforms. By making use of the FCT kernel, in practise, the descriptor becomes an FCT with respect to the band index for the cosine transforms corresponding to the wavelength resolved image in the sequence.

Following the rationale above, we commence by computing the distortion-invariant integral transform for each band in the image. To do this, we use Eq. (7.47) to estimate the distortion matrix with respect to a predefined reference. Here, we employ the peaks of the power spectrum and express the relation of the integral transforms for two locally planar image patches, i.e. the one corresponding to the reference and that for the object under study. We have done this following the notion that a blob-like shape composed of a single transcendental function on the image plane would produce two peaks in the Fourier domain. That is, we have set, as our reference, a moment generating function arising from a cosine on a plane perpendicular to the camera.

Let the peaks of the power spectrum for two locally planar object patches, **A** and **B**, be given by $\mathbf{U}_A$ and $\mathbf{U}_B$. Those for the reference **R** are $\mathbf{U}_R$. The affine distortion matrices are $\mathbb{T}_A$ and $\mathbb{T}_B$, respectively. As a result, the matrices $\mathbf{U}_A$, $\mathbf{U}_B$ and $\mathbf{U}_R$ are such that each of their columns corresponds to the x–y coordinates for one of the two peaks in the power spectrum. These relations are given by

$$\mathbf{U}_A = \left(\mathbb{T}_A^T\right)^{-1} \mathbf{U}_R, \qquad (7.48)$$

$$\mathbf{U}_B = \left(\mathbb{T}_B{}^T\right)^{-1}\mathbf{U}_R, \qquad (7.49)$$

where $\mathbb{T}_A : \mathbf{R} \Rightarrow \mathbf{A}$ and $\mathbb{T}_B : \mathbf{R} \Rightarrow \mathbf{B}$ are the affine coordinate transformation matrices of the planar surface patches under consideration in the spatial domain.

Note that this is reminiscent of the shape-from-texture approaches hinging on the use of the Fourier transform for the recovery of the local distortion matrix (Ribeiro and Hancock 2001). Nonetheless, in Ribeiro and Hancock (2001), the normal is recovered explicitly making use of the Fourier transform, whereas here we employ the integral transform and aim at relating the FCTs for the two locally planar patches with that of the reference. We can do this by making use of the composition operation given by

$$\mathbf{U}_B = \left(\mathbb{T}_A \mathbb{T}_B^{-1}\right)^T \mathbf{U}_A, \qquad (7.50)$$

$$\mathbf{U}_B = \Phi_{BA} \mathbf{U}_A, \qquad (7.51)$$

where $\Phi_{BA} = (\mathbb{T}_A \mathbb{T}_B^{-1})^T$ is the matrix representing the distortion of the power spectrum of $\mathbf{U}_A$ with respect to $\mathbf{U}_B$. Here, we use the shorthand notation $\mathbb{T}_A{}^T = \mathbf{U}_R \mathbf{U}_A{}^{-1}$ and $(\mathbb{T}_B{}^T)^{-1} = \mathbf{U}_B \mathbf{U}_R{}^{-1}$ to write

$$\Phi_{BA} = \left(\mathbf{U}_R \mathbf{U}_A^{-1}\right)\left(\mathbf{U}_B \mathbf{U}_R^{-1}\right). \qquad (7.52)$$

As a result, we fix a reference for every locally planar patch in order to compute the matrix Φ directly through the expression above. This contrasts with other methods in that, for our descriptor computation, we do not recover the principal components of the local distortion matrix, but rather construct a band-level matrix of the form

$$\mathbf{V} = \left[F(I_1)^* \mid F(I_2)^* \mid \ldots \mid F(I_{|\mathbb{I}|})^*\right], \qquad (7.53)$$

which is the concatenation of the affine invariant integral transforms $F(\cdot)^*$ for the band-resolved locally planar object surface patch $\mathbb{I}$ in the image. Moreover, we render the band-level integral transform invariant to affine transformations by making use of the reference peak matrix $\mathbf{U}_R$ such that the transform for the band indexed t is given by

$$F(I_R) = F(I_t)^* \Phi_t^{-1}, \qquad (7.54)$$

where Φ_t^{-1} is the matrix which maps the transform for the band corresponding to the wavelength λ to the transform $F(I_R)$ for the reference plane. Here, we use as reference the power spectrum given by two peaks rotated 45° about the upper left corner of the 2D FCT.

Note that, since we have derived our descriptor based on the properties of integral transforms and Hilbert spaces, each element of the matrix $\mathbf{V}$ can be considered as arising from the inner product of a set of orthonormal vectors. Moreover, from a harmonic analysis perspective, the elements of $\mathbf{V}$ are represented in terms of discrete wave functions, over an infinite number of elements (Katznelson 2004). This is analogous to the treatment given to time series in signal processing, where the

7.2 Spectrum Descriptors

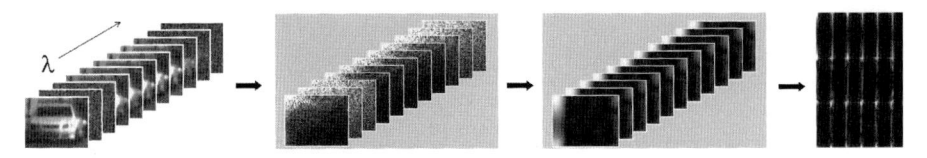

Fig. 7.6 *From left to right*: hyperspectral texture, the band-wise FCT, the distortion-invariant cosine transforms for every band in the image and the raster-scanned 3D matrix **V**

variance of the signal is described based on spectral density. Usually, the variance estimations are performed by using Fourier transform methods (Stein and Shakarchi 2003). Thus, we can make use of the discrete analogue of Eq. (7.27) to recover the kth coefficient for the image descriptor $\mathfrak{G}$, which becomes

$$\mathfrak{G}_k = F(\mathbf{V}) = \sum_{n=0}^{|\mathbb{I}|-1} F(I_n)^* K\left(\frac{\pi}{|\mathbb{I}|}\left(n+\frac{1}{2}\right),\left(k+\frac{1}{2}\right)\right), \quad (7.55)$$

where $|\mathfrak{G}| = |\mathbb{I}|$ and, for the FCT, the harmonic kernel above becomes

$$K\left(\frac{\pi}{|\mathbb{I}|}\left(n+\frac{1}{2}\right),\left(k+\frac{1}{2}\right)\right) = \cos\left(\frac{\pi}{|\mathbb{I}|}\left(n+\frac{1}{2}\right)\left(k+\frac{1}{2}\right)\right). \quad (7.56)$$

In Fig. 7.6, we illustrate the step sequence of the descriptor computation procedure. We depart from a series of bands in the image and compute the band-by-band FCT. With the band FCTs at hand, we apply the distortion correction approach presented in the previous sections to obtain a "power-aligned" series of cosine transforms that can be concatenated into **V**. The descriptor is then given by the cosine transform of **V** over the wavelength index. Note that the descriptor will be three dimensional (3D) in nature, with size $N_x \times N_y \times N_\lambda$, where N_x and N_y are the sizes of the locally planar object patches in the image lattice and N_λ is equivalent to the wavelength range for the hyperspectral image bands. In the figure, for purposes of visualisation, we have raster-scanned the descriptor to display a 2D matrix whose rows correspond to the wavelength indices of the hyperspectral image under study.

In Fig. 7.7 we use a texture plane to better illustrate the distortion correction operation at the band level. Moreover, in Fig. 7.8, we show a sample textured plane with affine distortion. In the figure, we have divided the distorted input texture into patches that are assumed to be locally planar. We then apply the FCT to each of these patches, represented in the form of a lattice on the input image in the left-hand panel. The corresponding power spectra are shown in the second column of the figure. Note that, as expected, the affine distortions produce a displacement on the power spectrum peaks. In the third panel, we show the power spectra after the matrix Φ has been recovered and multiplied so as to obtain the corrected FCTs given by $F(\cdot)^*$. The distortion-corrected textures in the spatial domain are shown in the rightmost panel in the figure. These have been obtained by applying the inverse cosine transform to the power spectra in the third column. Note that, from both the corrected power spectra and the inverse cosine transforms, we can conclude that the

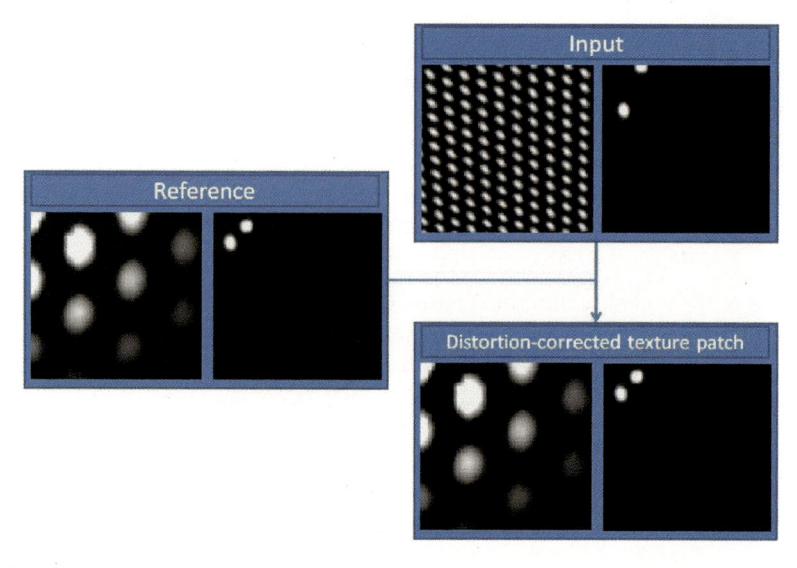

Fig. 7.7 Example of reference, input and distortion-corrected single-band textures. In the panels, the *left-hand image* shows the single-band reference texture, whereas the *right-hand panel* shows the power spectrum of the distorted and affine corrected FCT for the texture under study

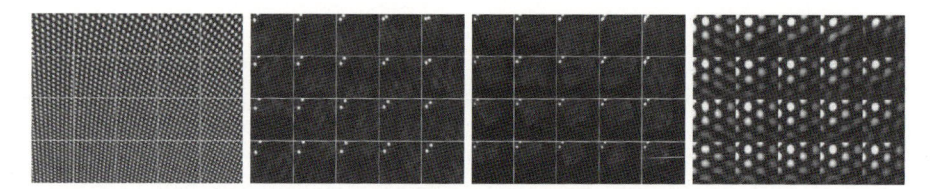

Fig. 7.8 *From left to right*: Affine distortion of a sample single-band image; FCT of the image patches in the *left-hand panel*, distortion-corrected power spectra for the FCTs in the *second panel* and inverse FCTs for the power spectrum in the third panel

correction operation can cope with large degrees of shear in the input texture-plane patches.

Finally, in Fig. 7.9, we show an example of the application of these affine invariant descriptors to pest detection. The process commences with an image of a canister with insects mixed with debris. The task is to identify the insects and separate them from the debris. From the pseudo-colour image at the top of the panel in Fig. 7.9, we can see that this is not a straightforward task. The next step is to recover the descriptors, which are centred at the SIFT (Lowe 2004) key points computed using the luminosity of the spectral image. Making use of these key points, the cosine transform above is used to compute the descriptors. This is possible due to the information compaction property mentioned earlier and the fact that computational methods for the efficient recovery of the FCT are readily available. These descriptors are then taken as input to a support vector classifier (Cristianini and Shawe-Taylor 2000) which has been trained beforehand.

Fig. 7.9 Sample result on pest detection

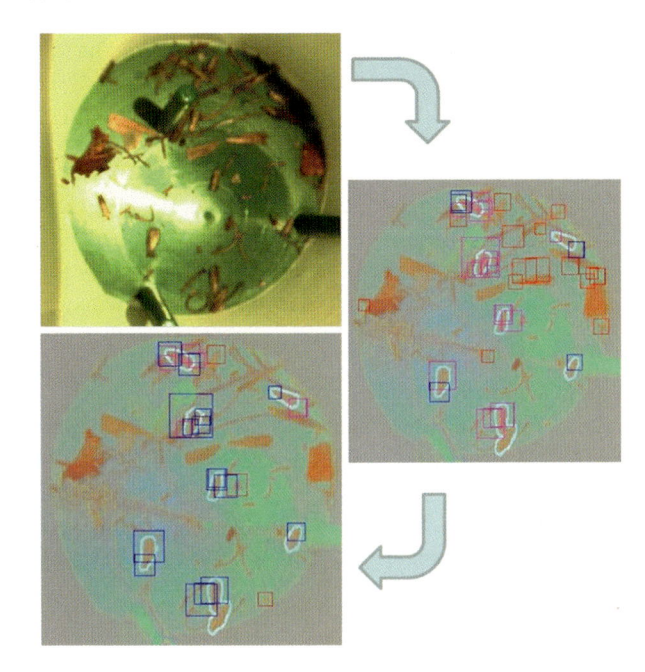

7.3 Automatic Absorption Detection

Hyperspectral imaging and material chemistry can be used for purposes of material identification, based on the notion that different materials have a characteristic response as a function of wavelength. A proof of concept is provided by Clark et al. (2003), who showed that an unknown spectrum can be identified efficiently by simply matching the dominant absorption feature in its reflectance spectrum with those present in a spectral library so as to recover the best fit. The reason is that reflections and absorptions are two complementary concepts of light behaviour. When light is incident on the surface of a material, some of it is reflected, some is absorbed and, in certain cases, some is transmitted.

The problem with absorptions is that they are less straightforward to recover compared to the irradiance measured by image sensors. Nevertheless, absorptions are closely related to the chemistry of the material; therefore, the presence of an absorption band at a certain spectral range is often a "signature" for chemicals and their concentrations. Thus, compared to reflectance-based approaches, the use of absorption features for material identification has a clear physical meaning. This principle has been successfully applied in systems such as the Tetracorder (Clark et al. 2003), which is a software tool for imaging spectroscopy that can be regarded as a collection of algorithms that permits the analysis of conditions in the spectra.

Despite the success of the Tetracorder system, it is still a semi-automatic one. A few typical spectra are selected for reference and their absorption bands are manually labelled in order to perform matching tasks against test spectra. This process involves non-trivial input of expert knowledge. The Tetracorder is based on spectral feature identification algorithms such as that in Clark et al. (1990), where a least-squares fit is used to match the spectrum under study to that in a reference library. As a result, the system performance is dependent on the selected reference spectral bands and purpose-labelled absorptions.

Therefore, it is not only theoretically important, but practically useful to recover absorption features automatically. This is especially relevant since a number of absorption features, such as that of water, chlorophyll and some gases find application in a wide variety of areas in geosciences. For instance, chlorophyll absorption (between approximately 0.4 and 0.7 μm) and leaf cell structure, which reflects near-infrared light, are the basis for the computation of the normalised difference vegetation index (NDVI) (Liang 2004). The NDVI is defined as the difference between the chlorophyll absorption and near-infrared (NIR) bands over their sum and has been employed to estimate biomass (Atkinson et al. 1994; Phinn et al. 1996).

Moreover, water absorption features and the NDVI are related to the leaf area index (LAI) (Roberts et al. 1998; Sims and Gamon 2003). This is understandable since vegetation is expected to have a high water content, whose absorption dominates the NIR band in plant spectral reflectance. In a related application to that of measuring and quantifying biomass, plant stress has been assessed using spectral absorption features. Adams et al. (1999) have used an approximation of the spectra based on three bands to measure chlorosis of leaves in stressed plants. Smith et al. (2004) have tested hyperspectral features in the red edge of the spectrum so as to detect plant stress caused by soil oxygen depletion, a common occurrence in natural gas leaks.

A number of methods have been proposed for the automatic detection of absorption bands. These range from the spectra derivative analysis (Lillesand and Ralph 2000) to scale space algorithms, such as Fingerprint (Piech and Piech 1987) and the maximum modulus wavelet transform (MMWT) (Hsu 2003). All these algorithms are based on the local analysis of the differential structure of the spectra and the subsequent recovery of absorption boundary points. The modelling of absorption features has attracted attention in the geoscience community. Sunshine et al. (1990) have proposed a modified Gaussian modelling (MGM) method to resolve spectra into their physical components.

7.3.1 Automatic Absorption Band Recovery

Derivative Analysis

We begin by reviewing derivative analysis (Butler and Hopkins 1970; Demetriades-Shah et al. 1990; Fell and Smith 1982; Lillesand and Ralph 2000), which is based on the extraction of interest points from the spectra making use of its derivatives with respect to wavelength index. Since an absorption band can be treated as a "dip" in the spectrum, we can locate these "dips" by recovering the spectral bands at which its second derivative changes sign or the first derivative is at an optimum, i.e. the

7.3 Automatic Absorption Detection

Fig. 7.10 (a) First and (b) second derivatives of a sample reflectance spectrum

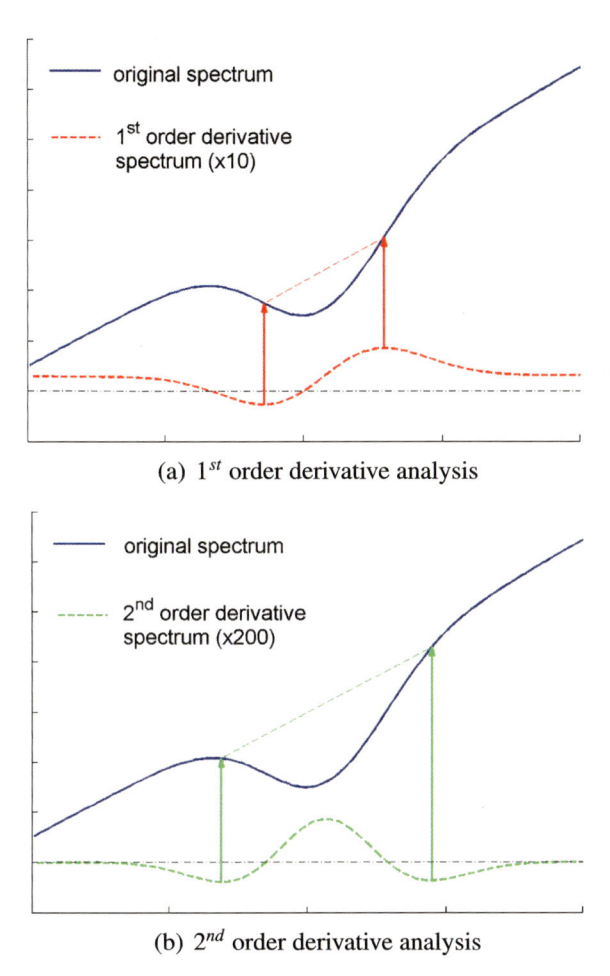

(a) 1st order derivative analysis

(b) 2nd order derivative analysis

inflection points of the spectrum. This can be achieved by making use of derivative analysis in two ways. The first of these is by taking the first-order derivative of the spectrum and finding the local maxima. The second route is by computing the second-order derivative and finding its zero-crossing points, i.e. where the trace of the second derivative changes sign.

An example of this is shown in Fig. 7.10, where the sample spectrum has been plotted using a solid curve. The first-order and second-order derivatives are those traces plotted using dashed lines. In Fig. 7.10, the sample spectrum has a single large absorption whose end points are given by the local minimum and maximum of the first-order derivative, as indicated by the dashed lines and upward-pointing arrows in the panel. This is equivalent to the recovery of the zero-crossing points in the second-order derivative of the example spectrum. For clarity, we have scaled the first- and second-order derivative traces by a factor of 10 and 200, respectively. In our plots, we have also shown the zero reference level making use of a dotted horizontal line. As shown in Fig. 7.10(a), the absorption band recovered by the first-order derivative analysis is quite narrow and does not cover the full range of the absorption feature. This can be resolved, in part, by computing the second-order derivative of the spectrum and, instead of making use of the zero-crossings for the absorption recovery process, by employing its local maxima as an alternative, as shown in Fig. 7.10(b). Here, the end points of the absorption band correspond to the local maxima of the second-order derivative of the spectrum, with a local minimum in the extreme point of the recovered absorption. It is clear that the absorption band recovered by the second-order derivative analysis covers a wider range than that corresponding to the first-order derivative analysis. This observation is an important one, since it constitutes the foundation for the MMWT method (Hsu 2003), which will be introduced later on.

The derivative analysis method, however, is extremely sensitive to noise, even after applying filtering and signal processing techniques as a preprocessing step. To illustrate this, we have added Gaussian noise to the example spectrum in Fig. 7.10. We apply derivative analysis to the noise-corrupted spectrum shown in Fig. 7.11. In Fig. 7.11(a) and (b), we show the recovered first- and second-order derivative traces, plotted in a similar fashion to that in Fig. 7.10.

From the panels, it is clear that derivative analysis is prone to error due to noise. In Fig. 7.11, the effect of the added Gaussian noise is significant, despite a small variance and a large signal-to-noise ratio (SNR). Moreover, from the traces, we can conclude that the second-order derivative is even more prone to noise corruption than the first-order one.

To overcome noise effects, Lillesand and Ralph (2000) proposed the use of Savitzky–Golay filters as a preprocessing step in hyperspectral derivative analysis. This approach aims at removing the noise before the spectrum is processed. Thus, the performance of the algorithm is dependent on the parameters of the filter bank. An example of this is shown in Fig. 7.12(a) and (b). Here, we have used a third-order Savitzky–Golay filter and exemplified the effects of tuning the filter in a non-optimal fashion. Note the presence of numerous small absorption features whose end points, indicated by the upward-pointing arrows, have been recovered making use of the local extrema of the first- and the second-order spectrum derivatives.

Scale Space Methods

Both Fingerprint (Piech and Piech 1987) and the maximum modulus wavelet transform (MMWT) (Hsu 2003) are scale space methods. The purpose of these methods is to analyse the signal under study at different scales so as to recover suitable operators for the task in hand from its scale space representation (Witkin 1983). In the case of spectral analysis, this scale space representation can be achieved via a convolution with Gaussian kernels of continuously increasing variances (Koenderink 1984). For purposes of absorption detection, the problems that concern us are twofold: firstly, the identification of absorption features in the spectra, and secondly, the localisation of the relevant absorption bands. For robust identification, we

7.3 Automatic Absorption Detection

Fig. 7.11 (**a**) First and (**b**) second derivative analysis on the spectrum yielded by adding noise to the reflectance spectrum in Fig. 7.10

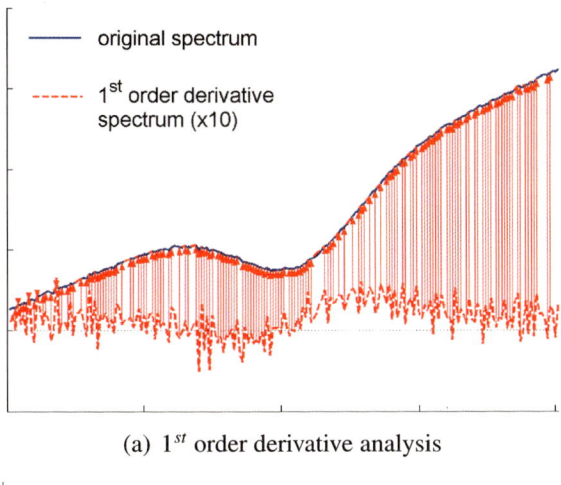

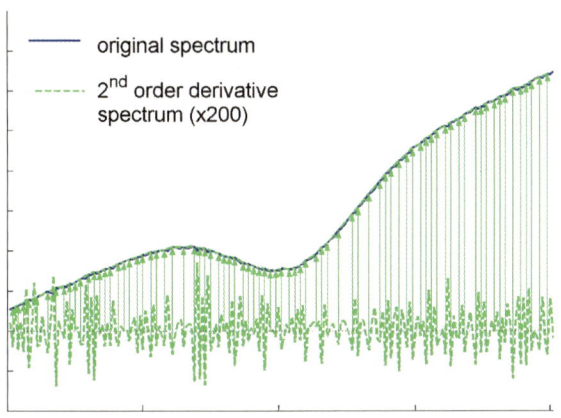

(a) 1st order derivative analysis

(b) 2nd order derivative analysis

aim at removing noise at a coarse scale. However, for accurate localisation, we make use of detailed information at finer scales. This leads to the general framework of scale space analysis: make use of the coarse scale to identify the absorption feature and locate the relevant bands by down-tracking them from coarse to fine scales.

Fingerprint Fingerprint (Piech and Piech 1987) is a method for recovering the absorption bands of the spectrum from its scale space image. The scale space image is a set of progressively smoothed traces computed from the original spectrum. As mentioned earlier, the smoothing of the spectrum is effected by convolving it with a Gaussian kernel with variance σ and mean μ as follows:

$$F\big(Q(\lambda), \sigma\big) = Q(\lambda) * G(\mu, \sigma), \tag{7.57}$$

where $G(\mu, \sigma) = \exp(-\frac{(\lambda-\mu)^2}{2\sigma^2})$ is a Gaussian kernel and $Q(\lambda)$ is the raw spectrum, i.e. the image radiance at a particular pixel. The scale space image then comprises all

Fig. 7.12 (a) first and (b) second derivative analysis on the noise-corrupted reflectance spectrum in Fig. 7.11 which has been smoothed using a Savitzky–Golay filter

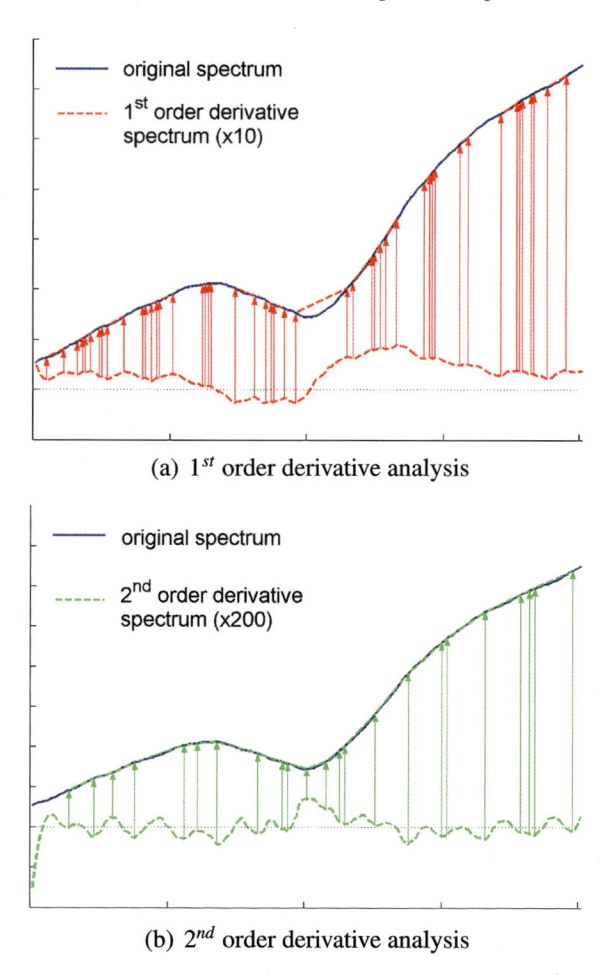

(a) 1st order derivative analysis

(b) 2nd order derivative analysis

the functions $F(Q(\lambda), \sigma)$ for a set of zero-mean Gaussian kernels with increasing values of σ. An example of the scale space for a sample spectrum is shown in Fig. 7.13. In the figure, we show the coarse-to-fine representation of the spectrum in a top-to-down fashion, i.e. the coarsest scale is represented by the topmost trace. As we can see, as the value of σ increases, i.e. the scale is coarser, the spectrum is smoother.

The Fingerprint method hinges on the rationale that inflection points which are stable across the scale-space representation of the spectrum should capture absorption features in a robust manner. The method recovers these inflection points, for each of the spectrum scale images, using first-order derivative analysis. There are many false inflection points at the fine scales due to signal noise. These false positives will be eliminated by making use of coarser scales. If an optimum is detected in nearby positions at adjacent scales, it is then likely to be either the beginning or the end of an absorption band. However, the inflection points recovered from the coarser

7.3 Automatic Absorption Detection

Fig. 7.13 A sample reflectance spectrum convolved with Gaussian kernels at different scales

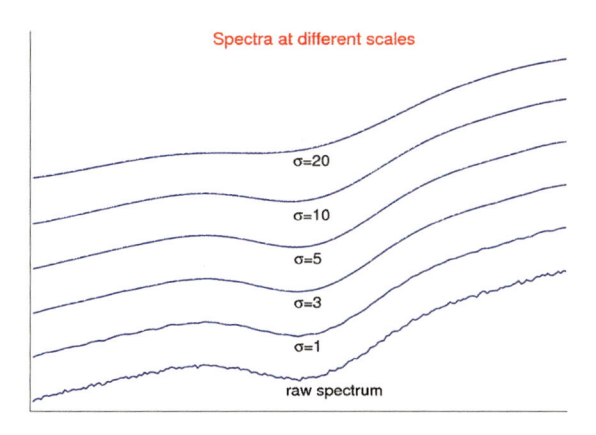

scales may be inaccurate, as the spectrum at these scales is often over-smoothed. Thus, the absorption end points are recovered by backtracking from coarse-to-fine scale images.

At this point, we note that the smoothing and differentiation steps can be computed efficiently by convolving the spectrum with the derivative of the Gaussian kernel. This is due to the fact that the convolution and the differentiation operations are commutative. This procedure can then be viewed as that of smoothing the spectrum with the Gaussian kernel $G(\mu, \sigma)$ and then taking its derivatives. We can formulate this by using the shorthand $x = \lambda - \mu$ as follows:

$$F\big(Q(\lambda), \sigma\big)\big|_{\sigma_i} = Q(\lambda) * G_x(\mu, \sigma), \tag{7.58}$$

where we have written $F(Q(\lambda), \sigma)|_{\sigma_i}$ to imply that the scale image is taken at the ith value of σ and

$$G_x(\mu, \sigma) = \frac{\partial G(\mu, \sigma)}{\partial x} = \frac{x}{\sigma} \exp\left(-\frac{x^2}{2\sigma^2}\right), \tag{7.59}$$

is the derivative of the Gaussian kernel with respect to the difference between the mean μ and the spectrum $Q(\lambda)$.

By finding the maxima and minima of $F(Q(\lambda), \sigma)$, we are then recovering the inflection points of the spectrum at the ith scale. This is the Fingerprint representation of the spectra. A Fingerprint example is shown in Fig. 7.14. The white traces in the panel indicate the local maxima of the first-order derivatives at different scales. The black lines indicate the local minima of the first-order derivative. An absorption band is recovered by a pair of minima-maxima lines. The minima traces indicate the beginning of the absorption, whereas the maxima evidence the end of the feature. Detection is accomplished at a large value of σ, i.e. a coarse scale. However, the end points of the absorption feature are given by the extrema at finer scales, which have been backtracked from the scale images corresponding to larger values of σ. In the figure, we have indicated the end points of the absorption feature using red vertical arrows.

Fig. 7.14 Absorption band recovered by backtracking across the scale space using the Fingerprint method

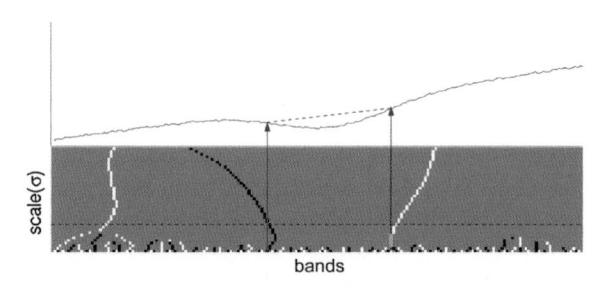

Figure 7.14 shows that Fingerprint is quite robust to noise-induced artifacts in the spectrum. Nonetheless, the recovered absorption appears narrower compared to the true feature. Moreover, the scale at which detection is made plays an important role in the identification of the absorption features. The variance σ at which the detection task takes place can be viewed as a cut-off value τ in the scale space (shown in the figure as a dashed horizontal line), which is often chosen empirically.

Maximum Modulus Wavelet Transform (MMWT) The Fingerprint method can be generalised to the maximum modulus wavelet transform (MMWT) (Hsu 2003) by making use of second-order derivatives. The idea is to apply a wavelet transform to the spectrum using the second-order derivatives of Gaussian (DoG) wavelets with continuously varying scale parameters. As in the Fingerprint approach, the absorption is then recovered using the extrema in the scale space spanned by the wavelet transforms. For the ith scale, the wavelet transform of the spectrum $Q(\lambda)$ is given by

$$F\big(Q(\lambda), \sigma\big)\big|_{\sigma_i} = \int Q(\lambda) \psi_{\mu,\sigma}(\lambda)\, d\lambda, \tag{7.60}$$

where $\psi_{\mu,\sigma}(\lambda) = \frac{1}{\sqrt{\sigma}} \psi(\frac{\lambda-\mu}{\sigma})$ is a scaled and translated Gaussian wavelet function.

Note that for the second-order DoG wavelet, Eq. (7.60) is equivalent to the second-order derivative analysis for fixed values of σ. This is due to the fact that the wavelet transform can be viewed, for the second-order DoG wavelet, as the smoothing and computation of the second-order spectrum derivative.

We are interested in the local maxima and minima of the scale space representation $F(Q(\lambda), \sigma)$. By relating the local maxima and minima at different scales to one another, we can obtain the positive and the negative minimal traces across the scale space. Each absorption is then characterised by end points which correspond to negative minimal traces with a positive maximal line in the middle point of the feature. As in the Fingerprint method, the detection is made at large values of σ and is dependent on a cut-off value. The localisation is achieved by backtracking the modulus maximal lines down to the finest scale.

An example of MMWT for absorption band detection is shown in Fig. 7.15. The white traces indicate the positive maximal lines, and the black traces show the negative minimal lines. The absorption feature is indicated using upward-pointing arrows, and the cut-off value is given by the horizontal dashed line. From the figure, we see that the absorption recovered by this method covers a larger range than that recovered by the Fingerprint algorithm.

Fig. 7.15 Absorption band recovered by backtracking across the scale space using the MMWT

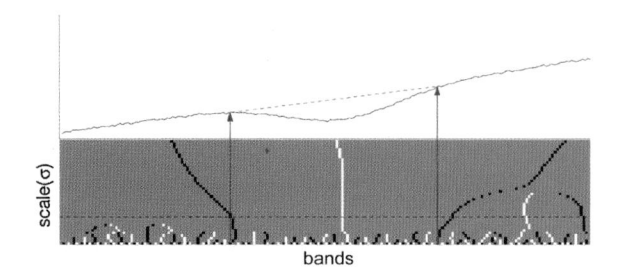

Uni-Modal Segmentation

All the methods for automatic absorption band recovery presented so far are based on the local analysis of the spectra derivatives, either in a single scale or a multiple-scale fashion. An alternative is to use the global shape of the spectrum.

If the global shape of the spectrum is convex, and a small region in it is concave, then this region will be regarded as an absorption. Thus, we can recover absorption features in a two-step fashion. First, we can model the shape of the spectrum as a convex hull with the continuum of the spectrum as an upper bound. Second, the absorption bands can then be viewed as local spectral segments which break the convexity at their corresponding local neighbourhoods. Viewed in this manner, the spectrum, as a function of the wavelength λ, can be treated as points distributed in a 2D space. The convex hull, which is the minimum convex polygon enclosing all these points, can be found using readily available algorithms. To illustrate this treatment, in the upper panels of Fig. 7.16(a) and (b) we show, in dashed lines, the continuum curve recovered using the convex hull for two sample spectra.

As shown in Fig. 7.16, the continuum of a spectrum is given by its convex hull so as to connect the spectrum local maxima. As a result, it can be viewed as an envelope that isolates the absorption bands in the reflectance spectra. In other words, the use of the convex hull splits the original spectrum into several spectral segments defined over local maxima. These segments assume a concave shape against the continuum.

After recovering the continuum, we can perform continuum removal on the spectrum. This is a standard procedure in remote sensing to isolate the local absorption feature from other photometric effects (Clark and Roush 1984). The removal operation is a simple normalisation of the spectrum $Q(\lambda)$ with respect to the continuum curve $C(\lambda)$, such that the continuum removed spectrum $r(\lambda)$ is given by

$$r(\lambda) = \frac{Q(\lambda)}{C(\lambda)}, \tag{7.61}$$

where, as before, λ is the wavelength variable.

Examples of continuum removed spectra are shown in the bottom panels of Fig. 7.16. The spectra shown here have been normalised to unity and inverted in order to turn absorptions into peaks. As we can see, the spectral segments in Fig. 7.16(a) are uni-modal, i.e. they only describe a single peak in the spectrum. In contrast, the segment in Fig. 7.16(b) is multi-modal with two peaks. These two peaks represent two absorption bands.

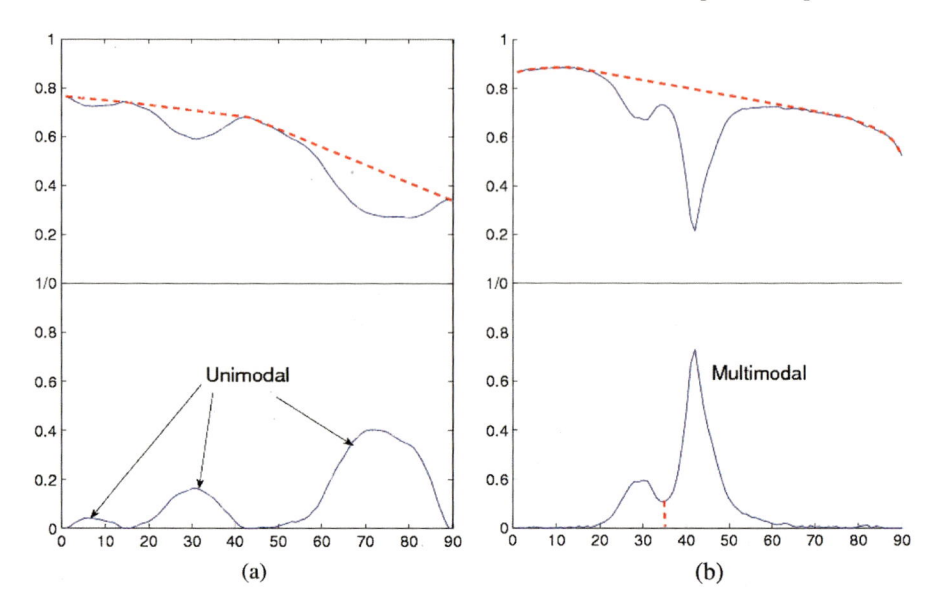

Fig. 7.16 *Top row*: sample reflectance spectra and their continuum; *Bottom row, left-hand panel*: continuum removal result for the uni-modal spectrum in the *top row*; *Bottom row, right-hand panel*: continuum removal result for the multi-modal spectrum in the *top row*

To take our analysis further and recover the absorption bands, we make use of uni-modal regression. The idea is to further split any number of segments into absorption bands, each of which satisfy uni-modal constraints via uni-modal regression (Stout 2002).

Uni-modal regression can be implemented through isotonic regressions. The purpose of an increasing/decreasing isotonic regression is to fit a monotonically increasing/decreasing piecewise smooth line to the original sequence so as to minimise the distance between them. Mathematically, denote by $\{y_i \mid i = 1, \ldots, n\}$ a sequence of spectral measurements of order n. The isotonic regression of the data is the set $\{\hat{y}_i \mid i = 1, \ldots, n\}$ that minimises

$$\sum_{i=1}^{n} |y_i - \hat{y}_i|^2,$$

s.t. $\hat{y}_1 \leq \cdots \leq \hat{y}_n$ for increasing isotonic regression, (7.62)

$\hat{y}_1 \geq \cdots \geq \hat{y}_n$ for decreasing isotonic regression.

In the equations above, we have used the L_2 Euclidean distance. Note that other distance measures can also be used. An advantage of using L_2 distance is that isotonic regression can be determined in $O(n)$ time by using the pool adjacent violators (PAV) algorithm (Ayer et al. 1955). PAV for increasing isotonic regression is defined as follows. For the spectral measurements $\{y_i \mid i = 1, \ldots, N\}$ ordered in decreasing

7.3 Automatic Absorption Detection

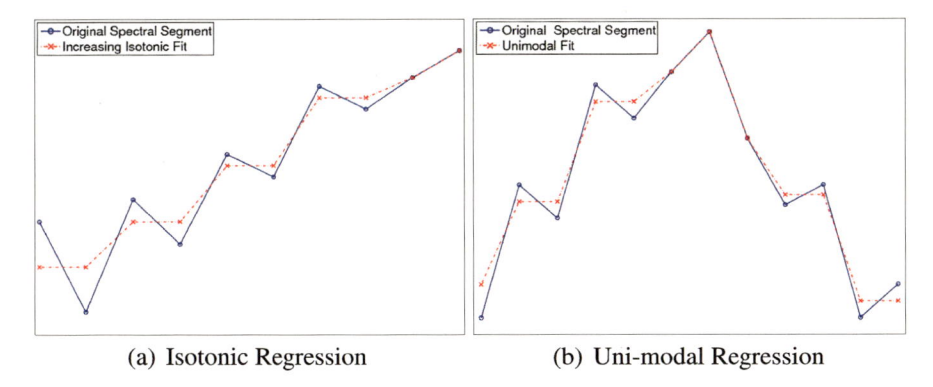

Fig. 7.17 Sample spectra and fitting results for isotonic (**a**) and uni-modal (**b**) regression

rank, i.e. $y_i \geq y_{i+1} \geq \cdots \geq y_j$, the PAV is given by

$$\hat{y}_k = \begin{cases} \frac{\sum_{k=i}^{j} y_k}{j-i+1}, & k \in [i, j], \\ y_k, & \text{otherwise.} \end{cases} \quad (7.63)$$

That is, each decreasing rank sequence is replaced by the mean value on the interval $[i, j]$, while non-decreasing values are preserved. PAV for decreasing isotonic regression can be defined likewise. In Fig. 7.17(a), we show an example of increasing isotonic regression. In the figure, the original signal has been plotted using a solid blue trace. The corresponding increasing isotonic regression result is given by the dash-dotted plot.

Uni-modal regression can be effected by locating a splitting point y_m in the sequence. Once y_m is at hand, we apply increasing isotonic regression to the left-hand segment of the sequence and decreasing isotonic regression to the right-hand segment. This is done to minimise the error function in Eq. (7.62) subject to the uni-modal constraint $\hat{y}_1 \leq \cdots \leq \hat{y}_m \geq \hat{y}_{m+1} \cdots \geq \hat{y}_N$. Examples of isotonic and uni-modal regression are shown in Fig. 7.17.

A straightforward way to implement uni-modal regression is to treat each point in the sequence as a candidate splitting point and apply increasing/decreasing isotonic regression to the left/right segments. The optimal splitting point is that whose sum of squared errors is minimum for the pair of isotonic regressions. Note that the approach above has a computational complexity of $O(n^2)$. An improved uni-modal regression algorithm is proposed in Stout (2002) that makes use of prefix isotonic regression, which is a method to determine the errors of isotonic regression for the interval $\{y_1, \ldots, y_i\}$ with $i = 1, \ldots, N$ in a single scan of the sequence. This is achieved via dynamic programming. A detailed review on prefix isotonic regression is beyond the scope of this book. Interested readers are referred to Stout (2002). By applying prefix isotonic regression to both the original and the inverted sequences, the optimal splitting point can be obtained with a complexity of $O(n)$.

With these ingredients, the step sequence for the absorption band detection algorithm is as follows.

1. Recover the convex hull over the spectrum using the Graham scan algorithm (Graham 1972).
2. Apply continuum removal and invert the continuum removed spectrum.
3. Perform uni-modal regression. The stopping criterion for the fitting procedure is given by the threshold ϵ. If the segment is not uni-modal, i.e. the fitting error is above the threshold ϵ, split the segment at the local minima.
4. Merge any two adjacent segments if the uni-modal constraint holds for their union. This is done, again, based on the threshold ϵ.
5. Interleave steps 3 and 4 until no further splitting and merging operations can be performed.

The main drawback of uni-modal regression is its tendency to generate false positives. In our aim of computation, this implies that the algorithm may yield many spurious, small absorptions when processing noisy spectra. To overcome this problem, a postprocessing step can be applied by eliminating the absorptions whose magnitude is negligible or those that cover a very small number of bands. Despite this, the main absorption features will still be well preserved by uni-modal segmentation.

Examples of absorption band detection for two example spectra are shown in Fig. 7.18. In contrast with the Fingerprint method, the absorption bands recovered by uni-modal segmentation are in better accordance with the actual absorptions in the spectra. Furthermore, the MMWT cannot recover all the absorptions in either of the spectra under study. The MMWT is less robust than the other two methods due to the use of second-order derivatives.

7.3.2 Complexity Analysis

In this section, we examine the computational and algorithmic complexity of the methods described above. Recall that the Fourier transform played an important role in the automatic absorption algorithms presented as alternatives to our uni-modal segmentation method. By making use of the fast Fourier transform (FFT), the Fourier transform can be computed in $\mathcal{O}(n \log n)$ floating-point operations, where n is the number of bands in the spectrum under study. Similarly, for a convolution mask of size k, multiplication in the frequency domain only takes $\mathcal{O}(kn)$ floating-point operations. Thus, the overall complexity is $O(kn) + 2\mathcal{O}(n \log n)$, which corresponds to the combined costs of the application of an FFT, its inverse and the multiplication operation in the frequency domain.

When using signal processing techniques in the frequency domain, the complexity of the derivative analysis method will be given by $O(kn) + 2\mathcal{O}(n \log n)$ plus the overhead introduced by the filtering algorithm or preprocessing step of choice. Note that Savitzky–Golay filters can be viewed as a finite impulse response filter (Oppenheim et al. 1999) governed by a least-squares fit to the spectrum, where the window size and the order of the fitting polynomial are parameters set by the user.

7.3 Automatic Absorption Detection

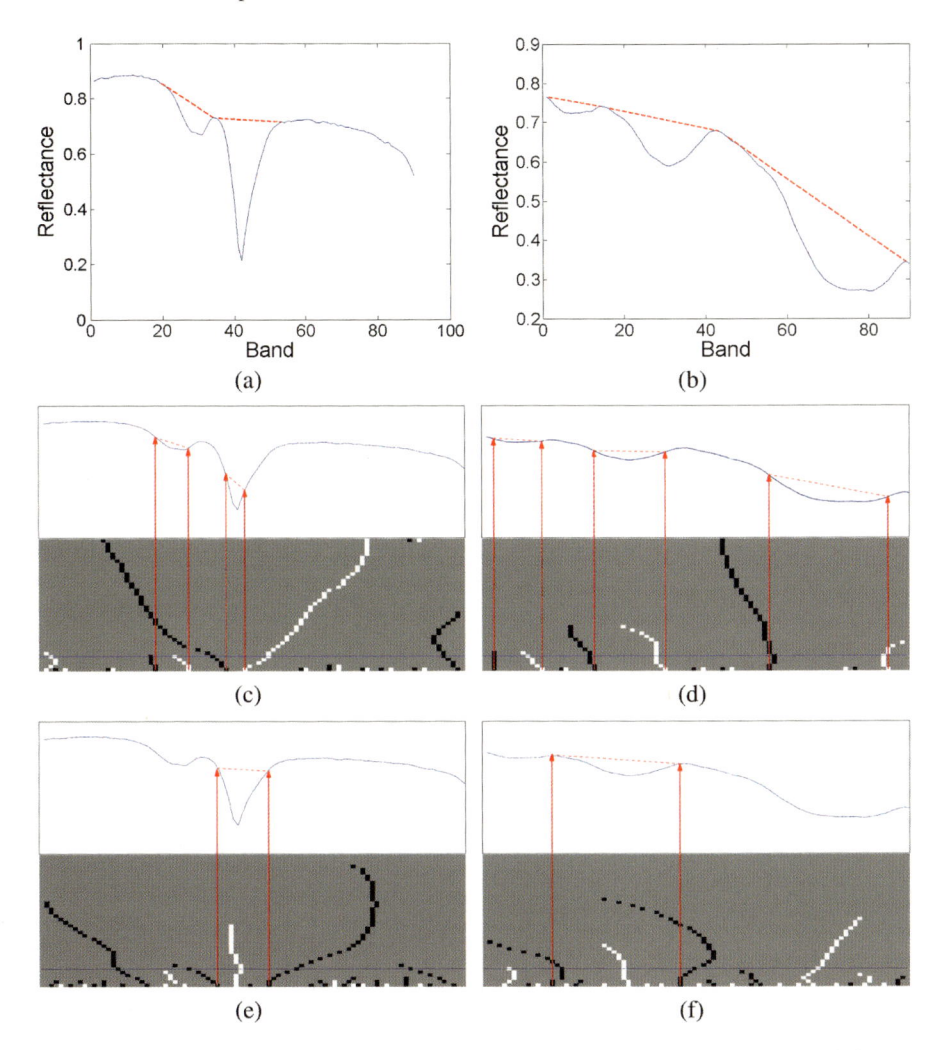

Fig. 7.18 Absorption band detection results. *First row*: sample reflectance spectra and their continuum. *Second row*: absorption bands yielded by the Fingerprint method. *Third row*: absorption bands yield by the MMWT

Scale space methods, i.e. Fingerprint and MMWT, are more computationally demanding due to the fact that first- and second-order derivatives must be computed for every scale. Here, the number of scales and their increments are parameters chosen by the user. In terms of complexity, for m scales, this is given by $m(O(kn) + 2\mathcal{O}(n \log n))$ plus the cost of the backtracking detection algorithm used to find the maxima and minima in the scale space. In our case, we have used a simple backtracking approach based on a nearest-neighbour search (Webb 2002) over the minima and maxima at every set of successive scales.

For uni-modal segmentation, the bottleneck is in the extraction of the convex hull and the uni-modal regression. Along these lines, the complexity for convex hull extraction is $O(n \log n)$, where n is the length of the spectrum. By using the implementation in Stout (2002), uni-modal regression can be achieved with a cost of $O(n \log n)$ floating-point operations. In fact, the cost of uni-modal regression is much reduced by the nature of our application, since we seldom have to apply it to the whole spectrum. This is due to the continuum removal operation where those bands which comprise the convex hull can be removed from further consideration at the uni-modal regression step.

In addition to the computational cost, we also examine the algorithmic complexity of these methods. This is important, because algorithmic complexity often determines the scalability of the method. Hence the simpler the algorithm, the more general situations to which it applies. On the other hand, the more complex the algorithm, the more accurate the underlying model. This is in agreement with the well-known Occam's razor theory (MacKay 2003), which states that models, and hence the algorithms they embody, should not be complicated beyond necessity. This complexity can, to some extent, be characterised by the number of parameters used in the algorithm.

For the derivative analysis, we use two parameters, i.e. the Savitzky–Golay polynomial order and window size. For scale space methods, both Fingerprint and MMWT require four parameters to be set by the user. These are the range defined by the initial and final values of the scale, i.e. σ_o and σ_m, the step of the scale increment, and the cut-off value τ. Despite being inefficient, the window size can be chosen to be the same length of the signal. Nonetheless, the choice of other parameters needs some scrutiny and will influence the performance. For example, in Fig. 7.18(c) and (d), the value of τ has been set to 5. If it had been chosen to be 10, then some absorption bands would not have been detected. However, $\tau = 10$ might be the right choice for other spectra. Similarly, the scale increment value is also important. In theory, it should be chosen as small as possible, but it is worth noting that the computational cost increases as does the number of scales m in the range $[\sigma_o, \sigma_m]$.

The uni-modal segmentation method, on the other hand, has only one parameter, i.e. the tolerance ϵ of the fitting error for the regression step. Hence, the performance of the uni-modal segmentation algorithm is easier to control, with only one parameter to tune. Furthermore, by normalising the spectrum to unity, the choice of the tolerance parameter is better constrained and the algorithm results are more robust to variations of ϵ.

7.3.3 Absorption Representation and Classification

As mentioned earlier, absorption bands have been modelled via parametric representations (Sunshine et al. 1990; Brown 2008). Here, an absorption band is represented by a parametric function after applying a predefined transformation to the original reflectance spectrum as a preprocessing step. Typical transformations comprise

7.3 Automatic Absorption Detection

Fig. 7.19 Plots of functions used for the parametric modelling of absorption bands

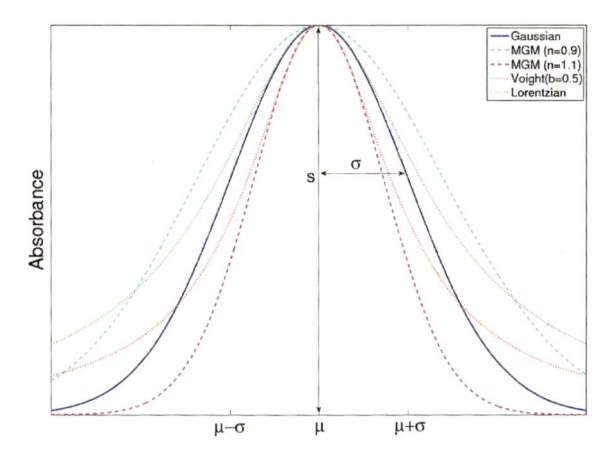

Table 7.1 Summary of the functions used for parametric modelling of absorption bands. In the table above, β and ϑ are user-defined parameters

Function	MGM (Sunshine et al. 1990)	Voight (Brown 2008)	Lorentzian (Brown 2008)
$g(Q(\lambda))$	$s \cdot \exp -\frac{(Q(\lambda)^\vartheta - \mu^\vartheta)^2}{2\sigma^2}$	$\frac{\upsilon}{[1+0.5\beta^2(Q(\lambda)-\mu)^2/\sigma^2]^{1/\beta^2}}$	$[\frac{2\upsilon\sigma^2}{2\sigma^2+(Q(\lambda)-\mu)^2}]$

converting wavelength to wave number in the x-axis and reflectance to negative log-reflectance (Brown 2008) or applying continuum removal to the log-reflectance spectrum plotted in the wave number space (Sunshine et al. 1990). The main purpose of these preprocessing steps is to convert reflectance to apparent absorption represented as a function of energy, as related to the wave number.

To represent the spectrum $Q(\lambda)$, a standard choice of the parametric function in the transformed spectrum space is the Gaussian function,

$$g(Q(\lambda)) = \upsilon \cdot \exp\left(-\frac{(Q(\lambda) - \mu)^2}{2\sigma^2}\right), \quad (7.64)$$

where υ is a proportionality constant and μ and σ are the centre and the bandwidth of the absorption band, respectively. Other choices include the modified Gaussian model (MGM) (Sunshine et al. 1990), the Lorentzian function (Brown 2008) and the Voight function (Brown 2008). These functions are plotted in Fig. 7.19 and summarised in Table 7.1.

Absorption band modelling techniques are specially useful in resolving overlapping absorptions. This is exemplified in Fig. 7.20(a), where three absorption bands are modelled as single Gaussian functions. Another advantage is their inherent relation to material chemistry. However, absorption modelling does not provide a direct means for matching different absorption features for purposes of material mapping and classification. A straightforward choice here is to use the parameters themselves as the input vectors to the mapping process. For instance, in the case of Gaussian functions, we can compare different absorptions by employing their central posi-

Fig. 7.20 Absorption band representations. (**a**) A parametric representation of the absorption bands in a sample reflectance spectrum. (**b**)–(**c**) Non-parametric representations of the absorption bands

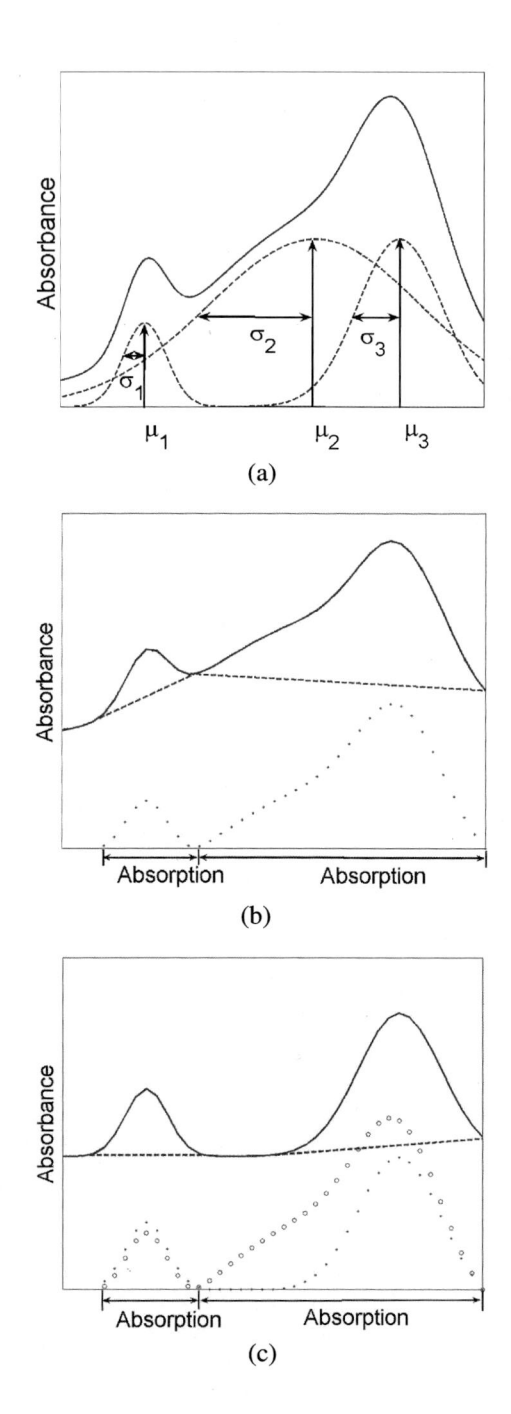

7.3 Automatic Absorption Detection

tions μ and bandwidths σ. Yet, this representation inevitably loses discriminant information due to the lower order statistics which it comprises.

Here we adopt a non-parametric representation of the absorptions by using the spectral values within the absorption bands as feature vectors. This enables us to discriminate between both separable and overlapping absorption bands without making use of lower order statistics. As a result, robust material mapping can be effected by discriminating between spectra with different absorption bands based on their shapes. To illustrate this, consider the continuum removed absorption spectrum shown in Fig. 7.20. As the overlapping bands in Fig. 7.20(b) vanish, the resulting spectrum becomes that shown in Fig. 7.20(c). Making use of the spectral band values of the continuum removed spectrum segment within the two remaining absorptions, it is still possible to discriminate between the two different spectra in the figures. Their non-parametric representations are plotted together in Fig. 7.20(c) using cyan and red dotted lines, respectively.

Another advantage of non-parametric absorption feature representation for material mapping is low computational cost. For classification, we only have to recover the absorption bands of reference spectra and store their spectral signatures for purposes of testing. Absorption modelling methods rely on curve fitting, which must be applied to each reference spectrum as well as unknown testing spectra. Therefore, parametric modelling of absorption bands for purposes of material mapping on large image sizes is potentially computationally expensive. In contrast, for a non-parametric representation, we can apply local analysis to the spectral range of interest as determined by the absorption bands detected from the reference spectra. This further reduces the computational complexity of the classification process.

We now proceed to illustrate the behaviour of the methods presented earlier on continuum removed spectra. In Fig. 7.21, we show example absorption detection results for some selected sample spectra from the USGS spectral library.[1] Example results for data acquired in-house making use of a spectrometer are shown in Fig. 7.22. From the figures, we can conclude that, especially for the mineral spectra in the USGS spectral library, the uni-modal segmentation algorithm achieves better results in terms of the area and separation of the recovered absorptions. Fingerprint and MMWT can also locate most of the absorption bands, but they are unable to recover the full absorption area, with the MMWT generally recovering a larger absorption bandwidth than the Fingerprint method. Scale space methods may not be able to recover the absorption feature in cases where the local spectral curve is not differentiable, which may be the case with two adjacent absorption bands. This is a common occurrence in many mineral spectra, as shown by the results on the alunite and the neodymium oxide spectra in Fig. 7.21.

The uni-modal segmentation algorithm can recover the full extent of the absorption feature. Nonetheless, it may fail to perform well in cases where the continuum is not convex-shaped. Examples of this are the results on Eucalyptus leaf powder shown in Fig. 7.22. Furthermore, the uni-modal assumption on the absorption features may hold for mixed absorptions, as in the case of the Navel leaf spectra shown

---

[1] http://speclab.cr.usgs.gov/spectral-lib.html.

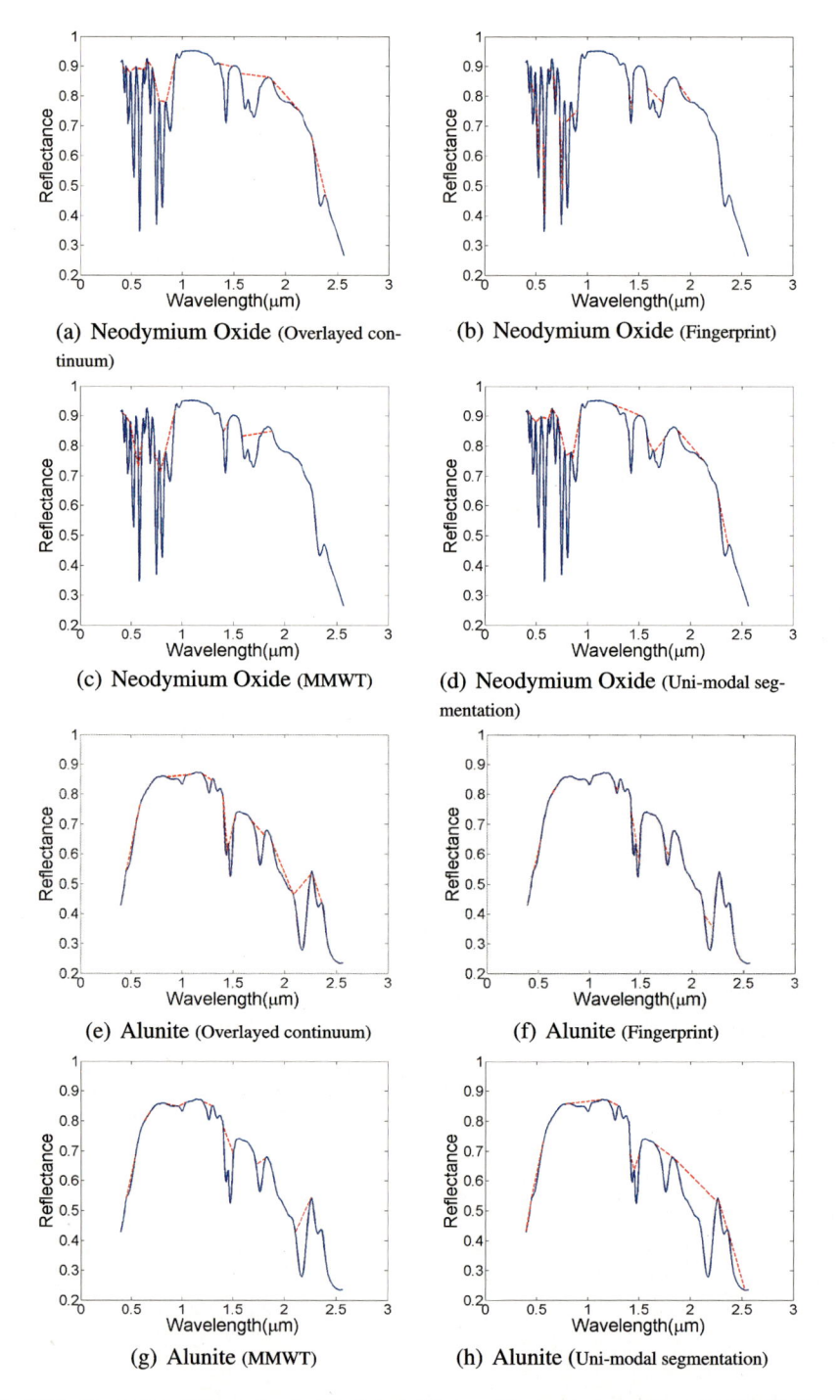

Fig. 7.21 Examples of absorption bands detected for sample spectra taken from the USGS library

7.3 Automatic Absorption Detection

(a) Eucalypt powder (Overlayed continuum)

(b) Eucalypt powder (Fingerprint)

(c) Eucalypt powder (MMWT)

(d) Eucalypt powder (Uni-modal segmentation)

(e) Navel leaf (Ground truth)

(f) Navel leaf (Overlayed continuum)

(g) Navel leaf (MMWT)

(h) Navel leaf (Uni_modal segmentation)

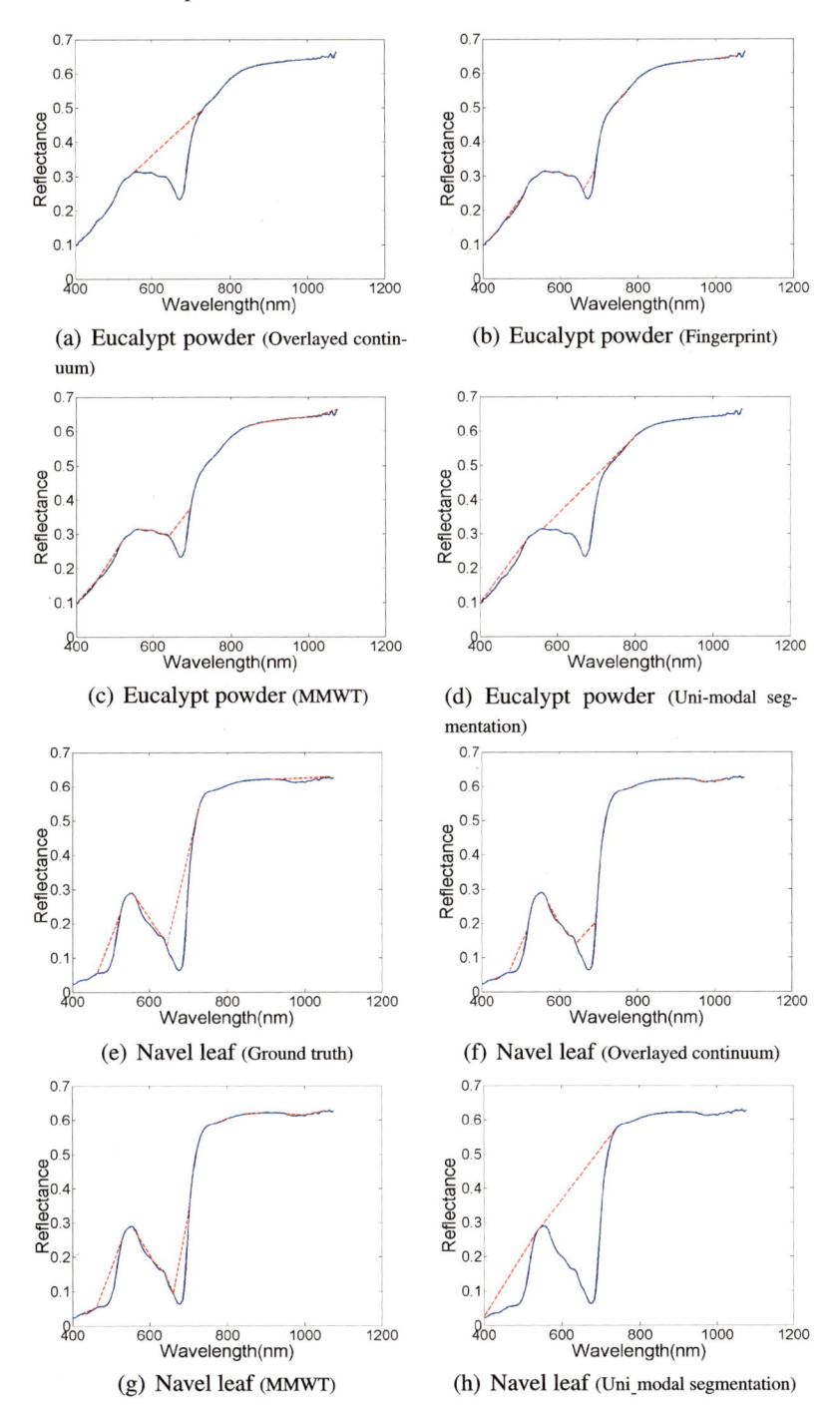

Fig. 7.22 Examples of absorption band detection results for spectroscopy data acquired by a spectrometer

in Fig. 7.22. Here, the two absorption features, one shallow segment around 600 nm and the deep band around 670 nm, were mistakenly detected as a single absorption by the uni-modal segmentation algorithm. The reason is that the absorption around 670 nm is too strong and it makes the shallow band less prominent. As a result, their mixture forms only one peak after continuum removal and cannot be separated simply based on the uni-modal constraint. From our comparison results, we can assert that there is no single method which consistently give optimal results for all the spectra in the datasets. In our experience, the uni-modal segmentation algorithm performs better for the mineral spectra in the USGS spectral library. This is due to the fact that, for mineral spectral data, the cusps and valleys in the spectra are more obvious. For organic material, like plants and leaves, whose absorption features are "shallow", the MMWT is, in general, a good option.

7.4 Notes

The use of some of the representations presented here permits interpolation at any point in the spectral domain via a set of numerically stable algorithms. This is particularly true for the splines and mixtures presented earlier, where spectra of dissimilar lengths can be compared on a consistent, continuous basis. This provides the added benefit of allowing operations such as absorption band detection to be effected on the continuous representation using tools from functional analysis rather than discrete approximation. Traditionally, absorption features are detected as signatures for chemical classification, through the analysis of the second derivative of the spectrum with respect to the wavelength (Butler and Hopkins 1970). Often the derivatives are evaluated using techniques similar to finite differences, which are inaccurate and noise-sensitive. With a continuous closed-form representation of reflectance spectra, derivative analysis can be performed analytically at any point in the spectral domain.

As opposed to other basis functions, splines possess a number of suitable properties for the representation of spectral data. Due to their local support property, noise corruption in the input data only affects a local spectral neighbourhood of the representation. As a result, the descriptor presented here is robust to noise and local perturbations in the spectra. Hence, this is an advantage over other choices of basis functions, such as Gaussian functions or wavelets (Angelopoulou et al. 2001). On the other hand, the flexibility to control the degree of the splines and the number of basis functions delivers a good trade-off between goodness of fit and representation complexity.

Finally, the combination of the spectral and spatial structure delivered by the imagery can provide a means to compact representations of the scene robust to affine transformations. This topic is somewhat under-researched. Note that the methods above employ statistics or signal processing techniques to "capture" the information in the spectra but make little use of the spatial information in the image. This contrasts with the use of image descriptors in computer vision, such as SIFT (Lowe 2004), spatial pyramids (Grauman and Darrell 2007) and maximally stable extremal regions (MSERs) (Matas et al. 2002).

References

Adams, M. L., Philpot, W. D., & Norvell, W. A. (1999). Yellowness index: an application of spectral second derivatives to estimate chlorosis of leaves in stressed vegetation. *International Journal of Remote Sensing, 20*(18), 3663–3675.
Angelopoulou, E. (2000). Objective colour from multispectral imaging. In *European conference on computer vision* (pp. 359–374).
Angelopoulou, E., Molana, R., & Daniilidis, K. (2001). Multispectral skin color modeling. In *Computer vision and pattern recognition* (pp. 635–642).
Asmussen, S., Binswanger, K., & Hojgaard, B. (2000). Rare events simulation for heavy-tailed distributions. *Bernoulli, 6*(2), 303–322.
Atkinson, P., Webster, R., & Curran, P. (1994). Cokriging with airborne mss imagery. *Remote Sensing of Environment, 50*, 335–345.
Ayer, M., Brunk, H. D., Ewing, G. M., Reid, W. T., & Silverman, E. (1955). An empirical distribution function for sampling with incomplete information. *Annals of Mathematical Statistics, 26*(4), 641–647.
Boser, B. E., Guyon, I., & Vapnik, V. (1992). A training algorithm for optimal margin classifiers. In *ACM conference on computational learning theory* (pp. 144–152).
Bracewell, R. N., Chang, K. Y., Jha, A. K., & Wang, Y. H. (1993). Affine theorem for two-dimensional Fourier transform. *Electronics Letters, 29*(3), 304.
Brown, A. J. (2008). Spectral curve fitting for automatic hyperspectral data analysis. *IEEE Transactions on Geoscience and Remote Sensing, 44*(6), 1601–1608.
Butler, W. L., & Hopkins, D. W. (1970). Higher derivative analysis of complex absorption spectra. *Photochemistry and Photobiology, 12*, 439–450.
Chang, C. q. I., Chakravarty, S., Chen, H. q. M., & Ouyang, Y. q. C. (2009). Spectral derivative feature coding for hyperspectral signature analysis. *Pattern Recognition, 42*(3), 395–408. doi:10.1016/j.patcog.2008.07.016. http://dx.doi.org/10.1016/j.patcog.2008.07.016.
Chum, O., Philbin, J., Sivic, J., Isard, M., & Zisserman, A. (2007). Total recall: automatic query expansion with a generative feature model for object retrieval. In *International conference on computer vision*.
Clark, R., & Roush, T. (1984). Reflectance spectroscopy: quantitative analysis techniques for remote sensing applications. *Journal of Geophysical Research, 89*, 6329–6340.
Clark, R., Swayze, G., Livo, K., Kokaly, R., Sutley, S., Dalton, J., McDougal, R., & Gent, C. (2003). Imaging spectroscopy: earth and planetary remote sensing with the usgs tetracorder and expert system. *Journal of Geophysical Research, 108*(5), 1–44.
Clark, R. N., Gallagher, A. J., & Swayze, G. A. (1990). Material absorption band depth mapping of imaging spectrometer data using a complete band shape least-squares fit with library reference spectra. In *Proceedings of the second airborne visible/infrared imaging spectrometer workshop* (pp. 176–186).
Cortes, C., & Vapnik, V. (1995). Support-vector networks. *Machine Learning, 20*(3), 273–297.
Cristianini, N., & Shawe-Taylor, J. (2000). *An introduction to support vector machines*. Cambridge: Cambridge University Press.
De Boor, C. (1978). *A practical guide to splines*. Berlin: Springer.
Demetriades-Shah, T. H., Steven, M. D., & Clark, J. A. (1990). High resolution derivatives spectra in remote sensing. *Remote Sensing of Environment, 33*, 55–64.
Doretto, G., Chiuso, A., Wu, Y., & Soatto, S. (2003). Dynamic textures. *International Journal of Computer Vision, 51*(2), 91–109.
Dror, R. O., Adelson, E. H., & Willsky, A. S. (2001). Recognition of surface reflectance properties from a single image under unknown real-world illumination. In *Proceedings of the IEEE workshop on identifying objects across variations in lighting*.
Fell, A. F., & Smith, G. (1982). Higher derivative methods in ultraviolet, visible and infrared spectrophotometry. *Analytical Proceedings, 22*, 28-33.

Fu, Z., Caelli, T., Liu, N., & Robles-Kelly, A. (2006a). Boosted band ratio feature selection for hyperspectral image classification. In *Proceedings of the international conference on pattern recognition* (Vol. 1, pp. 1059–1062).

Fu, Z., Tan, R., & Caelli, T. (2006b). Specular free spectral imaging using orthogonal subspace projection. In *Proceedings of the international conference on pattern recognition* (Vol. 1, pp. 812–815).

Fukunaga, K. (1990). *Introduction to statistical pattern recognition* (2nd ed.). New York: Academic Press.

Graham, R. L. (1972). An efficient algorithm for determining the convex hull of a finite planar set. *Information Processing Letters, 1*, 132–133.

Grauman, K., & Darrell, T. (2007). The pyramid match kernel: efficient learning with sets of features. *Journal of Machine Learning Research, 8*, 725–760.

Hsu, P. (2003). *Spectral feature extraction of hyperspectral images using wavelet transform*. PhD thesis, Department of Survey Engineering, National Cheng Kung University.

Huynh, C. P., & Robles-Kelly, A. (2008). A NURBS-based spectral reflectance descriptor with applications in computer vision and pattern recognition. In *IEEE conference on computer vision and pattern recognition*.

Jacobs, D. W., Belheumer, P. N., & Basri, R. (1998). Comparing images under variable illumination. In *Computer vision and pattern recognition* (pp. 610–617).

Jimenez, L., & Landgrebe, D. (1999). Hyperspectral data analysis and feature reduction via projection pursuit. *IEEE Transactions on Geoscience and Remote Sensing, 37*(6), 2653–2667.

Jolliffe, I. T. (2002). *Principal component analysis*. Berlin: Springer.

Katznelson, Y. (2004). *An introduction to harmonic analysis*. Cambridge: Cambridge University Press.

Khuwuthyakorn, P., Robles-Kelly, A., & Zhou, J. (2009). An affine invariant hyperspectral texture descriptor based upon heavy-tailed distributions and Fourier analysis. In *Joint IEEE international workshop on object tracking and classification in and beyond the visible spectrum* (pp. 112–119).

Koenderink, J. J. (1984). The structure of images. *Biological Cybernetics, 50*, 363–370.

Landgrebe, D. (2002). Hyperspectral image data analysis. *IEEE Signal Processing Magazine, 19*, 17–28.

Lansel, S., Parmar, M., & Wandell, B. A. (2009). Dictionaries for sparse representation and recovery of reflectances. In *SPIE proceedings: Vol. 7246. Computational imaging*. Bellingham: SPIE.

Lee, E. T. Y. (1989). Choosing nodes in parametric curve interpolation. *Computer Aided Design, 21*(6), 363–370.

Lenz, R., Latorre, P., & Meer, P. (2007). The hyperbolic geometry of illumination-induced chromaticity Changes. In *Computer vision and pattern recognition*.

Liang, S. (2004). *Quantitative remote sensing of land surfaces*. New York: Wiley-Interscience.

Lillesand, T., & Ralph, W. (2000). *Remote sensing and image interpretation* (4th ed.). New York: Wiley.

Lin, S., & Lee, S. W. (1997). Using chromaticity distributions and eigenspaces for pose, illumination, and specularity invariant 3D object recognition. In *Computer vision and pattern recognition* (pp. 426–431).

Lowe, D. (2004). Distinctive image features from scale-invariant keypoints. *International Journal of Computer Vision, 60*(2), 91–110.

MacKay, D. J. C. (2003). *Information theory, inference, and learning algorithms*. Cambridge: Cambridge University Press.

Maloney, L. T. (1986). Evaluation of linear models of surface spectral reflectance with small numbers of parameters. *Journal of the Optical Society of America A, 3*(10), 1673–1683.

Marimont, D. H., & Wandell, B. A. (1992). Linear models of surface and illuminant spectra. *Journal of the Optical Society of America A, 9*(11), 1905–1913.

Matas, J., Chum, O., Martin, U., & Pajdla, T. (2002). Robust wide baseline stereo from maximally stable extremal regions. In *Proceedings of the British machine vision conference* (pp. 384–393).

References

Nayar, S., & Bolle, R. (1996). Reflectance based object recognition. *International Journal of Computer Vision, 17*(3), 219–240.

Nilsback, M. E., & Zisserman, A. (2006). A visual vocabulary for flower classification. In *Computer vision and pattern recognition* (pp. 1447–1454).

Nister, D., & Stewenius, H. (2006). Scalable recognition with a vocabulary tree. In *Computer vision and pattern recognition* (pp. 2161–2168).

Oppenheim, A. V., Schafer, R. W., & Buck, J. R. (1999). *Discrete-time signal processing*. New York: Prentice Hall.

Phinn, S. R., Franklin, J., Stow, D., Hope, A., & Huenneke, L. (1996). Biomass distribution mapping in a semi-arid environment using airborne video imagery and spatial statistics. *Journal of Environmental Management, 46*(2), 139–165.

Piech, M., & Piech, K. (1987). Symbolic representation of hyperspectral data. *Applied Optics, 26*, 4018–4026.

Piegl, L., & Tiller, W. (1995). *The nurbs book*. Berlin: Springer.

Rao, K. R., & Yip, P. (1990). *Discrete cosine transform: algorithms, advantages, applications*. New York: Academic Press.

Ribeiro, E., & Hancock, E. R. (2001). Shape from periodic texture using the eigenvectors of local affine distoriton. *IEEE Transactions on Pattern Analysis and Machine Intelligence, 23*(12), 1459–1465.

Roberts, D. A., Brown, K., Green, R. O., Ustin, S. L., & Hinckley, T. (1998). Investigating the relationships between liquid water and leaf area in clonal populus. In *Summaries of the 7th annual JPL earth science workshop* (Vol. 1, pp. 335–344).

Shah, C., Watanachaturaporn, P., Arora, M., & Varshney, P. (2003). Some recent results on hyperspectral image classification. In *IEEE workshop on advances in techniques for analysis of remotely sensed data*.

Sims, D. A., & Gamon, J. A. (2003). Estimation of vegetation water content and photosynthetic tissue area from spectral reflectance: a comparison of indices based on liquid water and chlorophyll absorption features. *Remote Sensing of Environment, 84*, 526–537.

Sivic, J., Russell, B., Efros, A., Zisserman, A., & Freeman, W. (2005). Discovering objects and their location in images. In *International conference on computer vision* (pp. 370–377).

Slater, D., & Healey, G. (1997). Object recognition using invariant profiles. In *Computer vision and pattern recognition* (pp. 827–832).

Smith, W., Robles-Kelly, A., & Hancock, E. R. (2004). Skin reflectance modelling for face recognition. In *Proceedings of the international conference on pattern recognition* (pp. 210–213).

Sneddon, I. N. (1995). *Fourier transforms*. New York: Dover.

Stein, E., & Shakarchi, R. (2003). *Fourier analysis: an introduction*. Princeton: Princeton University Press.

Stokman, H. M. G., & Gevers, T. (1999). Detection and classification of hyper-spectral edges. In *British machine vision conference*.

Stout, Q. (2002). Optimal algorithms for unimodal regression. *Computing Science and Statistics, 32*.

Sunshine, J., Pieters, C. M., & Pratt, S. F. (1990). Deconvolution of mineral absorption bands: an improved approach. *Journal of Geophysical Research, 95*(B5), 6955–6966.

Tan, R. T., Nishino, K., & Ikeuchi, K. (2004). Separating reflection components based on chromaticity and noise analysis. *IEEE Transactions on Pattern Analysis and Machine Intelligence, 26*(10), 1373–1379.

Tiller, W. (1992). Knot-removal algorithms for NURBS curves and surfaces. *Computer Aided Design, 24*(8), 445–453.

Varma, M., & Ray, D. (2007). Learning the discriminative powerinvariance trade-off. In *International conference on computer vision*.

Vasconcelos, N. (2004). On the efficient evaluation of probabilistic similarity functions for image retrieval. *IEEE Transactions on Information Theory, 50*(7), 1482–1496.

Webb, A. (2002). *Statistical pattern recognition*. New York: Wiley.

Winder, S., & Brown, M. (2007). Learning local image descriptors. In *IEEE conference on computer vision and pattern recognition.*

Witkin, A. P. (1983). Scale-space filtering. In *Proceedings of the eighth international joint conference on artificial intelligence* (Vol. 2, pp. 1019–1022).

Yasuma, F., Mitsunaga, T., Iso, D., & Nayar, S. K. (2010). Generalized assorted pixel camera: post-capture control of resolution, dynamic range and spectrum. *IEEE Transactions on Image Processing, 19*(9), 2241–2253.

Young, G., & Householder, A. S. (1938). Discussion of a set of points in terms of their mutual distances. *Psychometrika, 3*, 19–22.

Young, N. (1988). *An introduction to Hilbert space.* Cambridge: Cambridge University Press.

Chapter 8
Material Discovery

For spectral image classification, each pixel is associated with a spectrum which can be viewed as an input vector in a high-dimensional space. Thus algorithms from statistical pattern recognition and machine learning have been adopted to perform pixel-level feature extraction and classification (Landgrebe 2002). These methods either directly use the complete spectra, or often make use of preprocessing and dimensionality reduction steps at input and attempt to recover statistically optimal solutions. Linear dimensionality reduction methods are based on the linear projection of the input data to a lower dimensional feature space. Typical methods include principal component analysis (PCA) (Jolliffe 2002), linear discriminant analysis (LDA) (Fukunaga 1990) and projection pursuit (Jimenez and Landgrebe 1999). Almost all linear feature extraction methods can be kernelised, resulting in kernel PCA (Schölkopf et al. 1999), kernel LDA (Mika et al. 1999) and kernel projection pursuit (Dundar and Landgrebe 2004). These methods exploit non-linear relations between different segments in the spectra by mapping the input data onto a high-dimensional space through different kernel functions (Scholkopf and Smola 2001).

From an alternative viewpoint, features of the absorption spectrum can be used as signatures for chemicals and their concentrations (Sunshine et al. 1990). Absorption and reflection are two complementary behaviours of the light incident on the material surface. Absorptions are inherently related to the material chemistry as well as other physical properties such as surface roughness (Hapke 1993). Therefore, the presence of an absorption at a certain spectral range is a "signature", which can be used for identification and recognition purposes.

8.1 Scenes in Terms of Materials

Up to this stage, we have elaborated on the invariance or representation problems related to imaging spectroscopy devoid of subpixel information. The use of imaging spectroscopy for scene analysis also permits the representation of a scene in terms of materials and their constituents. Note that the problem of decomposing a surface material into its primordial constitutive material compounds is a well-known setting in geoscience. This "unmixing" of the spectra is commonly stated as the problem of decomposing an input spectrum into relative portions of known spectra of end members. These end members are often provided as input in the form of a library and can correspond to minerals, chemical elements, organic matter, etc. The problem of unmixing applies to cases where a capability to provide subpixel detail is needed, such as geosciences, food quality assessment and process control. Moreover, unmixing is not exclusive to subpixel processing but can be viewed as a pattern recognition task related to soft clustering with known or unknown end members.

Note that a scene may comprise a number of naturally occurring materials such as wood, paint, etc. These materials vary from scene to scene, whereas their constituents are, in general, pure chemical compounds. In the case of scene analysis, the constitutive compounds or end members, as they are known in geosciences and mineralogy, are generally known. In the terrestrial setting, the materials have to be extracted from the scene. The challenge here is, hence, twofold. Firstly, it concerns the unknown scene material spectrum recovery, and secondly, it pertains to the unmixing of these material spectra based on a library of end members. This treatment allows one to avoid cumbersome labelling of the scene spectral signatures, which, at the moment, is mainly effected through expert intervention (Lennon et al. 2001). Automatic scene material extraction is often complicated due to the confounding factors introduced by illumination and the complex nature of real-world settings, where the number of materials in the scene is not known.

Furthermore, this task can be viewed as a blind source labelling problem which can be tackled in a number of ways. Here, we explore a probabilistic treatment of the problem where soft clustering is effected on the pixel reflectance spectra. This is done using deterministic annealing in a manner akin to clustering so as to recover groups of spectra with similar signatures. As the inference process converges, these groups can be further refined.

Thus, the problem becomes the following. Let a multispectral or hyperspectral image $\mathcal{I}$ consisting of wavelength indexed radiance spectra in the visible spectrum be the input. The aim of computation becomes the recovery of a set of scene materials and illuminants which are given in terms of end members in two libraries. The first of these accounts for chemical compounds which constitute the materials in the scene. These materials can then be unmixed into a set of man-made or naturally occurring compounds. The second library corresponds to "canonical" illuminant power spectra corresponding to light sources, such as the CIE standard illuminants or those illuminants proposed in Barnard et al. (2002). To illustrate this setting, in Fig. 8.1 we show an input hyperspectral image, in pseudo-colour, for which we have computed a scene material map. Two samples of these materials are then unmixed into a set of end member spectra. In the bottom and top panels, we show the scene material spectra and the end member spectra which correspond to the mixed components. In this figure, for clarity, we have omitted the unmixing step for the scene illuminants.

8.1 Scenes in Terms of Materials

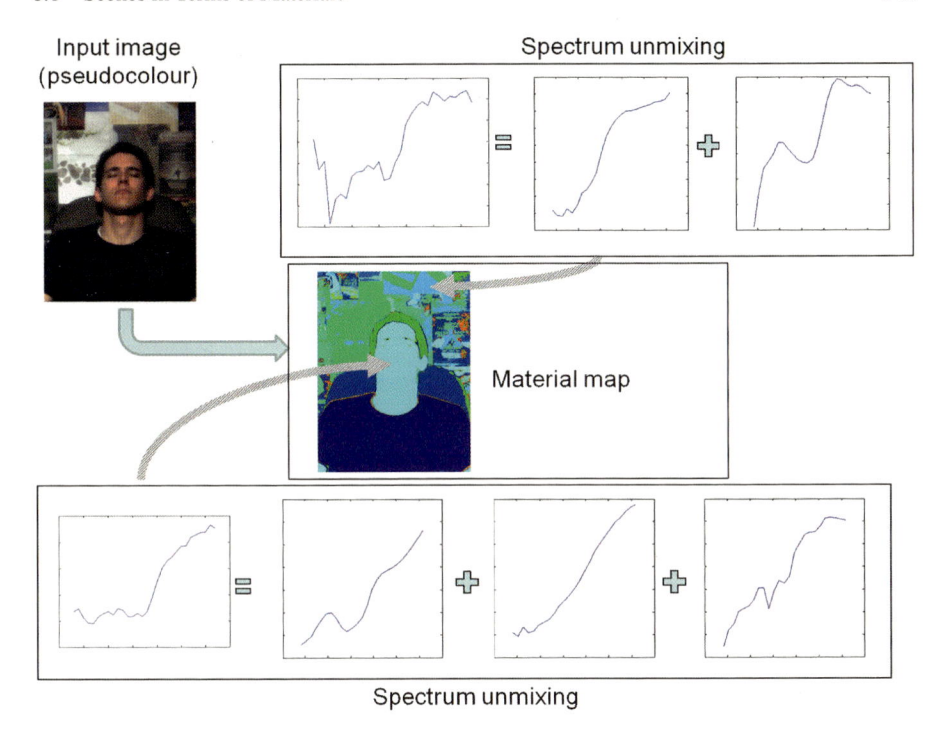

Fig. 8.1 Example material map and unmixing for the input image in the *upper left panel*. In the figure, we show the unmixing results for the spectra of two sample materials in the scene

To be more formal, note that we can express the image irradiance using the dichromatic model and the mixture coefficients $\rho_{m,u}$ and α_ℓ as follows:

$$I(u, \lambda_i) = L(\lambda_i)\big(g(u)S(u, \lambda_i) + k(u)\big)$$
$$= \left(\sum_{\ell=1}^{M} \alpha_\ell L_\ell(\lambda_i)\right)\left(g(u)\sum_{m=1}^{N} \rho_{m,u} S_m(\lambda_i) + k(u)\right), \quad (8.1)$$

where $S_m(\lambda_i)$ is the reflectance for the material indexed m at the ith wavelength indexed band and $L_\ell(\lambda_i)$ is the power spectrum of the ℓth illuminant in the scene.

Note that, based on the dichromatic model and the developments in previous chapters, we can process an input image so as to recover its illuminant, shading factor, specular coefficient and reflectance. Once the reflectance is at hand, the material and map may be computed and, if libraries for end members and standard illuminants are available, the scene materials and light may be unmixed. Viewed in this manner, the material reflectance S_m can be further expressed as

$$S_m(\cdot) = \mathbf{Qa}, \quad (8.2)$$

where $\mathbf{Q}$ is a matrix whose columns correspond to the end member spectra and $\mathbf{a}$ is a vector of abundance fractions, i.e. the contribution of each of the end member

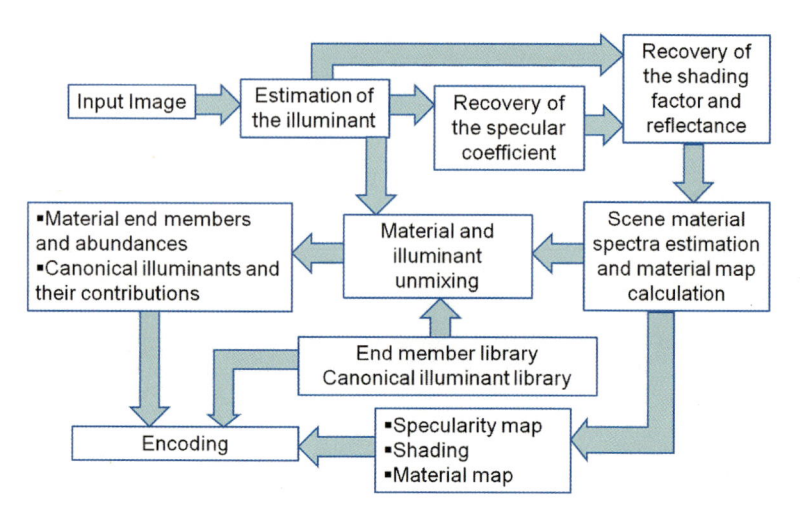

Fig. 8.2 Processing flow of a hyperspectral image based on the dichromatic model

spectra to $S_m(\cdot)$. In this equation we have written $S_m(\cdot)$ to imply that, in this case, the material reflectance is a vector whose elements are indexed to wavelength.

Similarly, for the ℓth illuminant in the scene, we can write

$$\mathcal{L}_\ell(\cdot) = \mathbf{L}\mathbf{b}, \tag{8.3}$$

where $\mathbf{L}$ is a matrix whose columns correspond to the standard illuminants and $\mathbf{b}$ is a vector of luminance fractions, i.e. the contribution of each of the illuminant power spectra to the scene illuminant $\mathcal{L}_\ell(\cdot)$. In the equation above, $\mathcal{L}_\ell(\cdot)$ implies that the scene illuminant is a vector whose elements are indexed to wavelength.

This processing flow is illustrated in Fig. 8.2. In the diagram, we take as input a hyperspectral or multispectral image and proceed to compute the illuminant as illustrated in Chap. 5. With the illuminant in hand, the shading factor and the specular coefficient are recovered. To do this, we can use specularity removal techniques or band subtraction. Another option is to recover the illuminant power spectrum, shading factor, specular coefficient and reflectance in a simultaneous operation using an optimisation approach like that presented in Sect. 6.2. Once the reflectance is available, the basis material reflectance spectra and the corresponding mixture coefficients are estimated (see the following sections). These material reflectance spectra and the scene illuminant can be further unmixed or encoded together with the shading and specularity information for material discovery, visualisation, shape analysis, etc. This encoding can be done using the representations presented in Chap. 7. Note that, if unmixing is effected, the encoding can be done on the end members and canonical illuminants rather than the reflectance or raw spectra. This prospectively can greatly reduce the storage required. It also permits the use of functional analysis on the representation basis, i.e. that used for the B-spline or the mixture, for purposes of derivative analysis, uni-modal regression, etc. It also allows for inter-

8.1.1 Imposing End Member Consistency

Note that decompositions like that in Eq. (8.1) have been used elsewhere (Drew and Finlayson 2007; Parkkinen et al. 1989; Judd et al. 1964). Their importance comes from the fact that, in practice, there may be variations in the actual reflectance of pixels belonging to the same material in the scene. This does not necessarily indicate a change of object material under consideration; it may also indicate small variations in composition. For instance, variations in the enamel of a mug do not change the fact that it is made of porcelain but can yield slight end member changes. Therefore, it would be rather undesirable to have objects partitioned or fragmented into inconsistent end members.

In this section, we aim at imposing material consistency between pixels sharing similar spectra in a scene. Instead of solving the problem at the pixel level, we extend the problem to material clusters, where the decomposition occurs per material cluster rather than per pixel.

Cost Function

We impose consistency of composition by recovering the mixture coefficients per material cluster from the estimated mean reflectance spectrum. To this end, we assign a partial membership $P(\omega(.)|u)$ of a material cluster with mean reflectance $S_m(.)$ to each pixel u in the image. Taking into account the material cluster set Ω, i.e. all the materials S_m in the scene, we view the recovery of the mixture coefficients as two successive optimisation problems. The first of these considers the clustering of image pixels based on their spectral reflectance. Here we employ an affinity metric $a(S_m(.), S(u, \cdot))$ to preselect k materials. This is done using the affinity between the pixel reflectance spectrum $S(u, \cdot)$ and the mean spectrum $S_m(.)$ for the mth material cluster. Mathematically, this affinity measure can be defined by their Euclidean angle

$$a\big(S_m(\cdot), S(u, \cdot)\big) = 1 - \frac{\langle S(u, \cdot), S_m(.)\rangle}{\|S(u, \cdot)\|\|S_m(.)\|}. \tag{8.4}$$

With this metric, the problem aims to minimise the total expected affinity for the entire image as

$$A_{\text{Total}} = \sum_{u \in \mathcal{I}, S_m(.) \in \Omega} P\big(S_m(.)|u\big) a\big(S_m(.), S(u, \cdot)\big), \tag{8.5}$$

subject to the law of total probability $\sum_{S_m(.) \in \Omega} P(S_m(.)|u) = 1 \ \forall u \in \mathcal{I}$.

Since the formulation in Eq. (8.5) often favours hard assignment of pixels to their closest materials, we restate the problem subject to the maximum entropy criterion (Jaynes 1957). The entropy of the material association probability distribution at each pixel is defined as

$$\mathcal{P} = -\sum_{u \in \mathcal{I}} \sum_{S_m(.) \in \Omega} P(S_m(.)|u) \log P(S_m(.)|u). \tag{8.6}$$

With the affinity metric in Eq. (8.4), the problem becomes that of finding a set of mean material reflectance spectra and a distribution of material association probabilities $P(S_m(.)|u)$ for each pixel u so as to minimise $A_{\text{Entropy}} = A_{\text{Total}} - \mathcal{L}$, where

$$\mathcal{L} = T\mathcal{P} + \sum_{u \in \mathcal{I}} \varrho(u) \left(\sum_{S_m(.) \in \Omega} P(S_m(.)|u) - 1 \right), \tag{8.7}$$

in which $T \geq 0$ and $\varrho(u)$ are Lagrange multipliers. Note that $T \geq 0$ weighs the level of randomness of the material association probabilities, whereas $\varrho(u)$ enforces the total probability constraint for every image pixel u.

The optimisation approach to the problem is somewhat similar to an annealing soft-clustering process. At the beginning, this process is initialised assuming all the image pixels are made of the same material. As the method progresses, the set Ω of materials grows. This, in essence, constitutes several "phase transitions", at which new materials arise from the existing ones. This phenomenon is due to the discrepancy in the affinity $a(S_m(.), S(u, \cdot))$ between the pixel reflectance $S(u, \cdot)$ and the material reflectance spectrum $S_m(.)$.

Material Cluster and End Member Proportion Recovery

We now derive the optimal set Ω of scene materials so as to minimise the cost function above. To do this, we compute the derivatives of A_{Entropy} with respect to the material reflectance spectrum $S_m(.)$ and equate them to zero, which yields

$$S_m(.) \propto \sum_{u \in \mathcal{I}} P(S_m(.)|u) \frac{S(u, \cdot)}{\|S(u, \cdot)\|}. \tag{8.8}$$

In Eq. (8.8), we require the probability $P(S_m(.)|u)$ to be available. To compute this probability, we employ deterministic annealing. A major advantage of the deterministic annealing approach is that it avoids being attracted to local minima. In addition, deterministic annealing converges faster than stochastic or simulated annealing (Kirkpatrick et al. 1983).

The deterministic annealing approach casts the Lagrangian multiplier T as a system temperature. At each phase of the annealing process, where the temperature T is kept constant, the algorithm proceeds as two interleaved minimisation steps in

8.1 Scenes in Terms of Materials 147

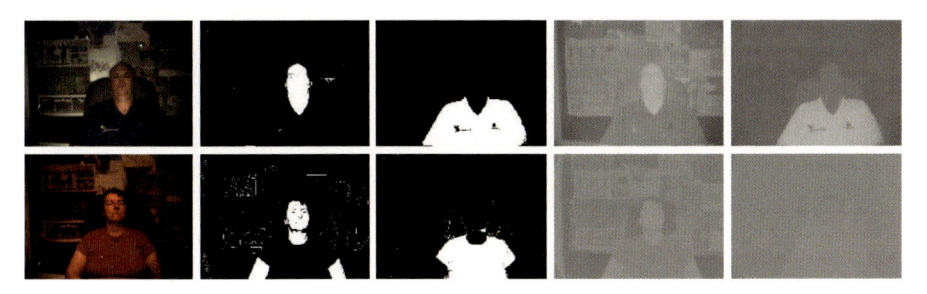

Fig. 8.3 Material maps estimated from a visible range image (*top row*) and a near-infrared range image (*bottom row*). *First column*: the input image in pseudo-colour. *Second and third columns*: the probability maps of skin and cloth produced by the method in Sect. 8.1.1. Fourth and fifth columns: the probability maps of skin and cloth produced by the Spectral Angle Mapper (Yuhas et al. 1992)

order to arrive at an equilibrium state. These two minimisation steps are performed alternately with respect to the material association probabilities and the end members.

For the recovery of the pixel-to-material association probabilities, we fix the material spectrum $S_m(.)$ and seek the probability distribution which minimises the cost function A_{Entropy}. This is achieved by setting the partial derivative of A_{Entropy} with respect to $P(S_m(.)|u)$ to zero. Since $\sum_{S_m(.) \in \Omega} P(S_m(.)|u) = 1$, it can be shown that the optimal material association probability for a fixed material set Ω is given by the Gibbs distribution,

$$P(S_m(.)|u) = \frac{\exp(\frac{-a(S_m(.), S(u, \cdot))}{T})}{\sum_{S_{m'}(.) \in \Omega} \exp(\frac{-a(S_{m'}(.), S(u, \cdot))}{T})} \quad \forall S_m(.), u. \tag{8.9}$$

In the second and third columns of Fig. 8.3, we show the probability maps of skin and cloth materials recovered by this method. In the fourth and fifth columns, we show the probability maps recovered by the Spectral Angle Mapper (Yuhas et al. 1992) commonly used in remote sensing. The two sample images shown have been captured under different illumination conditions, one in the visible and the other in the infrared spectrum. In the panels, the brightness of the probability maps is proportional to the association probability with the reference material.

Note that, from the figure, we can appreciate not only that the scene analysis setting is quite dissimilar to the remote sensing one, but also that, by retrieving the scene materials, even for very simple images, we open up the possibility of effecting segmentation and recognition simultaneously. This treatment is quite general in the sense that it is applicable to any number of wavelength indexed bands assuming no prior knowledge of the materials in the scene.

8.2 Material Unmixing

Hyperspectral data is acquired by high-spectral-resolution imaging sensors, containing hundreds of contiguous narrow spectral band images. Due to the low spatial resolution of the sensor, disparate substances may contribute to the spectrum for a single pixel, leading to the existence of "mixed" spectra in hyperspectral imagery. Hence, hyperspectral unmixing, which decomposes a mixed pixel into a collection of constituent spectra, or end members, and their corresponding fractional abundances is often employed to preprocess hyperspectral data (Keshava 2003). Many hyperspectral unmixing methods have been proposed in recent years. These include N-FINDR (Winter 1999), vertex component analysis (VCA) (Nascimento and Dias 2005), independent component analysis (ICA) (Wang and Chang 2006), alternating projected subgradients (APS) (Zymnis et al. 2007), minimum volume based algorithms (Craig 1994; Chan et al. 2009; Bioucas-Dias 2009; Ambikapathi et al. 2010) and flexible similarity measures (Chen et al. 2009).

Most of these methods assume a linear spectral mixture model for the unmixing process. If the number and signatures of end members are unknown, unmixing becomes a blind source separation (BSS) problem. This is compounded by the need to estimate the parameters of the mixing and/or filtering processes. It is impossible to uniquely estimate the original source signals and mixing matrix if no a priori knowledge is applied to the BSS. Various approaches have specific physical and statistical assumptions for modelling the unmixing process. For example, the assumption of source independence leads to the ICA method (Wang and Chang 2006), whereas the assumption of Markov random distribution of abundance leads to the spatial structure method in Jia and Qian (2007). Since the source signals are normally independent from one another in some specific frequencies, it can be assumed that the subcomponents of the sources are mutually independent, which leads to sub-band ICA (Qian and Wang 2010).

From the linear algebra point of view, BSS is a constrained matrix factorisation problem that has found numerous applications in feature and signal extraction (Ye and Li 2004). For general matrix factorisation problems, traditional matrix computation tools such as singular value decomposition (SVD), QR decomposition and LU factorisation can be used. However, these tools cannot be directly applied to hyperspectral unmixing because two constraints must be considered (Chang 2003). The first constraint is the non-negativity of both spectra and their fractional abundances. This is natural, as the contribution from end members should be larger than or equal to zero. Secondly, the additivity constraint over the fractional abundances has to be considered; this constraint guarantees that the addition of the proportional contribution from the end members matches the mixed observation.

Non-negative matrix factorisation (NMF) (Paatero 1997; Lee and Seung 1999), which decomposes the data into two non-negative matrices, is a natural solution to the non-negativity constraint (Pauca et al. 2006). From the data analysis point of view, NMF is very attractive because it usually provides a part-based representation of the data, making the decomposition matrices more intuitive and interpretable (Cichocki et al. 2009). However, the solution space of NMF is very large if no further constraints are considered. This added to the fact that the cost function is not convex, making the algorithm prone to noise corruption and computationally demanding.

To reduce the space of solutions, extensions of NMF including symmetric NMF, semi-NMF, non-smooth NMF and multi-layer NMF have been proposed (Cichocki et al. 2009). Researchers have also tried improving NMF-based unmixing methods by imposing further constraints (Pauca et al. 2006; Jia and Qian 2009; Huck et al. 2010). Donoho and Stodden (2004) analysed the assumptions required for NMF to generate unique solutions and lead to a well-defined answer. More recently, sparsity constraints have gained much attention, since they allow one to exploit the notion that most of the pixels are mixtures of only a few of the end members in the scene (Iordache et al. 2010; Thompson et al. 2009). This implies that a number of entries in the abundance matrix are zeros, which manifests itself as a large degree of sparsity.

Regularisation methods are usually utilised to define the sparsity constraint on the abundance of the end members. Along these lines, the L_0 regulariser accounts for the number of zero elements in an abundance matrix so as to yield the most sparse result given a cost function. However, the solution of the L_0 regulariser is an NP-hard optimisation problem that cannot be solved in practise. The L_2 regulariser, on the other hand, generates smooth but not sparse results (Berry et al. 2007). In general, the L_1 regulariser is the most popular choice for achieving sparsity of the abundance matrix (Hoyer 2004; Pascual-Montano et al. 2006; Zare and Gader 2008).

More recent work on this topic include the semi-supervised algorithms based on sparse regression in Iordache et al. (2010) and Thompson et al. (2009). These methods assume that the pixel signature can be expressed in the form of linear combinations of a number of pure spectral signatures known in advance, because the library of pure spectral signatures contains more possible sources than those actually present in the scene. Guo et al. (2009) also used a sparse regression model to estimate the abundances, while the end members were extracted using the N-FINDR algorithm. In Zare and Gader (2008), sparsity-promoting priors are used with an extension of the iterated constrained end member (ICE) algorithm (Berman et al. 2004) to determine the number of end members, in which the sparsity is achieved by a zero-mean Laplacian distribution akin to the L_1 regulariser.

8.2.1 Linear Spectral Mixture Model

In this section, we provide an overview of classical linear unmixing as a means to recover the composition of the scene materials and illuminants. Note that the concept of unmixing applies equally to both illuminant power and material reflectance spectra; hence, we elaborate on it in a general setting. The concept of unmixing hinges on representing the spectrum across N wavelength indexed bands based on K end member abundances. Linear unmixing can be, hence, expressed mathematically as follows:

$$\mathbf{x} = \mathbf{Q}\mathbf{a} + \mathbf{e}, \tag{8.10}$$

where **x** denotes an $N \times 1$ vector containing the spectra to be unmixed, **a** is a $K \times 1$ vector of abundance fractions for each end member, **e** is an $N \times 1$ vector of additive noise representing the measurement errors, and **Q** is an $N \times K$ non-negative spectral signature matrix whose columns correspond to an end member spectrum.

Using matrix notation, the mixing model above for the $|\mathcal{I}|$ pixels in the image can be rewritten as

$$\mathbf{X} = \mathbf{QA} + \mathbf{E}, \tag{8.11}$$

where the matrices $\mathbf{X} \in \mathbb{R}_+^{N \times |\mathcal{I}|}$, $\mathbf{A} \in \mathbb{R}_+^{K \times |\mathcal{I}|}$ and $\mathbf{E} \in \mathbb{R}^{N \times |\mathcal{I}|}$ represent, respectively, the hyperspectral data, the end member abundances and additive noise. Note that, in general, only **X** is known in advance, while the other two matrices, **A** and **Q** are our targets of computation. Moreover, from observation, we can see that the product in the first right-hand side term leads itself to a matrix factorisation problem.

8.2.2 Non-Negative Matrix Factorisation (NMF)

Non-negative matrix factorisation (NMF) has received considerable attention in the fields of pattern recognition and machine learning, where it leads to a "part-based" representation since it allows only additive combination of factors. Linear mixing models assume that the hyperspectral image is constituted of spectral signatures of end members with corresponding non-negative abundances. Therefore, the non-negativity of **A** and **Q** mentioned above is a natural property of the measured quantities in hyperspectral data. This non-negativity can replace the independence constraint used for BSS and exploited by methods such as ICA.

To obtain **A** and **Q**, NMF can be performed by minimising the difference between **X** and **QA** and enforcing non-negativity on **A** and **Q**. This difference is often measured using the Euclidean distance, relative entropy, or Kullback–Leibler divergence. The loss function for NMF based on the Euclidean distance is as follows:

$$\mathcal{C}(\mathbf{A}, \mathbf{Q}) = \frac{1}{2} \|\mathbf{X} - \mathbf{QA}\|_2^2. \tag{8.12}$$

Although there are numerous optimisation algorithms to estimate **A** and **Q**, it is difficult to obtain a globally optimal solution because of the non-convexity of $\mathcal{C}(\mathbf{A}, \mathbf{Q})$ with respect to both **A** and **Q**. Moreover, NMF is always utilised with other constraints, such as sparsity. This is due to the fact that NMF lacks a unique solution. This can be easily verified by considering $\mathbf{QA} = (\mathbf{QD})(\mathbf{D}^{-1}\mathbf{A})$ for any non-negative invertible matrix **D**.

Sparsity is also illustrated in compressive sensing (Donoho 2006). Likewise, sparsity is an intrinsic property of hyperspectral data. In most cases, the abundance distribution of any end member does not apply to the whole scene. This implies that the mixed pixel is usually the superposition of only a few end members, but not all those present in the scene. That is, for each end member, its abundance is localised with a degree of sparseness.

8.2 Material Unmixing

For this reason, we now consider NMF with a sparsity constraint as an objective function for our minimisation problem. This objective function is the combination of the reconstruction error and a sparsity measure as follows:

$$\mathcal{C}(\mathbf{A}, \mathbf{Q}) = \frac{1}{2}\|\mathbf{X} - \mathbf{Q}\mathbf{A}\|_2^2 + \gamma f(\mathbf{A}), \quad (8.13)$$

where $\gamma \in \mathbb{R}^+$ is a scalar that weights the contribution of the sparsity measure function $f(\cdot)$ of the matrix $\mathbf{Q}$, which is usually regarded as a regularisation term.

Many forms of regularisers exist to induce sparsity. In recent years, there has been an increasing interest in the L_1 regulariser, since it has a better asymptotic consistency than its L_2 counterpart. However, for spectral unmixing, the L_1 regulariser does not enforce the full additivity constraint. Finding new regularisers that yield sparse solutions while preserving the additivity constraint over the end members is an important problem in NMF-based unmixing methods.

Note that, in theory, all L_q norms for $0 < q < 1$ can be used for purposes of sparsity. More specifically, the L_q-NMF model for unmixing is given by

$$\mathcal{C}(\mathbf{A}, \mathbf{Q}) = \frac{1}{2}\|\mathbf{X} - \mathbf{Q}\mathbf{A}\|_2^2 + \gamma \|\mathbf{A}\|_q, \quad (8.14)$$

where

$$\|\mathbf{A}\|_q = \left(\sum_{k=1}^{K} \sum_{n=1}^{|\mathcal{I}|} a_n(k)^q \right)^{1/q}, \quad (8.15)$$

and $a_n(k)$ is the abundance fraction for the kth end member at the nth pixel in the image.

The multiplicative update rules to solve $\mathcal{C}(\cdot)$ with respect to $\mathbf{A}$ and $\mathbf{Q}$ are hence given by

$$\mathbf{Q} \leftarrow \mathbf{Q} .* \mathbf{X}\mathbf{A}^T ./ \mathbf{Q}\mathbf{A}\mathbf{A}^T, \quad (8.16)$$

$$\mathbf{A} \leftarrow \mathbf{A} .* \mathbf{Q}^T \mathbf{X} . / \left(\mathbf{Q}^T \mathbf{Q}\mathbf{A} + \frac{\gamma}{2}\mathbf{Q}^{q-1} \right). \quad (8.17)$$

where $.*$ and $./$ are the element-wise matrix multiplication and division.

8.2.3 Geometric Interpretation

As summarised in Bioucas-Dias and Plaza (2010), most linear spectral unmixing methods can be categorised into geometric and statistical algorithms. The geometrical methods are based on the connection between the linear spectral unmixing model and the convex geometry. In this sense, the regularised NMF is a geometric method. Here, we give the geometrical interpretation for NMF and show how it is related to some other algorithms.

In general, Eq. (8.10) is based on two main constraints.

(a)   (b)

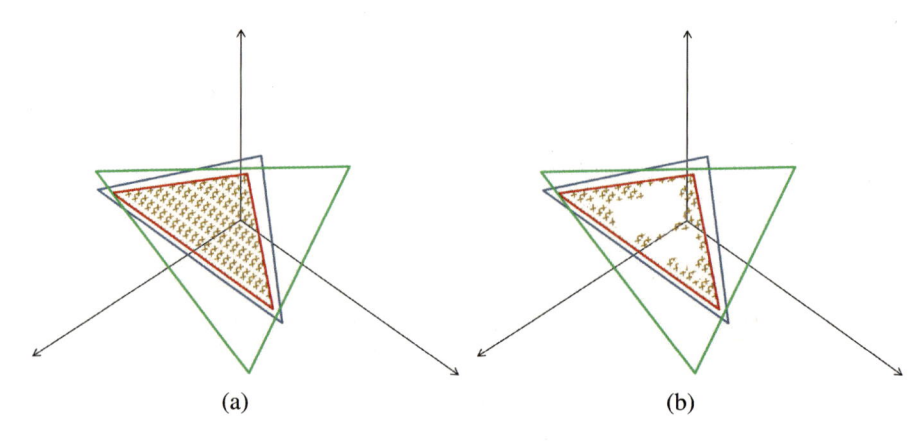

Fig. 8.4 Simplices for a uniform distribution of the spectra (**a**) and a sparse one (**b**)

- (A1) Non-negative constraint: **A** and **Q** are non-negative matrices.
- (A2) Full additivity constraint: the sum of entries in each column of **Q** is unity.

The first of these is met by NMF, with all the mixed or pure pixels laying on a positive simplicial cone that is given by

$$\text{Cone}(\mathbf{Q}) = \left\{ \mathbf{X} = \mathbf{QA} \mid \mathbf{A} \in \mathbb{R}_+^{K \times |\mathcal{I}|} \right\}. \tag{8.18}$$

If (A2) is met, all the mixed or pure pixels lie in an affine hull that is represented by

$$\text{Affine}(\mathbf{Q}) = \left\{ \mathbf{X} = \mathbf{QA} \mid \mathbf{1}_N^T \mathbf{A} = 1 \right\}. \tag{8.19}$$

This implies that, if both (A1) and (A2) are satisfied, all the mixed or pure pixels lie in a convex hull (simplex) that is represented by

$$\text{Simplex}(\mathbf{Q}) = \left\{ \mathbf{X} = \mathbf{QA} \mid \mathbf{A} \in \mathbb{R}_+^{K \times |\mathcal{I}|},\ \mathbf{1}_N^T \mathbf{A} = 1 \right\}. \tag{8.20}$$

Note that, if pure pixels exist in the scene, many geometrical unmixing algorithms can be used, such as PPI, N-FINDR and VCA. However, in the case when pure pixels do not exist, i.e., pixels are highly mixed, the above-mentioned constraints are not sufficient to produce a unique result. This can be observed in Fig. 8.4(a), in which three simplices meet the assumptions, but only one of them is accurate. Therefore, additional constraints have to be added to the linear spectral unmixing model, for example, those that enforce smoothness (Pauca et al. 2006) on **Q** and **A**, or low variance on the end member spectra (Huck et al. 2010).

In recent years, a very common practise has been to use minimum volume as an additional constraint. Methods such as ICE (Berman et al. 2004), MVC-NMF (Miao and Qi 2007) and SISAL (Bioucas-Dias 2009) follow Craig's unmixing criterion (Craig 1994) that the end member matrix **Q** and its spanned simplex should enclose the observed pixels and minimise its volume.

8.3 Material Classification Using Absorptions

The regulariser is closely linked to the minimum volume method through its geometric interpretation, as the sparsity represented by L_q enforces the volume of the simplex to be minimised. This is intuitive, because if the volume of the simplex becomes large, the observed pixels become highly mixed (less sparse). Here, we present a simple experiment to illustrate this phenomenon. In Fig. 8.4, we construct a simplex, which is represented by the red triangle (R) in Fig. 8.4(a). In the figure, 100 data points spanned by three vertices are generated with uniform distribution across in this simplex. Then, we construct two more simplices, the green (G) and blue triangles (B), which also contain all data points. Note that the volumes meet the following condition:

$$V_G > V_B > V_R.$$

Thus, the red simplex is the minimum one containing all the data points. The sum of the abundances of each data point in each simplex is unity. As a result, the L_1 regulariser cannot distinguish the volume of the simplices from one another. On the contrary, the $L_{1/2}$ regulariser yields a different average penalty cost of the data points for each simplex, using

$$\|\mathbf{A}\|_q = \frac{1}{|\mathcal{I}|}\left(\sum_{n=1}^{|\mathcal{I}|}\sum_{k=1}^{K} a_n(k)^q\right)^{1/q},$$

where $|\mathcal{I}| = 100$ and $q = 1/2$, such that

$$\|\mathbf{A}\|_{1/2}^{(V_R)} < \|\mathbf{A}\|_{1/2}^{(V_B)} < \|\mathbf{A}\|_{1/2}^{(V_G)}, \tag{8.21}$$

which has verified our claim that the minimum volume is related to the $L_{1/2}$ regulariser.

The same observation also applies to more clustered data. If we let the 100 data points drift towards the vertices of the red simplex, as shown in Fig. 8.4(b), then their real abundances are sparser than those corresponding to uniformly distributed data points in Fig. 8.4(a), with penalty costs fulfilling the relationship in inequality (8.21).

8.3 Material Classification Using Absorptions

Once the materials are in hand, classification methods can be applied for purposes of classifying the spectra. In this section, we examine two widely used approaches which employ absorptions to perform material classification.

8.3.1 Spectral Angle Mapper (SAM)

The Spectral Angle Mapper (SAM) (Boardman and Kruse 1994) is an automatic method for comparing image and library spectra. By taking the apparent reflectance at input, i.e. the true reflectance up to an unknown scaling factor, the correlation coefficient between two spectra can be used as a scale invariant similarity measure. The spectral angle (SA) between two absorption band segments is then defined as the inverse cosine of their correlation coefficient as follows:

$$\alpha(\mathbf{x}, \hat{\mathbf{x}}) = \arccos\left(\frac{\mathbf{x}^T \hat{\mathbf{x}}}{\|\mathbf{x}\|\|\hat{\mathbf{x}}\|}\right), \tag{8.22}$$

where $\mathbf{x}$ and $\hat{\mathbf{x}}$ are the vector forms of the absorption band segments under comparison.

SAM performs material mapping based on the SA. The smaller the SA, the more similar the two spectra under comparison are and the more likely the two materials correspond to one another. As a result, we can use SAM for comparison of local spectral segments and assign a likelihood value based on their similarity. The likelihood of the spectral segments $\mathbf{x}$ and $\hat{\mathbf{x}}$ falling into the interval $[\lambda_i, \lambda_j]$ becomes

$$f(\mathbf{x}, \hat{\mathbf{x}} | (\lambda_i, \lambda_j)) = \exp\left(\frac{-\alpha(\mathbf{x}, \hat{\mathbf{x}})^2}{2\sigma^2}\right), \tag{8.23}$$

where α is a constant and σ is a predefined variance parameter.

8.3.2 Spectral Feature Fitting (SFF)

The other method examined here, Spectral Feature Fitting (SFF) (Clark et al. 1990), is based on the local comparison of continuum removed spectra. Let us denote the continuum removed spectra of $\mathbf{x}$ and $\hat{\mathbf{x}}$ as $\mathbf{y}$ and $\hat{\mathbf{y}}$, respectively. SFF aims at matching the reference spectrum $\mathbf{y} = [y(\lambda_i), y(\lambda_{i+1}), \ldots, y(\lambda_j)]^T$ to the unknown candidate spectrum $\hat{\mathbf{y}} = [\hat{y}(\lambda_i), \hat{y}(\lambda_{i+1}), \ldots, \hat{y}(\lambda_j)]^T$ by recovering a scaling factor κ, which is determined by solving the following least-squares (LS) problem:

$$\min \sum_{\lambda_k \in [\lambda_i, \lambda_j]} (a + by(\lambda_k) - \hat{y}(\lambda_k))^2 \quad \text{and} \quad \kappa = \frac{1-b}{b}. \tag{8.24}$$

The solution to the above equation is given by

$$\begin{aligned} a &= \frac{1}{n}\left(\sum_{\lambda_k \in [\lambda_i, \lambda_j]} \hat{y}(\lambda_k) - b \sum_{\lambda_k \in [\lambda_i, \lambda_j]} y(\lambda_k)\right), \\ b &= \frac{\sum_{\lambda_k \in [\lambda_i, \lambda_j]} \hat{y}(\lambda_k) y(\lambda_k) - \frac{1}{n} \sum_{\lambda_k \in [\lambda_i, \lambda_j]} \hat{y}(\lambda_k) \sum_{\lambda_k \in [\lambda_i, \lambda_j]} y(\lambda_j)}{\sum_{\lambda_k \in [\lambda_i, \lambda_j]} y(\lambda_k)^2 - \frac{1}{n}(\sum_{\lambda_k \in [\lambda_i, \lambda_j]} y(\lambda_k))^2}, \end{aligned} \tag{8.25}$$

where n is the number of bands under comparison, i.e. $n = i - j + 1$, and k can be calculated from b by making use of Eq. (8.24).

8.3 Material Classification Using Absorptions

The likelihood or affinity of the two spectra compared using SFF is determined by two factors, the scale factor κ and the goodness of fit (GoF), i.e. the minimum value achieved by the LS problem in Eq. (8.24). The two spectra under comparison are deemed to be similar for large scale factors κ and small GoFs. With these ingredients, we can define the likelihood or fitting score for two local spectra segments $\mathbf{y}$ and $\hat{\mathbf{y}}$ being similar as follows:

$$f\left(\mathbf{y}, \hat{\mathbf{y}} \mid (\lambda_i, \lambda_j)\right) = \exp\left(\frac{-\text{GoF}(\mathbf{y}, \hat{\mathbf{y}})^2}{2\sigma^2 \kappa^2}\right). \tag{8.26}$$

8.3.3 Absorption Feature Based Material Classification

To perform material classification, we can use the values of $f(\mathbf{y}, \hat{\mathbf{y}}|(\lambda_i, \lambda_j))$ yielded by either SAM or SFF and compute the posterior probability of a reference spectrum $\mathbf{x}$ given a candidate spectrum $\hat{\mathbf{x}}$. We use the likelihood of the local spectral segments to recover a posterior probability that can be used for purposes of material classification. The posterior indicates whether a candidate spectrum is likely to correspond to a reference one. Therefore, classification becomes a Bayesian winner-take-all procedure (Duda and Hart 2000). The algorithm for the computation of the posterior probability is shown in Fig. 8.5. Note that, in calculating the posterior probabilities with respect to different reference spectra, we do not enforce a solution if the likelihood values are all very small. This is due to the fact that, if the posterior is close to zero, the input spectrum corresponds to a material not represented by any of the reference spectra. Therefore, we just set the posterior to zero for all classes.

To illustrate the practical application of absorption-based material classification, we have used a hyperspectral image captured by the AVIRIS sensor flown over Cuprite, Nevada in 1995. The region of interest has 400 samples, 350 lines and 220 bands covering the visible, near-infrared (NIR) and short wave infrared (SWIR) spectral ranges. For our material mapping experiment, we used the last 50 bands, from 1990.8 to 2479 nm in the SWIR spectral range, because of the higher signal-to-noise ratio (SNR) in the SWIR range. Radiometric calibration was performed using the ATmospheric REMoval program (ATREM) (Davis and Gao 1997) so as to convert radiance spectra to apparent reflectance spectra proportional to the true reflectance up to a yet unknown scaling factor. The image was then processed using the Empirical Flat Field Optimal Reflectance Transformation (EFFORT) (Boardman and Huntington 1996) to remove residual calibration errors and atmospheric effects.

The pseudo-colour image of three different bands, 2.1010 μm (red), 2.2008 μm (green) and 2.3402 μm (blue), is shown in Fig. 8.6(a). We used seven reference spectra of different minerals selected from the USGS spectral library for material mapping and classification. The material spectra employed were the playa surface, calcite, alunite, kaolinite, muscovite, silica and buddingtonite. In Fig. 8.6(b), we have plotted, in a bottom-up order, the end member spectra. In the figure, for clarity, we have offset the plots.

Input:	the set of reference spectra $\mathbf{x}_1, \ldots, \mathbf{x}_N$ and the hyperspectral image

- Training:
  Apply uni-modal segmentation to each reference spectrum and recover the absorption bands $\mathcal{C}^l = \{(\lambda_i^{(k)}, \lambda_j^{(k)}) \mid k = 1, \ldots, m(l)\}$, where $\mathcal{C}^l$ denotes the set of absorption bands for the ith spectrum, $\lambda_i^{(k)}$ and $\lambda_j^{(k)}$ refer to the first and the last channel for the kth absorption band of the lth reference spectrum and $m(l)$ denotes the total number of absorptions for the reference spectrum indexed l.
- Testing:
  For each candidate spectrum $\hat{\mathbf{x}}$ in the image
  - For each reference spectrum $\mathbf{x}_l$
    1. Extract the absorption features of the candidate spectrum.
    2. Compute the likelihood of the candidate spectrum $\hat{\mathbf{x}}$ being a match to the reference spectrum $\mathbf{x}_l$. The likelihood is given by

    $$L(\mathbf{x}_l; \hat{\mathbf{x}}) = \sum_k w_l^{(k)} f\left(\mathbf{x}_l, \hat{\mathbf{x}} \mid \left(\lambda_i^{(k)}, \lambda_j^{(k)}\right)\right), \qquad (8.27)$$

    where $w_l^{(k)}$ is the weight of the kth absorption band for reference spectrum $\mathbf{x}_l$. Here $w_l^{(k)}$ is proportional to the area of the absorption band such that $\sum_k w_l^{(k)} = 1$. Note that $f(\mathbf{x}_l, \hat{\mathbf{x}} \mid (\lambda_i^{(k)}, \lambda_j^{(k)}))$ is the likelihood for the absorption band indexed k, which can be calculated via Eqs. (8.23) or (8.26) for either SAM or SFF.

  - Compute the posterior probability for the candidate spectrum $\hat{\mathbf{x}}$ given the reference spectrum $\mathbf{x}_l$ using Bayes's rule. To do this, we assume equal prior probability for each reference spectrum

    $$P(\mathbf{x}_l \mid \hat{\mathbf{x}}) = \begin{cases} \frac{L(\mathbf{x}_l; \hat{\mathbf{x}})}{\sum_i L(\mathbf{x}_i; \hat{\mathbf{x}})}, & \max_i L(\mathbf{x}_i; \hat{\mathbf{x}}) > \varepsilon, \\ 0, & \text{otherwise.} \end{cases} \qquad (8.28)$$

Fig. 8.5 Algorithm for material mapping based on absorption bands

For purposes of illustrating the use of absorption features for pixel-level mineral mapping, we have performed experiments making use of the method in Kruse et al. (1993), the SFF method in (Clark et al. 1990) and the likelihood values yielded by Eq. (8.23). To ensure optimal performance, the parameters for each of the methods have been separately tuned. The corresponding results are displayed in Fig. 8.7, where different material types are denoted by different colours and the black pixels indicate that no matchings are found in the region.

In Fig. 8.7, SAM yields a high number of false positives for the playa surface. Furthermore, it misses the detail on mineral distributions across the scene. SFF and the likelihood yield much better results, with the latter gaining a margin of improve-

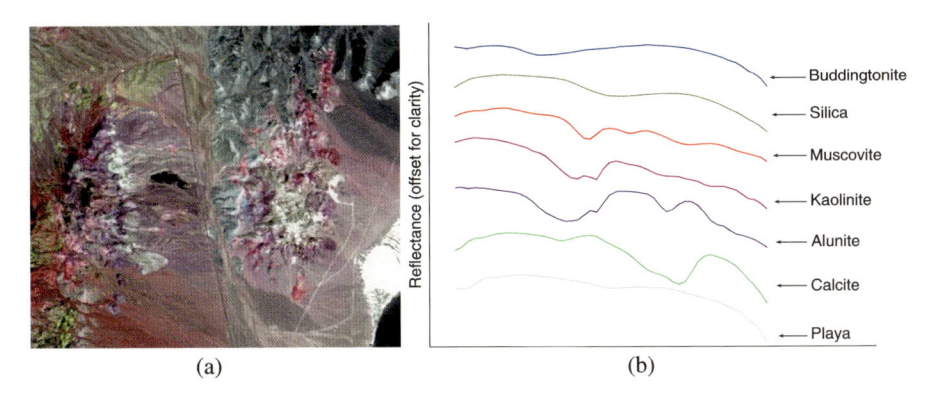

8.3 Material Classification Using Absorptions 157

Fig. 8.6 (**a**) Pseudo-colour image of Cuprite, Nevada. (**b**) End member reflectance spectra for a number of sample minerals in the scene

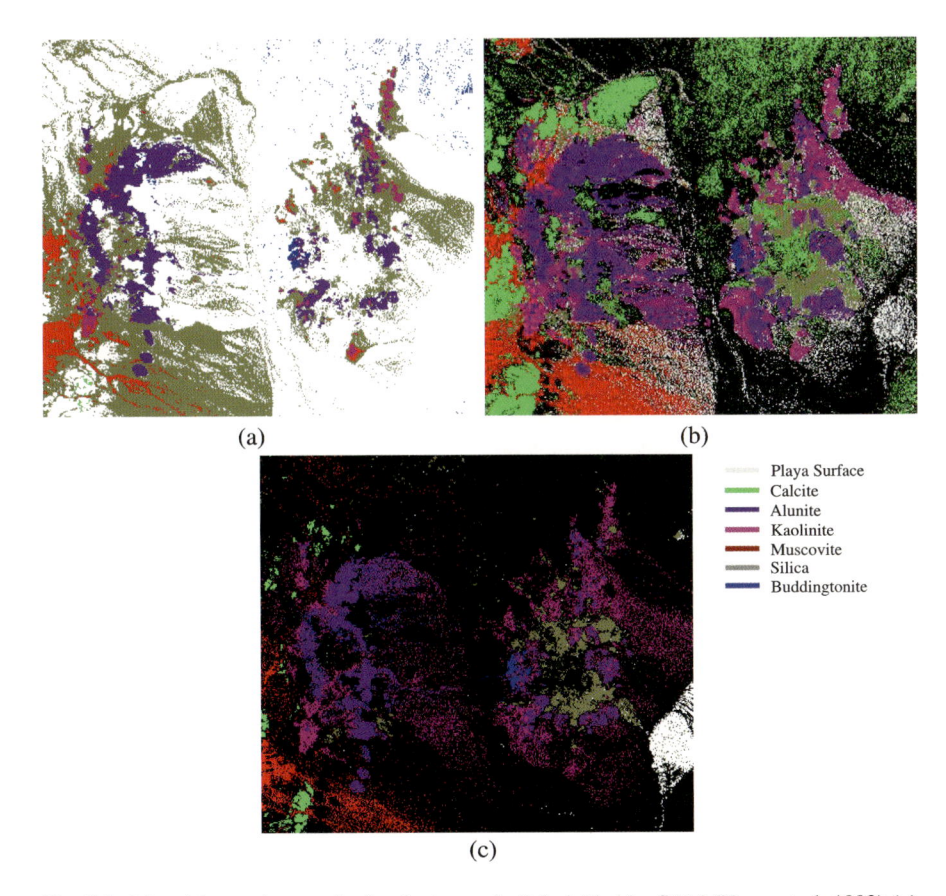

Fig. 8.7 Material mapping results for the image in 8.6 yielded by SAM (Kruse et al. 1993) (**a**), SFF (Clark et al. 1990) (**b**) and the likelihood obtained using Eq. (8.26) (**c**)

ment in performance. Nonetheless it exhibits better performance than the SAM. SFF maps a large area on the north-east corner of the scene to calcite, while most calcite is distributed in the south-west corner of the scene, as can be observed by a small green-coloured patch in Fig. 8.6(a). The dark green colour region on the north-east of the scene is actually formed by jarosite and various other minerals, which are not included in the reference spectra. In contrast with SAM and SFF, the likelihood can infer the calcite distribution more accurately by effectively rejecting false matching. Moreover, since the method is based on the local comparison of absorption bands, whose positions are known a priori from the reference spectra, we only have to extract the spectral segment from the bands of interest for each image spectrum and remove the continuum via Eq. (7.61), where $C(\lambda)$ here is simply a straight line joining the first and the last band in the absorption interval.

Note also that SFF is based on a global shape comparison between the continuum removed spectra and the candidate spectrum. In order to compare the image spectra, we have to apply continuum removal to both the candidate and the reference spectra. This implies an additional cost of $O(n \log n)$, where n is the number of bands, for each spectrum in the image before spectral matching. This computational cost of the matching operation is $O(n)$ for both SAM and SFF. For the likelihood method, the computational complexity is $O(n')$, where n' is the total number of spectral samples in the absorption band of the spectrum. Since n' is always less than or equal to n, the method is not only more accurate, but also much more efficient than traditional mapping methods for hyperspectral image analysis.

8.4 Statistical Learning and Band Selection

Statistical and absorption-based approaches are complementary in nature with respect to their respective strengths and limits. There has been little work, to our knowledge, linking the statistical analysis of the spectra with the domain knowledge of material physics and chemistry. In this section, we illustrate the use of statistical learning approaches for purposes of material identification based on absorption features. Moreover, the use of absorption features has made it feasible to train statistical classifiers on a small sample set with a marginal detriment on its generalisation properties. The reason is that absorptions are intrinsic to the object under study. The main advantage of this treatment is that, by using absorptions for purposes of identification and recognition, we can perform a localised analysis of the spectra to discriminate amongst different spectral segments.

8.4.1 Discriminant Absorption Feature Selection

It should be noted that some absorption bands may not be significant for purposes of classification. For example, some absorption bands are due to the presence of water and, hence, arise in a wide variety of materials. Furthermore, the spectra of the light source may, potentially, skew the observed absorptions. These "spurious" absorptions must be eliminated before classification can be undertaken.

This can be cast as a supervised feature selection problem. With all detected absorption bands at hand, only a few of the most discriminant bands are selected to form the final feature for classification. The other bands that do not contain sufficient discriminant information are not considered. Here, we focus on the binary setting for feature selection, where the training spectral samples are from two labelled classes. Extension to the multi-class problem is handled by making use of a pairwise fusion framework (discussed later). For absorption feature selection in a binary classification setting, the purpose is to identify those absorption bands for which the corresponding truncated absorption spectral segments from positive and negative labelled classes are best separated.

8.4.2 Canonical Correlation Analysis

For the supervised learning task, we are given a set of training spectra with known labels. The purpose is to infer a classifier in order to assign labels to the unknown testing spectra with minimum error. Let $\mathcal{R} = \{R(u, \lambda) \mid u = 1, \ldots, |\mathcal{R}|\}$ denote the set of training spectra and $Z = \{z_u \mid u = 1, \ldots, |\mathcal{R}|\}$ denote their corresponding labels, where $R(u, \lambda)$ and $z_u \in \{-1, 1\}$ represent the reflectance spectrum and the binary label for the training pixel u, respectively. We define $\mathbf{F}_u^j = [\hat{F}_u(\lambda_1^j), \ldots, \hat{F}_u(\lambda_2^j)]$ as the feature vector for the pixel u and the jth absorption band identified by the band edges $(\lambda_1^j, \lambda_2^j)$. The feature vector is computed from the truncated spectrum within the absorption band based on the representation scheme defined in Eq. (6.32). Hence, each training spectrum has m feature vectors, where m is the total number of absorption bands recovered by the absorption detection algorithm discussed in the previous chapter. Our feature selection scheme requires that the pairs of feature vectors being compared have equal length. This is not necessarily the case for absorption features, since the length of the recovered absorptions may differ greatly.

An option here is to re-evaluate the absorptions at a set length using NURBS as presented earlier in the book. Another alternative is the application of canonical correlation analysis (CCA) as a preprocessing step to align the feature vectors to allow for feature selection. This aims at reducing the computational cost of the method and allowing the comparison of spectral segments of different lengths on an equal basis.

Let $\mathcal{F}_i = \{\mathbf{F}_u^i \mid u = 1, \ldots, |\mathcal{R}|\}$ denote the set of absorption features obtained from the ith absorption band in the training set. We perform CCA to discover the correlations between features extracted from any two absorption bands $\mathcal{F}_i$ and $\mathcal{F}_j$. CCA recovers two basis vectors a and b for $\mathcal{F}_i$ and $\mathcal{F}_j$, respectively, such that the one-dimensional projections of the feature vectors in the two sets along the direction of the corresponding basis are maximally correlated. Here, we can apply CCA to each pair of absorption feature vectors and recover their respective bases. Hence, CCA is performed $\binom{m}{2}$ times where m is the total number of absorption bands detected. Nevertheless, each CCA operation is similar to one another. Therefore in the following text, we use $\mathcal{X}$ and $\mathcal{Y}$ to represent feature sets $\mathcal{F}_i$ and $\mathcal{F}_j$ and denote the corresponding multi-dimensional random variables by X and Y. The problem for seeking the optimal projection directions can be formulated as follows:

$$\max_{a,b} \rho(X, Y) = \frac{a^T \Sigma_{XY} b}{(a^T \Sigma_{XX} a)^{1/2} (b^T \Sigma_{YY} b)^{1/2}},$$

$$\mu_X = E_X[\mathbf{x}], \quad \mu_Y = E_Y[\mathbf{y}],$$

$$\Sigma_{XX} = E\left[(\mathbf{x} - \mu_X)(\mathbf{x} - \mu_X)^T\right], \quad (8.29)$$

$$\Sigma_{YY} = E\left[(\mathbf{y} - \mu_Y)(\mathbf{y} - \mu_Y)^T\right],$$

$$\Sigma_{XY} = E\left[(\mathbf{x} - \mu_X)(\mathbf{y} - \mu_Y)^T\right],$$

where $E[\cdot]$ is the expectation operator, μ_X and μ_Y are the mean vectors for the random variables X and Y, Σ_{XY} is the cross-covariance matrix between X and Y and Σ_{XX} and Σ_{YY} are the covariance matrices for X and Y, respectively. For the paired datasets $\mathcal{X}$ and $\mathcal{Y}$, we can estimate the above terms using the sample mean vectors and covariance matrices given by

$$\hat{\mu}_{\mathcal{X}} = \frac{1}{N} \sum_{i=1}^{N} \mathbf{x}_i, \quad \hat{\mu}_{\mathcal{Y}} = \frac{1}{N} \sum_{i=1}^{N} \mathbf{y}_i,$$

$$\hat{\Sigma}_{\mathcal{X}\mathcal{X}} = \frac{1}{N} \sum_{i=1}^{N} (\mathbf{x}_i - \hat{\mu}_{\mathcal{X}})(\mathbf{x}_i - \hat{\mu}_{\mathcal{X}})^T,$$

$$\hat{\Sigma}_{\mathcal{Y}\mathcal{Y}} = \frac{1}{N} \sum_{i=1}^{N} (\mathbf{y}_i - \hat{\mu}_{\mathcal{Y}})(\mathbf{y}_i - \hat{\mu}_{\mathcal{Y}})^T, \quad (8.30)$$

$$\hat{\Sigma}_{\mathcal{X}\mathcal{Y}} = \frac{1}{N} \sum_{i=1}^{N} (\mathbf{x}_i - \hat{\mu}_{\mathcal{X}})(\mathbf{y}_i - \hat{\mu}_{\mathcal{Y}})^T,$$

where $\mathbf{x}_i$ and $\mathbf{y}_i$ denote the ith element in $\mathcal{X}$ and $\mathcal{Y}$, respectively, and $N = |\mathcal{X}| = |\mathcal{Y}|$ is the total number of training samples in both datasets.

Let $a' = \hat{\Sigma}_{\mathcal{X}\mathcal{X}}^{1/2} a$ and $b' = \hat{\Sigma}_{\mathcal{Y}\mathcal{Y}}^{1/2} b$. With this change of variables and the use of the plug-in estimators defined in Eq. (8.30), the CCA formulation in Eq. (8.29) can be re-expressed as

$$\max_{a',b'} a'^T K b' = a'^T \hat{\Sigma}_{\mathcal{X}\mathcal{X}}^{-1/2} \hat{\Sigma}_{\mathcal{X}\mathcal{Y}} \hat{\Sigma}_{\mathcal{Y}\mathcal{Y}}^{-1/2} b',$$

$$\text{s.t.} \quad a'^T a' = 1 \quad \text{and} \quad b'^T b' = 1, \quad (8.31)$$

which can be solved via singular value decomposition (SVD), where the maximum canonical correlation is given by the largest singular value of matrix $K = \hat{\Sigma}_{XX}^{-1/2} \hat{\Sigma}_{XY} \hat{\Sigma}_{YY}^{-1/2}$, and the corresponding bases a' and b' are given by the left and right singular vectors corresponding to the largest singular value.

8.4.3 Feature Selection and Entropy

The previous CCA step provides subspace-projected scalar features for each pair of absorption features $\mathcal{F}^i$ and $\mathcal{F}^j$. Let $\mathbf{a}^T \mathcal{F}^i = \{\mathbf{a}^T \mathbf{F}_u \mid \mathbf{F}_u \in \mathcal{F}^i\}$ and $\mathbf{b}^T \mathcal{F}^j = \{\mathbf{b}^T \mathbf{F}_u \mid \mathbf{F}_u \in \mathcal{F}^j\}$ denote the set of scalar feature values obtained by applying CCA for the ith and jth absorption bands, respectively; we further define $\mathcal{X}^i = \{\mathbf{x}_u \in \mathbf{a}^T \mathcal{F}^i \mid z_u = 1\}$ for the subset of features from the ith absorption with positive labels and $\mathcal{Y}^j = \{\mathbf{x}_u \in \mathbf{b}^T \mathcal{F}^j \mid z_u = -1\}$ for the subset from the jth absorption with negative labels. The feature selection task can be viewed as that of selecting absorption bands with the maximum discrimination between positive and negative spectral samples. This can be achieved by solving the following optimisation problem:

$$\max_\alpha \{\pi\} = \max_\alpha \left\{ \sum_{i,j} D_\alpha \left(P(\mathcal{X}^i) \parallel P(\mathcal{Y}^j) \right) \right\}, \tag{8.32}$$

where $P(\cdot)$ denotes the probability density function and $D_\alpha(\cdot)$ refers to the divergence measure parameterised by α. The divergence $D_\alpha(\cdot)$ can be viewed as a measure of the separability between the two probability distributions $P(\mathcal{X}^i)$ and $P(\mathcal{Y}^j)$. The purpose here is to find the parameter α that maximises the total pairwise separability.

Now we discuss the selection of the divergence measure D_α in Eq. (8.32). To this end, we make use of entropy, which is a measure of uncertainty for random variables. In information theory, Renyi (1960) proposed the concept of α-entropy, which is a family of parametric functions defined over the random variable X with probability distribution function (PDF) $P(\mathcal{X}^i)$ parameterised by the scalar variable α as follows:

$$H_\alpha(\mathcal{X}^i) = \frac{1}{1-\alpha} \log \left(\int P(\mathcal{X}^i)^\alpha \, d\lambda \right). \tag{8.33}$$

Note that so far, and hereafter, we view $\mathcal{X}^i$ and $\mathcal{Y}^j$ as random variables with PDFs $P(\mathcal{X}^i)$ and $P(\mathcal{Y}^j)$. This is important since we can characterise the dissimilarity between class conditional distributions of projected absorption features, where the divergence measures are usually used to model the separability between different classes. This has a number of advantages. Firstly, it permits the feature selection task to be cast as an optimisation setting as shown above. Secondly, we can select multiple bands simultaneously, recovering the optimal solution for the combination of absorption bands under study. Thus, here, we make use of the Rényi α-divergence

(Renyi 1960), defined as

$$D_\alpha(P(\mathcal{X}^i)\|P(\mathcal{Y}^j)) = \frac{1}{\alpha-1}\log\left(\int P(\mathcal{X}^i)^\alpha P(\mathcal{Y}^j)^{1-\alpha}\,d\lambda\right), \quad (8.34)$$

to recover the most discriminant absorption bands. We then obtain a specific form of the optimisation problem by implementing the α-divergence $D_\alpha(\cdot\|\cdot)$ for each $\mathcal{X}^i$ and $\mathcal{Y}^j$ in Eq. (8.32).

Here, we take a two-step process to solve Eq. (8.32) so as to recover the most discriminative features. Firstly, we recover the parameter α for each pair so as to maximise the objective function. Secondly, we select the absorption bands that yield the larger divergences given the recovered α-parameters. To do this, we commence by using the shorthand $\gamma = \frac{1}{\alpha-1}$ and write the α-divergence above as follows:

$$D_\alpha(P(\mathcal{X}^i)\|P(\mathcal{Y}^j)) = \gamma\log\left(\int P(\mathcal{X}^i)\left(\frac{P(\mathcal{X}^i)}{P(\mathcal{Y}^j)}\right)^{\frac{1}{\gamma}}\,d\lambda\right). \quad (8.35)$$

The change of variable above is important since it allows the algebraic manipulation necessary to use the lower bound of the α-divergence for optimisation purposes. To this end, note that for $\gamma > 0$, we have $\alpha > 1$. Further, it is straightforward to see that the α-divergence is monotonically increasing with respect to α. As a result, by using the log-divergence, we can write

$$\mathcal{D}_\alpha(P(\mathcal{X}^i)\|P(\mathcal{Y}^j)) = \log(D_\alpha(P(\mathcal{X}^i)\|P(\mathcal{Y}^j)))$$
$$= \log(\gamma) + \log\left(\log\left(\int P(\mathcal{X}^i)\left(\frac{P(\mathcal{X}^i)}{P(\mathcal{Y}^j)}\right)^{\frac{1}{\gamma}}\,d\lambda\right)\right). \quad (8.36)$$

Note that the first term on the left-hand side of the expression above does not depend on either of the random variables $\mathcal{X}^i$ or $\mathcal{Y}^j$, i.e. our subspace-projected absorption bands. Thus, we remove it from further consideration and focus our attention on the second term. Since the logarithm is a monotonic function, maximising the log-divergence $\mathcal{D}_\alpha(P(\mathcal{X}^i)\|P(\mathcal{Y}^j))$ implies maximising

$$\hat{\mathcal{D}}_\alpha(P(\mathcal{X}^i)\|P(\mathcal{Y}^j)) = \log\left(\int P(\mathcal{X}^i)\left(\frac{P(\mathcal{X}^i)}{P(\mathcal{Y}^j)}\right)^{\frac{1}{\gamma}}\,d\lambda\right). \quad (8.37)$$

To turn the problem of optimising over the log-divergence above into one that can be solved algebraically, we use the following relation presented in Alencar and Assis (1998):

$$\int P(\mathcal{X}^i)\log\left(\frac{P(\mathcal{X}^i)}{P(\mathcal{Y}^j)}\right)^{\frac{1}{\gamma}}\,d\lambda = \frac{1}{\gamma}KL(P(\mathcal{X}^i)\|P(\mathcal{Y}^j))$$
$$\leq \log\left(\int P(\mathcal{X}^i)\left(\frac{P(\mathcal{X}^i)}{P(\mathcal{Y}^j)}\right)^{\frac{1}{\gamma}}\,d\lambda\right), \quad (8.38)$$

8.4 Statistical Learning and Band Selection

where $KL(P(\mathcal{X}^i)\|P(\mathcal{Y}^j))$ is the Kullback–Leibler (KL) divergence between the subspace-projected absorption bands given by

$$KL(P(\mathcal{X}^i)\|P(\mathcal{Y}^j)) = \int P(\mathcal{X}^i)\log\frac{P(\mathcal{X}^i)}{P(\mathcal{Y}^j)}\,d\lambda. \tag{8.39}$$

The expression above is important since it provides a lower bound for $\hat{\mathcal{D}}_\alpha(P(\mathcal{X}^i)\|P(\mathcal{Y}^j))$. This permits using the lower bound of the log-divergence as an alternative to $\hat{\mathcal{D}}_\alpha(P(\mathcal{X}^i)\|P(\mathcal{Y}^j))$. Further, note that the KL divergence is, in general, not symmetric, i.e. $KL(P(\mathcal{X}^i)\|P(\mathcal{Y}^j)) \neq KL(P(\mathcal{Y}^j)\|P(\mathcal{X}^i))$. To obtain a symmetric log-divergence, we can employ the average distance over the two possible KL divergences for the two distributions. This yields Jeffrey's divergence (Jeffreys 1946), which can be viewed as an expected value over the KL divergences for those random variables whose distributions are exponential in nature.

Thus, we can use, as an alternative to π in Eq. (8.32), the quantity

$$\hat{\pi} = \sum_{i,j\in\Gamma}\frac{1}{2\gamma}\left(KL(P(\mathcal{X}^i)\|P(\mathcal{Y}^j)) + KL(P(\mathcal{Y}^j)\|P(\mathcal{X}^i))\right)$$

$$= \sum_{i,j\in\Gamma}\frac{1}{\gamma}\mathcal{KL}(P(\mathcal{X}^i), P(\mathcal{Y}^j)), \tag{8.40}$$

where $\mathcal{KL}(P(\mathcal{X}^i), P(\mathcal{Y}^j)) = \frac{1}{2}(KL(P(\mathcal{X}^i)\|P(\mathcal{Y}^j)) + KL(P(\mathcal{Y}^j)\|P(\mathcal{X}^i)))$ is Jeffrey's divergence, i.e. the symmetrised KL divergence. The summation is taken over the set of all pairs $\Gamma = \{(i, j) \mid i, j \in [1, \ldots, m]$ and $i < j\}$ with m being the total number of absorption bands detected.

Moreover, we can write the expression above in compact form. We do this by constructing a matrix $\mathbf{D}$ of order $|\Gamma|$ whose element $D_{i,j}$ indexed i, j is given by the symmetrised KL divergence between every pair of subspace-projected absorption segments, i.e. $D_{i,j} = \mathcal{KL}(P(\mathcal{X}^i), P(\mathcal{Y}^j))$. Thus, $D_{i,j}$ is defined as the divergence between the positively labelled subset of subspace-projected features for the ith absorption band and negatively labelled subset of projected features for the jth absorption band, as delivered at output by the CCA processing.

By expressing $\frac{1}{\gamma}$ as the product of two other variables ϕ_i, ϕ_j for each random-variable pair $\mathcal{X}^i, \mathcal{Y}^j$ in Eq. (8.40), we can reformulate the optimisation problem defined in Eq. (8.32) as follows:

$$\max\{\hat{\pi}\} = \max_\phi\left\{\sum_{i,j}\phi_i D_{i,j}\phi_j\right\} = \max_\phi\{\phi^T\mathbf{D}\phi\}. \tag{8.41}$$

From inspection, it is clear that maximising $\hat{\pi}$ can be posed as an eigenvalue problem, and that ϕ is an eigenvector of $\mathbf{D}$ whose corresponding eigenvalue is $\hat{\pi}$. Furthermore, the expression above is reminiscent of the Rayleigh quotient. Note that the elements of the vector ϕ are required to be real and positive. This is a consequence of the fact that $\gamma = \frac{1}{\alpha-1}$ and that α is bounded between -1 and unity.

Consequently, the coefficients of the eigenvector ϕ are always non-negative. Since the elements of the matrix $\mathbf{D}$ are non-negative, it follows that the quantity $\phi^T \mathbf{D} \phi$ should be positive. Hence, the set of solutions reduces itself to one that corresponds to a constant $\hat{\pi} > 0$. We also require the coefficients of the vector ϕ to be linearly independent of the all-ones vector $\mathbf{e} = [1, 1, \ldots, 1]^T$.

With these observations in mind, we focus on proving the existence of a vector ϕ, linearly independent from $\mathbf{e}$, and demonstrate that this vector is unique. To this end, we use the Perron–Frobenius theorem (Varga 2000). This concerns the proof of existence regarding the eigenvalue $\hat{\pi} = \max_{i=1,2,\ldots,|\Gamma|}\{\pi_i\}$ of a primitive, real, non-negative, symmetric matrix $\mathbf{D}$, and the uniqueness of the corresponding eigenvector ϕ. The Perron–Frobenius theorem states that the eigenvalue $\hat{\pi} > 0$ has a multiplicity of one. Moreover, the coefficients of the corresponding eigenvector ϕ are all positive and the eigenvector is unique. As a result, the remaining eigenvectors of W have at least one negative and one positive coefficient. If $\mathbf{D}$ is substochastic, ϕ is also known to be linearly independent of the all-ones vector $\mathbf{e}$. As a result, the leading eigenvector of $\mathbf{D}$ is the unique solution of Eq. (8.41).

Thus, we can recover the optimum α for each pair of subspace-projected absorption bands $\mathcal{X}^i$ and $\mathcal{Y}^j$ by building the matrix $\mathbf{D}$ and computing its leading eigenvector ϕ. Once the eigenvector ϕ is in hand, we can recover the α parameter by using the relationship $\alpha = 1 + \frac{1}{\gamma} = 1 + \phi_i \phi_j$ presented earlier. Once the α parameter has been recovered, the pairs X, Y corresponding to the largest divergences $\hat{\mathcal{D}}_\alpha(P(\mathcal{X}^i) \| P(\mathcal{Y}^j))$ can be selected for purposes of classification.

8.4.4 Material Identification via the α-Entropy

We now elaborate further on the algorithm for material identification. The step sequence of the algorithm is presented in Fig. 8.8. The idea underpinning the algorithm here is that the optimal absorption band should best discriminate between positive and negative spectral classes. This can be achieved by maximising the α-divergence between distributions of corresponding spectral features.

Note that, for every candidate absorption band, each spectral segment can be seen as a feature vector in the N-dimensional space, where the dimensionality N is given by the number of bands in the absorption spectral segment. As the number of bands may be different for every pair of candidate absorptions, the computed KL divergence may be biased towards features with a greater number of bands. This may occur regardless of the separability of the two classes and is due to the fact that distributions in higher dimensional spaces are, in general, more scattered than those defined in lower dimensions. To balance this effect, we apply CCA to each of pair of candidate absorption bands as a preprocessing step for the algorithm above. Note that, since we have at our disposal training data, the sample covariance and cross-covariance matrices used by the CCA can be computed directly. Thus, CCA will project all the spectral segments for every candidate absorption band onto a one-dimensional space. We then compute the KL divergence between the distributions

8.4 Statistical Learning and Band Selection

Given the set of training spectral features $\mathcal{R} = \{r_1(\lambda), \ldots, r_{|\mathcal{R}|}(\lambda)\}$ and the label set $\mathcal{Z} = \{z_u\}$, where $z_u \in \{-1, 1\}$ is the label for the spectrum $r_u(\lambda)$, we compute the set of tuples $\{(\lambda_1^1, \lambda_2^1), \ldots, (\lambda_1^m, \lambda_2^m)\}$ corresponding to the extreme spectral values for the m absorption bands recovered by our algorithm, as described in the previous section. For each tuple $(\lambda_1^j, \lambda_2^j)$, do

- Extract the spectral bands for the recovered absorption segments and transform them to invariant feature representation $\hat{F}_u(\lambda_i)$ as shown in Sect. 8.4.2.
- For each pair of absorption features, apply CCA to the feature vectors and project them along the direction given by the maximum canonical correlation.
- Estimate the symmetrised KL divergence values for the projected feature elements via Eq. (8.42).
- With the symmetrised KL divergences, construct the matrix $\mathbf{D}$ and recover its leading eigenpair. Use the corresponding eigenvector to recover the values of α that maximise the divergence between classes. Use the m absorption bands which correspond to the m-largest α-divergence values.

Fig. 8.8 Discriminant absorption band selection algorithm

of these subspace-projected one-dimensional features and select the most discriminative absorptions.

Since CCA is a correlation-based technique, in the algorithm above we have assumed the positive and negative sample set distributions to be Gaussian in nature. This is a reasonable assumption due to the uni-modality of the absorption bands. Since linear projection preserves Gaussian distributions, the subspace-projected feature elements also satisfy Gaussian distributions whose KL divergence can be computed in closed form. Let $\hat{\mu}_\mathcal{X}$ and $\hat{\sigma}_\mathcal{X}^2$ be the sample mean and variance of the subspace-projected feature set $\mathcal{X}$, and $\hat{\mu}_\mathcal{Y}$ and $\hat{\sigma}_\mathcal{Y}^2$ be the sample mean and variance of $\mathcal{Y}$. The symmetrised KL divergence between distributions $P(\mathcal{X})$ and $P(\mathcal{Y})$ is then given by

$$\mathcal{KL}(P(\mathcal{X}) \| P(\mathcal{Y})) = \frac{(\hat{\mu}_\mathcal{X} - \hat{\mu}_\mathcal{Y})^2 (\hat{\sigma}_\mathcal{X}^2 + \hat{\sigma}_\mathcal{Y}^2)}{2\hat{\sigma}_\mathcal{X}^2 \hat{\sigma}_\mathcal{Y}^2} + \frac{\hat{\sigma}_\mathcal{X}^2}{2\hat{\sigma}_\mathcal{Y}^2} + \frac{\hat{\sigma}_\mathcal{Y}^2}{2\hat{\sigma}_\mathcal{X}^2} - 1. \quad (8.42)$$

In the expression above, for brevity, we have omitted the superscripts for the subspace-projected features $\mathcal{X}$ and $\mathcal{Y}$ without further disambiguation. The KL divergences estimated from the above equation are then used to compute the matrix elements $D_{i,j}$ in Eq. (8.41), where $D_{i,j} = \mathcal{KL}(P(\mathcal{X}^i), P(\mathcal{Y}^j))$ captures the divergence between positively labelled subspace-projected features from the ith absorption band $\mathcal{X}^i$ and negatively labelled projected features from the jth absorption band $\mathcal{Y}^j$.

Finally, we employ a support vector machine (SVM) (Cristianini and Shawe-Taylor 2000) for the classification of absorption features. To this end we can take the truncated absorption spectral segment from each absorption band making use of the feature representation. The truncated absorption feature segments, scaled by

Training

Given a set $\Gamma = \{\omega_1, \omega_2, \ldots, \omega_{|\Omega|}\}$ of materials to identify

- Obtain positive samples for every pair of materials ω_i and ω_j in Ω.
- Perform absorption feature extraction and selection for every pair of classes ω_i and ω_j.
- Train the SVM classifier over the selected features.

Testing

- For any novel test spectrum, apply each of the trained binary classifiers above.
- Combine the binary classification results and assign the spectrum to the class that maximises the total posterior probability given by

$$\xi = \arg\max_i \sum_{j=1, j\neq i}^{N} Q_{i,j}, \qquad (8.45)$$

where $Q_{i,j}$ is the decision value output of the classifier for the pair of classes indexed i and j in Ω. $Q_{i,j}$ is anti-symmetric, i.e. $Q_{i,j} = -Q_{j,i}$ for any $i \neq j$.

Fig. 8.9 Pairwise classifier fusion method

their corresponding eigenvalues of D in Eq. (8.41), are then concatenated to form the final absorption feature vector to be used for SVM classification. To solve multi-class classification problems, we can adopt a pairwise fusion framework (Hastie and Tibshirani 1998). The idea is to build a binary classifier for any pair of two classes, where the final classification result is obtained by combining the output of all binary classifiers. Instead of voting on hard decisions, we aggregate the decision value output produced by each component classifier and assign the testing data to the class with the highest aggregation value. This can be viewed as a generalisation to majority voting. The procedure of the fusion algorithm is presented in Fig. 8.9.

The use of a pairwise classifier fusion framework has two major advantages over a single classifier directly applied to the multi-class classification task. Firstly, it is often easier to separate two classes and combine the binary classification results than finding a multi-class classifier that can distinguish all classes. Secondly and, more importantly, the use of a pairwise fusion framework allows us to use different sets of absorption features for every pair of classes. In our case, we can use those absorption bands which best distinguish between different classes without being constrained to a unique band or feature that is applied to all material classes under study.

Note that, despite our focus on classification problems, pairwise classification can also be used for the more complicated problem of supervised spectral unmixing (Chang 2006). In classification, each pixel in the image is assumed to have a pure spectrum and is assigned to a single material class. This may not be the case for imaging spectroscopy data, where image pixel spectra may correspond to multiple material types due to the variability in material composition and restriction in spatial resolution. We note that the SVM classifier can also produce probability values

8.4 Statistical Learning and Band Selection

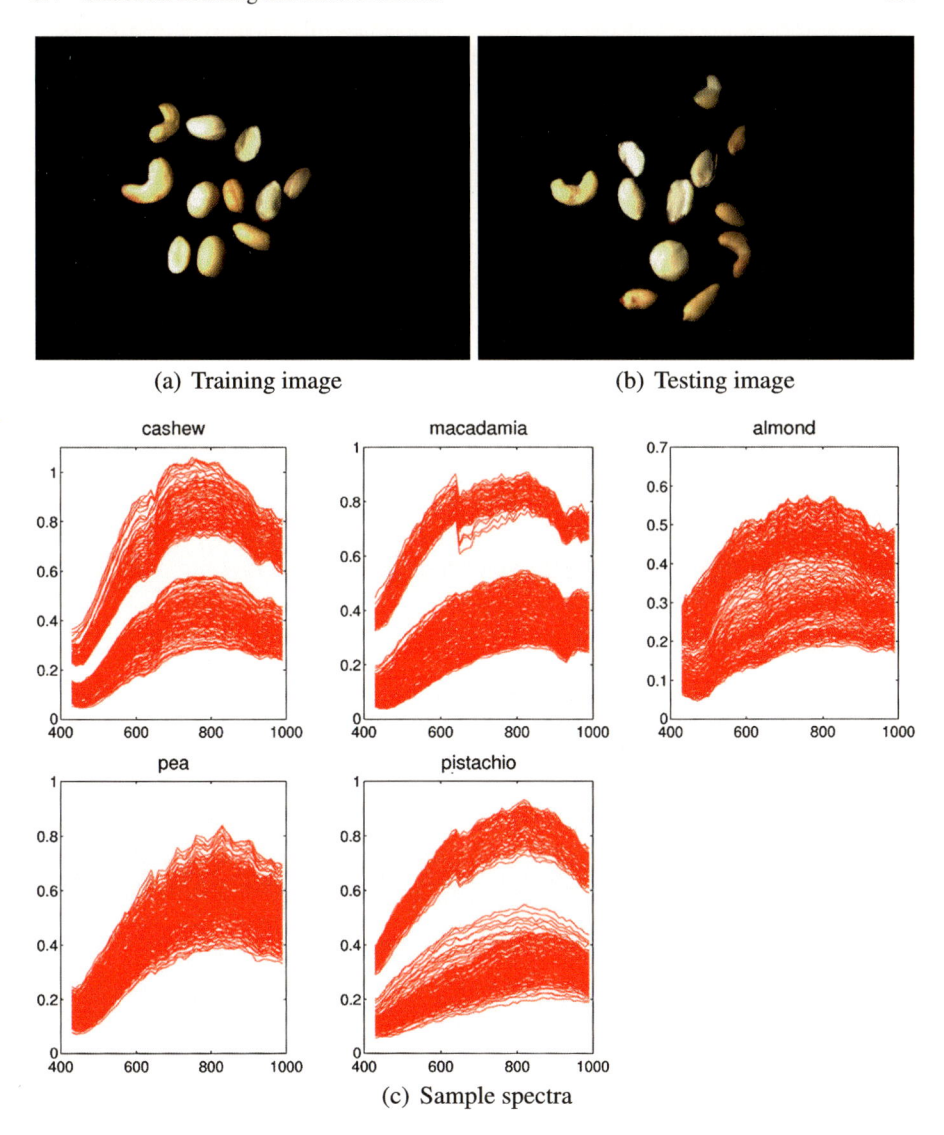

Fig. 8.10 *Top row*: pseudo-colour images of assorted nuts used for training (*left-hand panel*) and testing (*right-hand panel*). *Bottom rows*: spectra for the nut types in the top row

at output for each label class (Platt 2000). We can exploit these probabilistic output values from each binary classifier and aggregate them via the right-hand side of Eq. (8.45), which is the final probability for class i (up to a scaling factor) corresponding to the mixing coefficient of the ith material class. The only difference is that now each $Q_{i,j}$ is a probability value and $Q_{j,i} = 1 - Q_{i,j}$.

In Fig. 8.10, we show the pseudo-colour images for the two hyperspectral images along with the example spectra of a set of nuts being classified. The problem is to classify these five nut types, i.e. cashew, macadamia, almond, peanut and pistachio.

For purposes of training, we have manually labelled foreground regions according to nut type. In Fig. 8.10(c) we have plotted 100 randomly selected pixel spectra for each type of nut. From the plots, it is not obvious how the spectra of different nuts can be employed for recognition purposes. Indeed, this is a very challenging classification task due to the high degree of similarity between the spectra of different types of nuts. Moreover, the intensity information of the nut spectra is not reliable due to the variation of the illuminant across the imagery.

Figure 8.11 shows example classification maps for the two images in Fig. 8.10 yielded by the absorption features as compared to the raw image data. In the figure, we have used a random training sample with 50 pixels for each nut type taken from the reference image in the left-hand column. The right-hand column presents the results on the novel image. The ground truth maps and the results yielded by the discriminative features are compared with the raw spectra. From the figure, it is clear that the selected features recover a more plausible material map, i.e. one which is a closer match to the ground truth, than that delivered by the use of raw spectra.

We now conclude the section by examining the features used for classification. Since we select features of different lengths for each pairwise SVM classifier, we have applied kernel principal component analysis (KPCA) (Schölkopf et al. 1999) to the kernel matrix induced by the feature vectors and kept the first two principal components for visualisation purposes. The distributions of the KPCA projections for the absorption features extracted for each pair of classes are plotted in the left-hand column of Fig. 8.12. In the figure, we have used a random training set of 100 spectra for each nut type. For purposes of comparison, we also plot the top two principal components recovered by KPCA from the raw spectra in the left-hand column of the figure. For brevity, we only show the feature projection maps for nut-type pairs corresponding to cashew against macadamia, almond, peanut and pistachio. In all the panels, the blue dots represent feature projections for the cashews, whereas the green dots correspond to the other nut types. From the feature projections, we can appreciate that the absorption features have better discriminative power than the raw spectra. This is exemplified by the large overlap between feature projections from different labels in the KPCA map of raw features, where classes, i.e. nut-type pairs, are scattered. In contrast, the plots corresponding to our method show better linearly separable mappings, which can then be employed by a large margin classifier like the SVM for improved classification performance.

8.5 Notes

In the chapter, we have treated scenes as being composed of material which can then be unmixed for purposes of recovering their end members. These end members and materials may be used for recognition and classification. Note that, in practice, object classification and image categorisation techniques (Nister and Stewenius 2006;

8.5 Notes

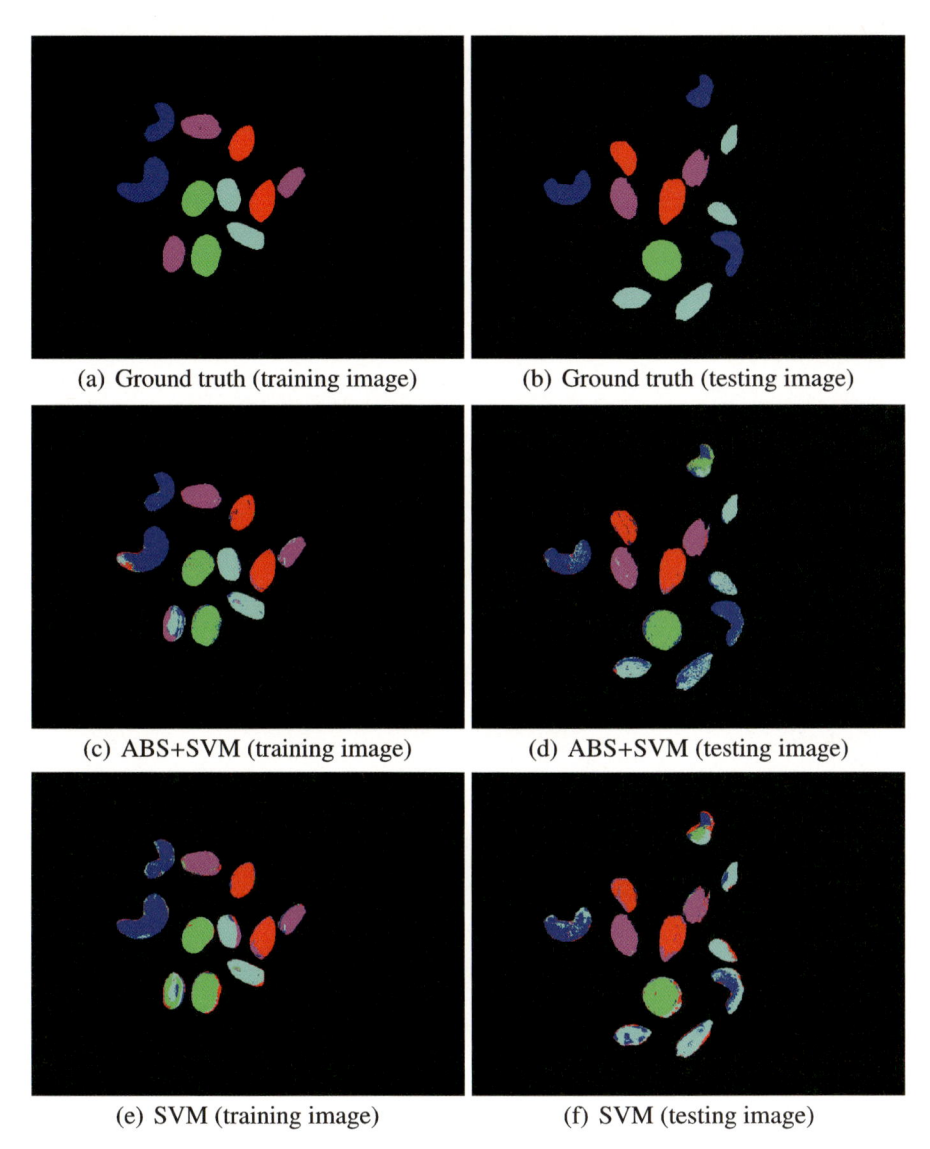

(a) Ground truth (training image)

(b) Ground truth (testing image)

(c) ABS+SVM (training image)

(d) ABS+SVM (testing image)

(e) SVM (training image)

(f) SVM (testing image)

Fig. 8.11 Material mapping results for the nut images in Fig. 8.10. *Top row*: ground truth material maps (cashew: *blue*, macadamia: *green*, almond: *red*, peanut: *cyan*, pistachio: *magenta*); *Middle row*: material maps delivered using the discriminative absorption features as applied to an SVM classifier; *Bottom row*: material maps yielded by applying an SVM classifier on the raw spectra

Sivic et al. 2005; Chum et al. 2007) are based on the comparison of image features between those in a query image and those corresponding to the images in a dataset. This is usually based on a codebook that serves as a means to recover the closest match through classification. Furthermore, descriptors are often combined so as to assign a different importance to each of them in order to maximise performance.

(a) cashew (green) vs. macadamia (blue)

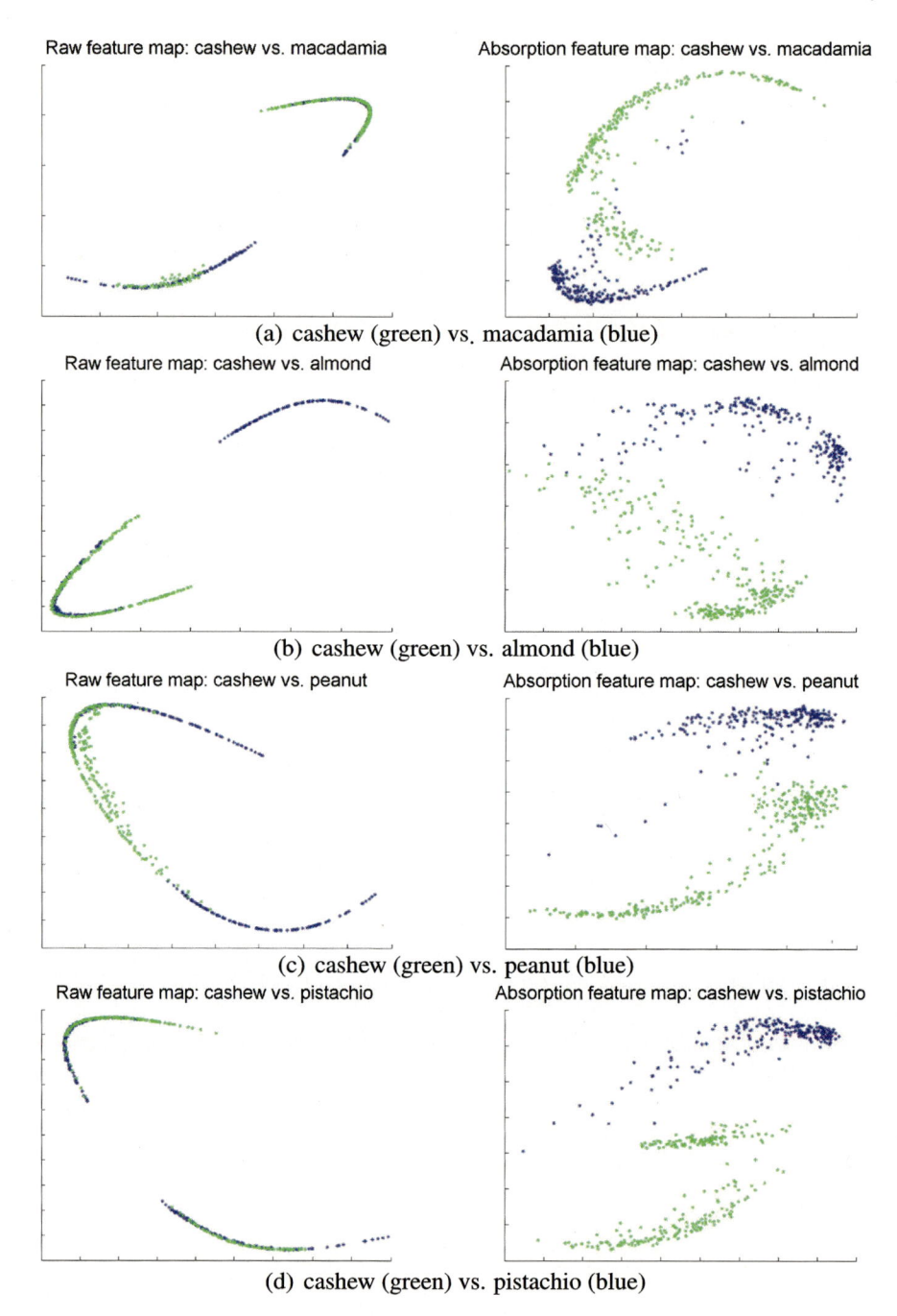

(b) cashew (green) vs. almond (blue)

(c) cashew (green) vs. peanut (blue)

(d) cashew (green) vs. pistachio (blue)

Fig. 8.12 Kernel principal component analysis (KPCA) embeddings for sample nut-type pairs used in our experiments. *Left-hand column*: KPCA embeddings for the raw spectra; *Right-hand column*: KPCA projections for the selected absorption features

Earlier, we elaborated on the use of features in imaging spectroscopy. These materials can be used to compute high-level features which can be further selected and combined making use of supervised learning techniques over a training set of known object categories or classes.

This may involve recovering either the optimal descriptors directly (Winder and Brown 2007) or their optimal combination (Nilsback and Zisserman 2006; Varma and Ray 2007) in order to maximise performance and minimise the variance across the dataset with respect to the classifier output. In any case, the classifier output is expected to be dependent on the image representation and the similarity measure employed to categorise the images (Vasconcelos 2004). These image features arise not only from the local descriptors but also from spectral signatures which capture the composition of objects in terms of materials and their quantities in the scene. To this end, feature selection can be adapted to a spatio-spectral model. This would allow us to apply mixtures of spatio-spectral features to the imagery for purposes of recognition.

Moreover, methods to handle the classification of objects composed of multiple end members can be improved by casting the feature combination and selection problem into a pairwise classification framework. Note that, here, selection can also be applied to the spectra. Further, by viewing the spectra as high-dimensional vector fields, one can also apply kernel methods to incorporate machine learning techniques into the feature recovery process. Other possibilities comprise the use of attributed graph-based representations, where pixels, object parts or super-pixels can be viewed as nodes. In this manner, the information delivered by their spectra can then be used to recover an attribute set for the graph.

References

Alencar, M. S., & Assis, F. M. (1998). A relation between the Renyi distance of order α and the variational distance. In *International telecommunications symposium* (Vol. I, pp. 242–244).

Ambikapathi, A., Chan, T., Ma, W., & Chi, C. (2010). A robust alternating volume maximization algorithm for endmember extraction in hyperspectral images. In *Proceedings of the IEEE workshop on hyperspectral image and signal processing: evolution in remote sensing* (pp. 1–4).

Barnard, K., Martin, L., Funt, B., & Coath, A. (2002). A data set for colour research. *Color Research and Application*, 27(3), 147–151.

Berman, M., Kiiveri, H., Lagerstrom, R., Ernst, A., Dunne, R., & Huntington, J. F. (2004). ICE: a statistical approach to identifying endmembers in hyperspectral images. *IEEE Transactions on Geoscience and Remote Sensing*, 42(10), 2085–2095.

Berry, M. W., Browne, M., Langville, A. N., Pauca, V. P., & Plemmons, R. J. (2007). Algorithms and applications for approximate nonnegative matrix factorization. *Computational Statistics & Data Analysis*, 52(1), 155–173.

Bioucas-Dias, J. (2009). A variable splitting augmented Lagrangian approach to linear spectral unmixing. In *Proceedings of the IEEE workshop on hyperspectral image and signal processing: evolution in remote sensing* (pp. 1–4).

Bioucas-Dias, J., & Plaza, A. (2010). Hyperspectral unmixing: Geometrical, statistical and sparse regression-based approaches. In *Society of Photo-Optical Instrumentation Engineers (SPIE) conference series: Vol. 7830. Image and signal processing for remote sensing XVI*.

Boardman, J. W., & Huntington, J. F. (1996). Mineral mapping with 1995 AVIRIS data. In *Summaries of the sixth annual JPL airborne research science workshop* (pp. 9–11).

Boardman, J. W., & Kruse, F. A. (1994). *Automated spectral analysis: a geological example using AVIRIS data, northern Grapevine Mountains, Nevada.*

Chan, T.-H., Chi, C.-Y., Huang, Y.-M., & Ma, W.-K. (2009). A convex analysis-based minimum-volume enclosing simplex algorithm for hyperspectral unmixing. *IEEE Transactions on Signal Processing, 47*(11), 4418–4432.

Chang, C.-I. (2003). *Hyperspectral imaging: techniques for spectral detection and classification.* Dordrecht: Kluwer Academic/Plenum.

Chang, C.-I. (2006). *Hyperspectral imaging: techniques for spectral detection and classification.* Berlin: Springer.

Chen, J., Jia, X., Yang, W., & Matsushita, B. (2009). Generalization of subpixel analysis for hyperspectral data with flexibility in spectral similarity measures. *IEEE Transactions on Geoscience and Remote Sensing, 47*(7), 2165–2171.

Chum, O., Philbin, J., Sivic, J., Isard, M., & Zisserman, A. (2007). Total recall: automatic query expansion with a generative feature model for object retrieval. In *International conference on computer vision.*

Cichocki, A., Zdunek, R., Phan, A. H., & Amari, S. (2009). *Nonnegative matrix and tensor factorizations: applications to exploratory multi-way data analysis and blind source separation.* New York: Wiley.

Clark, R. N., Gallagher, A. J., & Swayze, G. (1990). Material absorption band depth mapping of imaging spectrometer data using a complete band shape least-squares fit with library reference spectra. In *Proceedings of the second airborne visible/infrared imaging spectrometer workshop* (pp. 176–186).

Craig, M. D. (1994). Minimum-volume transforms for remotely sensed data. *IEEE Transactions on Geoscience and Remote Sensing, 32*(1), 99–109.

Cristianini, N., & Shawe-Taylor, J. (2000). *An introduction to support vector machines.* Cambridge: Cambridge University Press.

Davis, B.-C., & Gao, C. O. (1997). Development of a line-by-line-based atmosphere removal algorithm for airborne and spaceborne imaging spectrometers. In *Proceedings of SPIE: Vol. 3118. Imaging Spectrometry III* (pp. 132–141).

Donoho, D., & Stodden, V. (2004). When does non-negative matrix factorization give a correct decomposition into parts? In *Neural information processing systems.* Cambridge: MIT Press.

Donoho, D. (2006). Compressed sensing. *IEEE Transactions on Information Theory, 52*(4), 1289–1306.

Drew, M. S., & Finlayson, G. D. (2007). Analytic solution for separating spectra into illumination and surface reflectance components. *Journal of the Optical Society of America A, 24*(2), 294–303.

Duda, R. O., & Hart, P. E. (2000). *Pattern classification.* New York: Wiley.

Dundar, M., & Landgrebe, D. (2004). Toward an optimal supervised classifier for the analysis of hyperspectral data. *IEEE Transactions on Geoscience and Remote Sensing, 42*(1), 271–277.

Fukunaga, K. (1990). *Introduction to statistical pattern recognition* (2nd ed.). New York: Academic Press.

Guo, Z., Wittman, T., & Osher, S. (2009). L1 unmixing and its application to hyperspectral image enhancement. In S. S. Shen & P. E. Lewis (Eds.), *Algorithms and technologies for multispectral, hyperspectral, and ultraspectral imagery* (Vol. 7334, p. 73341M).

Hapke, B. (1993). *Theory of reflectance and emittance spectroscopy.* Cambridge: Cambridge University Press.

Hastie, T., & Tibshirani, R. (1998). Classification by pairwise coupling. In *Advances in neural information processing systems* (Vol. 10).

Hoyer, P. O. (2004). Non-negative matrix factorization with sparseness constraints. *Journal of Machine Learning Research, 5*, 1457–1469.

Huck, A., Guillaume, M., & Blanc-Talon, J. (2010). Minimum dispersion constrained nonnegative matrix factorization to unmix hyperspectral data. *IEEE Transactions on Geoscience and Remote*

Sensing, 48(6), 2590–2602.

Iordache, M. D., Plaza, A., & Bioucas-Dias, J. (2010). Recent developments in sparse hyperspectral unmixing. In *IEEE international geoscience and remote sensing symposium*.

Jaynes, E. T. (1957). Information theory and statistical mechanics. *Physical Review, 106*(4), 620–630.

Jeffreys, H. (1946). An invariant form for the prior probability in estimation problems. In *Proceedings of the royal society A* (Vol. 186, p. 453).

Jia, S., & Qian, Y. (2007). Spectral and spatial complexity-based hyperspectral unmixing. *IEEE Transactions on Geoscience and Remote Sensing, 45*(12), 3867–3879.

Jia, S., & Qian, Y. (2009). Constrained nonnegative matrix factorization for hyperspectral unmixing. *IEEE Transactions on Geoscience and Remote Sensing, 47*(1), 161–173.

Jimenez, L., & Landgrebe, D. (1999). Hyperspectral data analysis and feature reduction via projection pursuit. *IEEE Transactions on Geoscience and Remote Sensing, 37*(6), 2653–2667.

Jolliffe, I. (2002). *Principal component analysis*. Berlin: Springer.

Judd, D. B., Macadam, D. L., Wyszecki, G., Budde, H. W., Condit, H. R., Henderson, S. T., & Simonds, J. L. (1964). Spectral distribution of typical daylight as a function of correlated color temperature. *Journal of the Optical Society of America, 54*(8), 1031–1036.

Keshava, N. (2003). A survey of spectral unmixing algorithms. *The Lincoln Laboratory Journal, 14*(1), 55–78.

Kirkpatrick, S., Gelatt, C. D., & Vecchi, M. P. (1983). Optimization by simulated annealing. *Science, 220*(4598), 671–680.

Kruse, F. A., Lefkoff, A. B., & Dietz, J. B. (1993). Expert system-based mineral mapping in northern Death Valley, California/Nevada using the airborne visible/infrared imaging spectrometer (AVIRIS). *Remote Sensing of Environment, 44*, 309–336.

Landgrebe, D. (2002). Hyperspectral image data analysis. *IEEE Signal Processing Magazine, 19*, 17–28.

Lee, D. D., & Seung, H. S. (1999). Learning the parts of objects by non-negative matrix factorization. *Nature, 401*, 788–791.

Lennon, M., Mercier, G., Mouchot, M. C., & Hubert-moy, L. (2001). Spectral unmixing of hyperspectral images with the independent component analysis and wavelet packets. In *Proceedings of the international geoscience and remote sensing symposium*.

Miao, L. D., & Qi, H. R. (2007). Endmember extraction from highly mixed data using minimum volume constrained nonnegative matrix factorization. *IEEE Transactions on Geoscience and Remote Sensing, 45*(3), 765–777.

Mika, S., Ratsch, G., Weston, J., Scholkopf, B., & Muller, K. (1999). Fisher discriminant analysis with kernels. In *IEEE neural networks for signal processing workshop* (pp. 41–48).

Nascimento, J. M. P., & Dias, J. M. B. (2005). Vertex component analysis: a fast algorithm to unmix hyperspectral data. *IEEE Transactions on Geoscience and Remote Sensing, 43*(4), 898–910.

Nilsback, M. E., & Zisserman, A. (2006). A visual vocabulary for flower classification. In *Computer vision and pattern recognition* (Vol. II, pp. 1447–1454).

Nister, D., & Stewenius, H. (2006). Scalable recognition with a vocabulary tree. In *Computer vision and pattern recognition* (Vol. II, pp. 2161–2168).

Paatero, P. (1997). Least squares formulation of robust non-negative factor analysis. *Chemometrics and Intelligent Laboratory Systems, 37*(1), 23–35.

Parkkinen, J., Hallikainen, J., & Jaaskelainen, T. (1989). Characteristic spectra of Munsell colors. *Journal of the Optical Society of America A, 6*(2), 318–322.

Pascual-Montano, A., Carazo, J. M., Kochi, K., Lehmann, D., & P.-Marqui, R. D. (2006). Nonsmooth nonnegative matrix factorization (nsNMF). *IEEE Transactions on Pattern Analysis and Machine Intelligence, 28*(3):403–415.

Pauca, V. P., Piper, J., & Plemmons, R. J. (2006). Nonnegative matrix factorization for spectral data analysis. *Linear Algebra and Its Applications, 416*(1), 29–47.

Platt, J. (2000). Probabilistic outputs for support vector machines and comparison to regularized likelihood methods. In *Advances in large learning a discriminative margin classifiers* (pp. 61–74).

Qian, Y., & Wang, Q. (2010). Noise-robust subband decomposition blind signal separation for hyperspectral unmixing. In *IEEE international geoscience and remote sensing symposium*.

Renyi, A. (1960). On measures of information and entropy. In *4th Berkeley symposium on mathematics, statistics and probability* (pp. 547–561).

Scholkopf, B., & Smola, A. J. (2001). *Learning with kernels: support vector machines, regularization, optimization, and beyond*. Cambridge: MIT Press.

Schölkopf, B., Smola, A. J., & Müller, K.-R. (1999). Kernel principal component analysis. In *Advances in kernel methods: support vector learning* (pp. 327–352).

Sivic, J., Russell, B., Efros, A., Zisserman, A., & Freeman, W. (2005). Discovering objects and their location in images. In *International conference on computer vision* (Vol. I, pp. 370–377).

Sunshine, J., Pieters, C. M., & Pratt, S. F. (1990). Deconvolution of mineral absorption bands: an improved approach. *Journal of Geophysical Research*, *95*(B5), 6955–6966.

Thompson, D., Castao, R., & Gilmore, M. (2009). Sparse superpixel unmixing for exploratory analysis of CRISM hyperspectral images. In *Proceedings of the IEEE workshop on hyperspectral image and signal processing: evolution in remote sensing* (pp. 1–4).

Varga, R. S. (2000). *Matrix iterative analysis* (2nd ed.). Berlin: Springer.

Varma, M., & Ray, D. (2007). Learning the discriminative powerinvariance trade-off. In *International conference on computer vision*.

Vasconcelos, N. (2004). On the efficient evaluation of probabilistic similarity functions for image retrieval. *IEEE Transactions on Information Theory*, *50*(7), 1482–1496.

Wang, J., & Chang, C.-I. (2006). Applications of independent component analysis in endmember extraction and abundance quantification for hyperspectral imagery. *IEEE Transactions on Geoscience and Remote Sensing*, *44*(9), 2601–2616.

Winder, S., & Brown, M. (2007). Learning local image descriptors. In *IEEE conference on computer vision and pattern recognition*.

Winter, M. E. (1999). N-FINDR: An algorithm for fast autonomous spectral end-member determination in hyperspectral data. In *Proceedings of the SPIE conference on imaging spectrometry V* (pp. 266–275).

Ye, J., & Li, Q. (2004). LDA/QR: An efficient and effective dimension reduction algorithm and its theoretical foundation. *Pattern Recognition*, *37*, 851–854.

Yuhas, R. H., Goetz, A. F. H., & Boardman, J. W. (1992). Discrimination among semiarid landscape endmembers using the spectral angle mapper SAM algorithm. In *Summaries of the third annual JPL airborne geoscience workshop* (pp. 147–149).

Zare, A., & Gader, P. (2008). Hyperspectral band selection and endmember detection using sparsity promoting priors. *IEEE Geoscience and Remote Sensing Letters*, *5*(2), 256–260.

Zymnis, A., Kim, S., Skaf, J., Parente, M., & Boyd, S. (2007). Hyperspectral image unmixing via alternating projected subgradients. In *Proceedings of the asilomar conference on signals, systems, and computers* (pp. 1164–1168).

Chapter 9
Reflection Geometry

A long-standing problem in computer vision is the recovery of reflection geometry including illuminant direction and position and object shape from image reflectance. To solve this problem, it is important to understand how the interaction between the surface orientation, the light source and viewing direction and the material properties of the surface under study influence the observed reflectance. In Chap. 4, we have described reflectance models that encode the relationship between these factors. In this chapter, we examine frameworks and methods for recovering the illuminant and shape from monochromatic, trichromatic and spectral images captured from a single view.

In the early days of computer vision, the shape and source from shading work aimed at recovering the illuminant direction and the geometry of smoothly curved surfaces with homogeneous reflecting properties using the variation in image irradiance, or *shading* (Brooks and Horn 1985; Horn and Brooks 1986). Classical approaches to shape from shading (Ikeuchi and Horn 1981; Horn and Brooks 1986; Zheng and Chellappa 1991) often rely on Lambertian, i.e. diffuse, object reflectance assumptions. Other shape from shading methods use a reflectance model to capture both the shape and photometric invariants of non-Lambertian surfaces. Along these lines, perhaps the work of Beckmann on modelling the reflectance of smooth and rough surfaces is one of the earliest approaches that employed a physical theory of the light wave scattering process (Beckmann and Spizzichino 1963). An extension of this work is the model proposed by Vernold and Harvey (1998). Torrance and Sparrow (1967) have developed a physically realistic account of specular reflectance that models the angular distribution of surface micro-facets.

For complex non-Lambertian surfaces, the shape recovery is not a straightforward task and is confronted with the complex physical models of the real world, including the dependence of the imagery on wavelength, and the reflection, transmission, scattering and refraction of light through the boundary between transmission media and object surfaces. The recovery of shape from a single view by solving an image irradiance equation based on a reflectance model can lead to a generalisation of classical methods in shape from shading and photometric stereo. Moreover, imaging spectroscopy methods are quite general in nature and can be applied in a straightforward manner to monochromatic or trichromatic imagery by specifying the wavelengths at which images are to be captured.

Imaging spectroscopy opens up research questions on the recovery of source, shape and wavelength indexed photometric parameters from multispectral and hyperspectral images. This recovery problem can be viewed as a more complex version of the traditional source and shape from shading problem, with additional features such as multiple wavelengths, spatially varying material albedo, unconstrained shape and complex reflection mechanism. Bearing this in mind, we intend to pose the problem in a unified view of the previous work in the areas of shape from shading and photometric stereo for grey-scale, trichromatic, multispectral and hyperspectral images.

The chapter begins with a general reflectance model in Sect. 9.1 that encompasses the components and features of a wide range of reflectance models, including those based on the Fresnel reflection theory. In Sect. 9.2, we show links between this general model and well-known reflectance models in the literature. With this general reflectance model at hand, we cast the problem of recovering shape and photometric invariants as that of solving the general image irradiance equation. The formulation of this equation in Sect. 9.3 allows for a general optimisation framework based on the calculus of variations. We also point out the relationship between this framework and existing work in shape from shading and photometric stereo. Lastly, in Sect. 9.4, we review existing techniques for estimating the illuminant direction. The literature in this area is highly relevant to the recovery of shape and photometric parameters, as the illuminant direction governs the shading of image irradiance. Many of these methods can be employed as a preceding step to shape recovery, while others involve the simultaneous recovery of source, shape and material reflectance properties.

9.1 General Reflectance Model

The reflectance models in this chapter are commonly defined with respect to a local reference coordinate system. Specifically, the reflectance at each surface point is defined in a coordinate system whose origin is located at the point and whose z-axis is aligned to the local surface normal $\vec{N}$. The incident light direction $\vec{L}$ is defined by the zenith and azimuth angles θ_i and ϕ_i, respectively. Accordingly, the zenith and azimuth angles of the outgoing direction $\vec{V}$ are θ_s and ϕ_s. For simplicity, we assume that the incident light is always in the x–z plane, i.e. $\phi_i = \pi$. Alternatively, reflectance models can be parameterised with another set of angles, making use of the halfway vector $\vec{H} = \vec{L} + \vec{V}$, which is the sum of the unit-length vectors in the light and viewing directions. Note that the reflection geometry can be equivalently represented by the angular difference θ_d between $\vec{L}$ and $\vec{H}$, the half angle θ_h between $\vec{N}$ and $\vec{H}$ and the incident angle θ_i. This geometry is illustrated in Fig. 9.1. Before further formalism, in Fig. 9.2, we summarise the commonly used symbols in this chapter for reference purposes.

The observed surface reflectance is the result of the interaction of the illumination, the shape of the surface and the material characteristics. Therefore, reflectance

9.1 General Reflectance Model

Fig. 9.1 Reflection geometry. $\vec{L}$, $\vec{V}$ and $\vec{N}$ are the light source direction, viewing direction and the local surface normal, respectively. $\vec{H}$ is the halfway vector. The polar coordinates of these vectors are defined with respect to a local coordinate system whose origin is located at the point under study and where the z-axis aligns with the surface normal

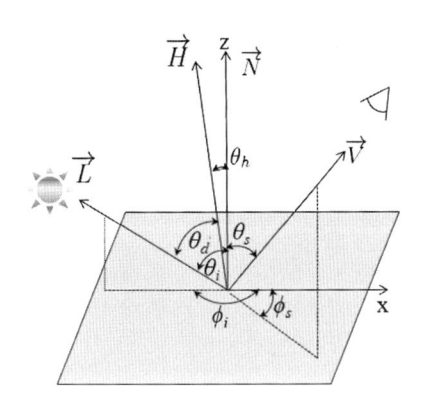

Notation	Description
$\vec{L} = (L_x, L_y, L_z)^T$	Illuminant direction.
$\vec{N} = (N_x, N_y, N_z)^T$	Surface normal vector.
$\vec{V}$	Viewing direction.
$\vec{H}$	Halfway vector (bisector) between $\vec{L}$ and $\vec{V}$.
ϕ_L and θ_L	Tilt (azimuth) and slant (zenith) angle of $\vec{L}$.
θ_i	Incident angle between $\vec{L}$ and $\vec{N}$.
θ_h	Half-angle between $\vec{N}$ and $\vec{H}$.
θ_d	Angle between $\vec{L}$ and $\vec{H}$.
θ_s	Reflection angle between $\vec{N}$ and $\vec{V}$.
$f(u, \lambda)$	General reflectance function at a pixel u and a wavelength λ.
$f^{\text{diff}}(u, \lambda)$	The diffuse component of the general reflectance function.
$f^{\text{spec}}(u, \lambda)$	The specular component of the general reflectance function.
$\Theta(u)$	Reflection angles at a pixel u.
$\Omega(u)$	Photogrammetric variables at a pixel u.
$\eta(u, \lambda)$	Refractive index at a pixel u and a wavelength λ.
$p(u), q(u)$	Horizontal and vertical surface gradients at a pixel u.
$F(\theta, \eta)$	Fresnel reflection coefficient with incident angle θ and refractive index η.
$\rho(u, \lambda)$	Total diffuse albedo at a pixel u and wavelength λ.
σ	Standard deviation of the height variation.
τ	Surface correlation length.
$D(\vartheta)$	The distribution of the angle ϑ between the micro-facet normal and the mean surface normal.
$\mathbb{E}_\mathcal{I}(\cdot)$ and $\text{Var}_\mathcal{I}(\cdot)$	Expected value and variance of the argument over the domain $\mathcal{I}$.

Fig. 9.2 Commonly used symbols in Chap. 9

can be expressed as a function of the parameters related to these factors. The reflectance models described in Chap. 4 generally involve three sets of variables. The first set of parameters is the reflection angle variables $\Theta(u)$, which describe the relative angles between the incident light, viewing and local surface normal directions.

With the viewing direction fixed to the z-axis, these angles are related to the illumination direction $\vec{L}$ and the surface normal $\vec{N}(u)$ corresponding to each pixel u. The second set consists of the wavelength-dependent material refractive index $\eta(u, \lambda)$ for the pixel u and wavelength λ, and other wavelength-dependent material properties such as the material albedo, which we denote as $\Upsilon(u, \lambda)$. Lastly, we relate reflectance to the set of wavelength-independent photogrammetric variables $\Omega(u)$, such as the local micro-facet slope and its distribution.

It has been a well-known approach in computer vision and computer graphics to model the scene radiance as a linear combination of two reflection components, namely specular and diffuse. This idea was first proposed by Shafer through the dichromatic reflection model (Shafer 1985). Formally, a general reflectance model is expressed as

$$f(u, \lambda) = w_d(u) f^{\text{diff}}(u, \lambda) + w_s(u) f^{\text{spec}}(u, \lambda). \tag{9.1}$$

Here, f denotes total reflectance at pixel location u and wavelength λ. The first term on the right-hand side represents the diff use component in which f^{diff} stands for the diffuse reflectance and w_{diff} denotes its weight. Similarly, the second term describes the specular component in which f^{spec} denotes specular reflectance and w_s denotes the weight of the specular term.

Moreover, the modelling of the diffuse and specular components in Eq. (9.1) can be formulated in a number of ways. In Chap. 4, we covered a range of reflectance models; a number of them solely represent the diffuse component, others are a linear combination of a diffuse and a specular component. In general, each of these components can be expressed as a product of a purely geometric term that is wavelength independent and a wavelength-dependent term that involves either the material albedo for diffuse reflection or the Fresnel reflection coefficients for specular reflection.

The first term explains the influence of the surface orientation, illuminant direction, viewing direction, surface roughness and the distribution of the micro-facet slope on the intensity of the reflected radiance. The relative angles between the surface orientation, illuminant direction and viewing direction affect the foreshortening of the surface patch observed from both the light source and the viewing position. Meanwhile, a number of reflectance models attribute the shape and distribution of specular lobes to the distribution of the micro-facet slope. As a measure of surface roughness, this statistic affects two kinds of attenuation of reflected light called masking and shadowing. Masking is the interception of flux reflected from a micro-facet by its adjoining facets. Shadowing is the case in which the reflecting facet is partially illuminated due to the blockage of nearby facets.

Material albedo is often used to quantify the fraction of diffusely reflected light at different wavelengths. To model specular reflection, the Fresnel reflection coefficients account for the reflection, transmission and refraction of light through the boundary between different object media. These coefficients are dependent on the incident angle of light and the material refractive index (Hecht 2002).

With the parameter sets defined above, the first term can be written as a function $\Lambda(\cdot)$ of the reflection angles $\Theta(u)$ and the photogrammetric variables $\Omega(u)$. In addition, the second term is treated as a function $\Gamma(\cdot)$ of the angles $\Theta(u)$, the refractive index $\eta(u, \lambda)$ and other wavelength-dependent material properties $\Upsilon(u, \lambda)$. As a result, the general reflectance equation at a surface location u and wavelength λ can be written as a product of these two functions:

$$f(u, \lambda) = \Lambda\big(\Theta(u), \Omega(u)\big) \Gamma\big(\Theta(u), \eta(u, \lambda), \Upsilon(u, \lambda)\big). \tag{9.2}$$

9.2 Parametric Form of Specific Reflectance Models

In this section, we show how the modelling of diffuse and specular reflection by various reflectance models conforms to the general model described in Sect. 9.1 by drawing the correspondences between their components. Later on, in Table 9.1, we summarise these correspondences, which provide an explicit link between the formulation in Eq. (9.2) and reflectance models in the computer vision and computer graphics literature. This is important, since it provides a proof of concept that a general process of recovering reflectance model parameters can be applied to each specific reflectance model.

In the following, we elaborate further on the correspondences between a number of reflectance models reviewed in Chap. 4 and the general reflectance model above. Here, we briefly mention the elements of each model again using the same notation as in the original formulation. The reader is referred to Chap. 4 for a detailed description of the models.

9.2.1 Dichromatic Reflection Model

In Sect. 4.2.2, we described the dichromatic model as a linear combination of a diffuse Lambertian component and a specular component having the same colour as the illuminant. The reflectance derived from the dichromatic reflection model is

$$f_{\text{dichromatic}}(u, \lambda) = g(u) S(u, \lambda) + k(u). \tag{9.3}$$

Since the shading factor is the cosine of the incident angle, i.e. $g(u) = \cos\theta_i$, the diffuse component can be cast as an instance of the general model by setting $\Theta = \{\theta_i\}$, $\Omega = \emptyset$, $\Upsilon = \{S(u, \lambda)\}$. The dichromatic model can then be separated into two factors conforming to the general reflectance model as $\Lambda(\Theta(u), \Omega(u)) = \cos\theta_i$ and $\Gamma(\Theta(u), \eta(u, \lambda), \Upsilon(u, \lambda)) = S(u, \lambda)$. Meanwhile, the specular component is regarded as a specialisation of the general model when $\Theta = \emptyset$, $\Omega = k(u)$ and $\Upsilon = \emptyset$.

9.2.2 Wolff Model

In Sect. 4.2.3, we presented a diffuse reflectance model proposed by Wolff (1994) for dielectric objects. Wolff viewed the diffuse reflection process as a combination of internal scattering and refraction at the material-air boundary. Therefore, the model is explained by both Snell's law of refraction and the Fresnel attenuation factor. As before, the diffuse subsurface scattering, as defined by the Wolff model, is rewritten as

$$f_W(u, \lambda) = \rho(u, \lambda) \cos\theta_i \left[1 - F(\theta_i, \eta(u, \lambda))\right] \left[1 - F\left(\theta'_s, \frac{1}{\eta(u, \lambda)}\right)\right]$$
$$= \rho(u, \lambda) \cos\theta_i \left[1 - F(\theta_i, \eta(u, \lambda))\right]\left[1 - F(\theta_s, \eta(u, \lambda))\right]. \quad (9.4)$$

Here, we derive the equality $F(\theta'_s, \frac{1}{\eta(u,\lambda)}) = F(\theta_s, \eta(u, \lambda))$ from Snell's law of refraction, i.e. $\theta'_s = \arcsin(\sin(\theta_s)/n(\lambda))$.

We observe that the Wolff model conforms to the general reflectance model described in Sect. 9.1. The specialisation is realised when the parameter sets of the general model are instantiated as $\Theta = \{\theta_i, \theta_s\}$, $\Omega = \emptyset$ and $\Upsilon = \{\rho(u, \lambda)\}$. Based on these correspondences, the factors of the reflectance model are $\Lambda(\Theta(u), \Omega(u)) = \cos\theta_i$ and $\Gamma(\Theta(u), \eta(u, \lambda), \Upsilon(u, \lambda)) = \rho(u, \lambda)[1 - F(\theta_i, \eta(u, \lambda))][1 - F(\theta_s, \eta(u, \lambda))]$.

9.2.3 Beckmann–Kirchhoff Model

Recall that in Sect. 4.1.1, the Beckmann–Kirchhoff model (Beckmann and Spizzichino 1963) predicts the scattered light power from a surface point as a summation of a diffuse scattering component and a specular component. The general formula for the diffuse scattering component involves an infinite sum and is computationally intractable. Instead, we model the surface correlation length with a Gaussian distribution and focus on the case where the surface is very rough. These assumptions allow the following approximation of the diffuse scattering component:

$$f_{BK}^{\text{diff}}(u, \lambda) = \frac{\pi \tau^2 G_{BK}^2}{A\sigma^2 v_z^2} \exp\left(-\frac{\tau^2 v_{xy}^2}{4\sigma^2 v_z^2}\right), \quad (9.5)$$

where we have maintained the same notation as in Chap. 4.

In Eq. (9.5), σ is the standard deviation of the height variation with respect to the mean surface level, and the surface correlation length τ is the correlation length between the micro-surface extrema. Note that the surface slope parameter $\frac{\sigma}{\tau}$ controls the scattering behaviour for various degrees of roughness. Therefore, it is sufficient to estimate $\frac{\sigma}{\tau}$ from reflectance data rather than σ and τ separately. In other words, this is equivalent to estimating the parameter $m \triangleq (\frac{\sigma}{\tau})^2$, which is the square inverse of the surface slope.

9.2 Parametric Form of Specific Reflectance Models

Now we rewrite Eq. (9.5) in terms of the parameter subsets of the general reflectance model in Eq. (9.2). To this end, we make use of the equalities $v_x = k(\sin\theta_i - \sin\theta_s \cos\phi_s)$, $v_y = -k\sin\theta_s \sin\phi_s$, $v_{xy}^2 = v_x^2 + v_y^2$ and $v_z = -k(\cos\theta_i + \cos\theta_s)$. When substituting these expressions into Eq. (9.5), we obtain

$$f_{\text{BK}}^{\text{diff}}(u, \lambda) = \frac{\lambda^2 m}{16\pi A \cos^2\theta_i \cos^4\theta_h} \exp\left(-\frac{m}{4}\tan^2\theta_h\right). \tag{9.6}$$

In Eq. (9.6), we can treat the area of a surface patch A as a constant. From this expression, we observe that the diffuse component of the Beckmann–Kirchhoff model is an instance of the general reflectance model by grouping the parameters as $\Theta = \{\theta_i, \theta_h\}$, $\Omega = \{m\}$, $\Upsilon = \{\lambda\}$, and separating the factors of the model as $\Lambda(\Theta(u), \Omega(u)) = \frac{m}{\cos^2\theta_i \cos^4\theta_h} e^{-\frac{m}{4}\tan^2\theta_h}$ and $\Gamma(\Theta(u), \eta(u, \lambda), \Upsilon(u, \lambda)) = \lambda^2$.

On the other hand, the specular component is fully expanded as

$$f_{\text{BK}}^{\text{spec}}(u, \lambda) = \frac{1}{4\pi\sigma^2} \exp\left(-\frac{\theta_h^2}{\sigma^2}\right) \exp\left(-\left(\frac{2\pi\sigma}{\lambda}(\cos\theta_i + \cos\theta_s)\right)^2\right) F(\theta_i, \eta(u, \lambda)), \tag{9.7}$$

due to the following formulations of the roughness parameter $g(\cdot)$ and the magnitude P_0 of the specular component:

$$g(\theta_i, \theta_s, \lambda, \sigma) = \left(\frac{2\pi\sigma}{\lambda}(\cos\theta_i + \cos\theta_s)\right)^2,$$

$$P_0(\theta_h, \sigma) = \frac{1}{\sqrt{2\pi}\sigma} \exp\left(-\frac{\theta_h^2}{2\sigma^2}\right). \tag{9.8}$$

The above formulation of the specular component also falls under the general reflectance model in Eq. (9.2). We note the correspondence between these two models by grouping the parameters into the sets $\Theta = \{\theta_i, \theta_h, \theta_s\}$, $\Omega = \{\sigma\}$ and $\Upsilon = \{\lambda\}$, and separating the model into two factors without constants as $\Lambda(\Theta(u), \Omega(u)) = \frac{1}{\sigma^2}\exp(-\frac{\theta_h^2}{\sigma^2})$ and $\Gamma(\Theta(u), \eta(u, \lambda), \Upsilon(u, \lambda)) = \exp(-(\frac{2\pi\sigma}{\lambda}(\cos\theta_i + \cos\theta_s))^2) F(\theta_i, \eta(u, \lambda))$.

9.2.4 Vernold–Harvey Model

As described in Sect. 4.1.2, the Vernold–Harvey model is a variant of the Beckmann–Kirchhoff model with the modification of the original geometric attenuation factor G_{BK}^2 to $G_{\text{VH}}^2 = \cos\theta_i$. This modification results in a change in the diffuse component, whereas the specular component remains unchanged. When the surface is rough and the surface correlation length is modelled with a Gaussian function, the diffuse component is parameterised with respect to the reflection angles as

$$f_{\text{VH}}^{\text{spec}}(u,\lambda) = \frac{\pi \tau^2 \cos\theta_i}{A\sigma^2 v_z^2} \exp\left(-\frac{\tau^2 v_{xy}^2}{4\sigma^2 v_z^2}\right)$$

$$= \frac{\lambda^2 m \cos\theta_i}{16\pi A \cos^2\theta_d \cos^2\theta_h} \exp\left(-\frac{m}{4}\tan^2\theta_h\right). \quad (9.9)$$

We note that the model in Eq. (9.9) and the general reflectance model in Eq. (9.2) are equivalent when $\Theta = \{\theta_i, \theta_h, \theta_d\}$, $\Omega = \{m\}$ and $\Upsilon = \{\lambda\}$, where the model can be separated into two factors given by $\Lambda(\Theta(u), \Omega(u)) = \frac{m\cos\theta_i}{16\pi A \cos^2\theta_d \cos^2\theta_h} \times \exp(-\frac{m}{4}\tan^2\theta_h)$ and $\Gamma(\Theta(u), \eta(u,\lambda), \Upsilon(u,\lambda)) = \lambda^2$.

9.2.5 Torrance–Sparrow Model

In Sect. 4.2.5, we introduced the Torrance–Sparrow reflectance model to explain the off-specular spike observed on rough surfaces. In Sect. 9.2.1, we have already shown that the diffuse Lambertian model is a special instance of the general model. Therefore, we now focus on showing the correspondence between the specular component of this model and the general model.

Recall from Sect. 4.2.5 that the Torrance–Sparrow model includes the following specular component:

$$f_{\text{TS}}^{\text{spec}}(u,\lambda) = F\bigl(\theta_i, \eta(u,\lambda)\bigr) \frac{G_{\text{TS}}(\theta_{ip}, \theta_{sp})}{\cos\theta_s} D(\vartheta). \quad (9.10)$$

We establish correspondences between the parameter sets of this component and those of the general model as $\Theta = \{\theta_i, \theta_s, \theta_{ip}, \theta_{sp}\}$, $\Omega = \{\vartheta, \sigma_\vartheta\}$ and $\Upsilon = \emptyset$, where σ_ϑ is the standard deviation of the Gaussian distribution $D(\vartheta)$ of the angle ϑ between the micro-facet normal and the mean surface normal. This leads to the following factorisation of the model: $\Lambda(\Theta(u), \Omega(u)) = \frac{G_{\text{TS}}(\theta_{ip},\theta_{sp})}{\cos\theta_s} D(\vartheta)$ and $\Gamma(\Theta(u), \eta(u,\lambda), \Upsilon(u,\lambda)) = F(\theta_i, \eta(u,\lambda))$.

9.2.6 Cook–Torrance Model

In Sect. 4.2.6, we have shown that the Cook–Torrance Model differs from the Torrance–Sparrow model mainly in the formulation of the geometric attenuation term. Hence, we only focus on mapping the specular term to the general reflectance model. The correspondences between the parameters of the two models are as follows: $\Theta = \{\theta_i, \theta_s, \phi_i, \phi_s, \theta_h\}$, $\Omega = \{m\}$ and $\Upsilon = \emptyset$. Here, m is the roughness of the surface.

Table 9.1 Parameter correspondence between the diffuse and specular components of several reflectance models and the model described in Sect. 9.1. For brevity, we have omitted the pixel location u and the wavelength λ from the notation

Model	Θ	Ω	Υ	$\Gamma(\Theta,\eta,\Upsilon)$	$\Lambda(\Theta,\Omega)$
Lambert	θ_i		S	S	$\cos\theta_i$
Dichromatic (specular)		k			k
Wolff	θ_i, θ_s		ρ	$\rho[1-F(\theta_i,\eta)]$ $\times[1-F(\theta_s,\eta)]$	$\cos\theta_i$
Beckmann–Kirchhoff (diffuse)	θ_i, θ_h	m	λ	λ^2	$\frac{m}{\cos^2\theta_i \cos^4\theta_h} e^{-\frac{m}{4}\tan^2\theta_h}$
Beckmann–Kirchhoff (specular)	$\theta_i, \theta_h, \theta_s$	σ	λ	$F(\theta_i,\eta)$	$e^{-(\frac{2\pi\sigma}{\lambda}(\cos\theta_i+\cos\theta_s))^2}$ $\times \frac{1}{\sigma^2} e^{-\frac{\theta_h^2}{\sigma^2}}$
Vernold–Harvey (specular)	$\theta_i, \theta_d, \theta_h$	m	λ	$\lambda^2 F(\theta_i,\eta)$	$\frac{m\cos\theta_i}{\cos^2\theta_d \cos^2\theta_h} e^{-\frac{m}{4}\tan^2\theta_h}$
Torrance–Sparrow (specular)	$\theta_i, \theta_s, \theta_{ip}, \theta_{sp}$	$\vartheta, \sigma_\vartheta$		$F(\theta_i,\eta)$	$\frac{G_{\text{TS}}(\theta_{ip},\theta_{sp})}{\cos\theta_s} D(\vartheta)$
Cook–Torrance (specular)	$\theta_i, \theta_s, \phi_i \phi_s, \theta_h$	m		$F(\theta_i,\eta)$	$\frac{G_{\text{CT}}(\theta_i,\theta_s,\phi_i,\phi_s)}{\cos\theta_i \cos\theta_s} D(\theta_h)$

The factorisation of the Cook–Torrance model conforming to the general reflectance model is $\Lambda(\Theta(u), \Omega(u)) = \frac{G_{\text{CT}}(\theta_i,\theta_s,\phi_i,\phi_s)}{\cos\theta_i \cos\theta_s} D(\theta_h)$ and $\Gamma(\Theta(u), \eta(u,\lambda), \Upsilon(u,\lambda)) = F(\theta_i, \eta(u,\lambda))$.

We summarise the correspondences between the specific reflectance models mentioned in this section and the general reflectance models in Table 9.1. In the table, we list the parameter subsets and the factors of the diffuse and the specular components of each specific model.

9.3 A Variational Approach to Shape Recovery

Having introduced the general reflectance model in Sect. 9.1, we now formulate the problem of recovering the object shape and photometric invariants from spectral images. The formulation rests on an image irradiance equation based on the general reflectance model, which is a generalisation of the shape from shading and photometric stereo problems. As a result, the problem amounts to finding the parameters of the general model that yield the closest rendering result to the input image irradiance. This general problem is applicable to the range of reflectance models described above. Subsequently, we present a generic optimisation framework to tackle this problem. The advantage of such a general framework is that the general solution can be specialised for each specific reflectance model. We also relate the proposed optimisation approach to classical solutions in shape from shading and photometric stereo.

9.3.1 General Irradiance Equation

Assume that we have a set of M spectral images $\mathcal{I}_1, \mathcal{I}_2, \ldots, \mathcal{I}_M$ as input, where each image $\mathcal{I}_l$ is taken under a different illuminant direction. Let the wavelengths at which these images are sampled be $\lambda \in \{\lambda_1, \ldots, \lambda_K\}$. In addition, all the images are observed from the same viewpoint.

Normally, spectral cameras acquire irradiance impinging on the image sensor, which is proportional to the scene radiance. This radiometric quantity is a product of the illumination spectrum and the scene reflectance. Further, the illumination of the scene can be estimated from a single spectral image using a method such as that in Huynh and Robles-Kelly (2010) or measured with a white calibration target. We have discussed several methods of estimating the illuminant power spectrum in Sect. 5.3. Subsequently, we can obtain the spectral reflectance $R_l(u, \lambda)$ at every pixel u and sample wavelength λ in the lth image after being normalised by the respective illumination power spectrum.

From the given set of reflectance images, we aim to recover the model parameters satisfying Eq. (9.2). To solve this problem, we commence by noting that the reflection angles depend on the illumination direction $\vec{L}$, the viewing direction $\vec{V}$ and the surface orientation $\vec{N}$. While the surface normal and viewing direction are fixed for all the input images, the angles $\Theta_l(u)$ at every pixel site in the lth image vary with respect to the illumination direction. On the other hand, the parameters $\eta(u, \lambda)$ and $\Omega(u)$ are invariant to illuminant power and direction.

While the surface reflectance is modelled at each surface point using an associated local coordinate system, the three-dimensional orientation and depth of the surface under study are defined with respect to a global coordinate system. In this system, the origin is located at the viewpoint, the positive z-axis is in the opposite direction to the line of sight and the positive x-axis points toward the right-hand side of the field of view.

In the formulation herein, we assume that the image is formed by orthographic image projection. With no perspective transformation between the world coordinates and the image coordinates, the surface depth can be treated as a function $z(x, y)$ of the image coordinates $u = (x, y)$, up to a scaling factor. Then the surface normal at the surface point corresponding to the pixel $u = (x, y)$ is, by definition, $\vec{N(u)} = [-p(u), -q(u), 1]^T$, where $p(u)$ and $q(u)$ are the surface gradients, i.e. $p(u) = \frac{\partial z(x,y)}{\partial x} = z_x$ and $q(u) = \frac{\partial z(x,y)}{\partial y} = z_y$.

Thus, given a light source direction and a viewer direction represented by unit vectors $\vec{L}$ and $\vec{V}$ respectively, the reflection angles can be expressed in terms of the surface gradients as

$$\cos\theta_i = \frac{-L_x p(u) - L_y q(u) + L_z}{\sqrt{p(u)^2 + q(u)^2 + 1}},$$

$$\cos\theta_d = \frac{\vec{L} \cdot \vec{H}}{\|\vec{H}\|},$$

9.3 A Variational Approach to Shape Recovery

$$\cos\theta_h = \frac{-H_x p(u) - H_y q(u) + H_z}{\|\vec{H}\|},$$

where the coordinates of the light and halfway vectors $\vec{L} = (L_x, L_y, L_z)^T$ and $\vec{H} = (H_x, H_y, H_z)^T$ are known given the viewing direction. This means that the reflection angles $\Theta(u)$ only depend on the light direction and local surface gradients $p(u)$ and $q(u)$.

Therefore, with known illuminant directions, we can re-parameterise the general reflectance Eq. (9.2) with respect to the surface gradients $p(u)$, $q(u)$ and index it with respect to the image number l as follows:

$$f_l(u, \lambda) = \Phi_l\big(p(u), q(u), \eta(u, \lambda), \Upsilon(u, \lambda)\big)\Psi_l\big(p(u), q(u), \Omega(u)\big).$$

This representation offers the advantage of replacing the reflection angle variables with the surface gradients. These are invariant with respect to the illuminant direction. More importantly, this formulation implies that the number of geometric variables is two per pixel site u, which, in turn constrains the number of reflectance equations needed to recover the surface shape. With the new parameterisation, we aim to solve the following system of equations:

$$f_l(u, \lambda) = R_l(u, \lambda), \quad \forall l = 1, \ldots, M. \tag{9.11}$$

Assuming that the number of pixels in each image is N, the system above consists of $M \times N \times K$ equations with $(|\Omega| + K + 2) \times N$ independent variables. These include K wavelength-dependent refractive indices at each pixel and the number of micro-facet scattering variables $|\Omega|$. Without further constraints, this system is well defined if and only if the number of equations at least equals the number of variables. In other words, the problem is only solvable given at least $M \geq (|\Omega| + K + 2)/K$ images. For all the reflectance models mentioned in Sect. 9.2, this number is lower-bounded at 2. In summary, this is the theoretical lower bound on the number of illuminants needed in order to recover the surface shape and photometric invariants when the illuminant directions are known.

Note that this lower bound is consistent with the classical photometric stereo methods for grey-scale images with the Lambertian model, where $K = 1$ and $|\Omega| = 0$. In addition, the formulation above can be viewed as a generalisation of Woodham's photometric stereo problem (Woodham 1980) for arbitrary reflectance models.

9.3.2 Cost Functional

When only a single image is at hand, the shape and photometric parameter recovery becomes ill-posed. In this case, we resort to constraints based on implicit assumptions on the spectral and local surface variations. Specifically, when the surface under study is smooth, we can enforce the surface integrability constraint, which has been utilised in shape from shading by several authors (Frankot and Chellappa 1988; Horn and Brooks 1986). This constraint states that the cross-partial derivatives of $p(u)$ and $q(u)$ must be equal, i.e. $p_y(u) = q_x(u)$. Furthermore, one can assume smoothness on the spectral variation of the refractive index. This assumption implies that the surface scattering characteristics should vary smoothly across the spatial domain. As we will show later in this section, these assumptions permit the recovery of the shape and photometric parameters by making use of a line-search approach.

To take our analysis further, we commence by noting that the parameters satisfying the reflectance equations are the minimisers of the following cost function:

$$C = \sum_{l=1}^{M} \int_{\mathcal{I}} \int_{\mathcal{W}} \left(R_l(u, \lambda) - f_l(u, \lambda)\right)^2 du\, d\lambda + \alpha \int_{\mathcal{I}} \left(\frac{\partial p(u)}{\partial y} - \frac{\partial q(u)}{\partial x}\right)^2 du$$
$$+ \beta \int_{\mathcal{I}} \int_{\mathcal{W}} \left(\frac{\partial \eta(u, \lambda)}{\partial \lambda}\right)^2 du\, d\lambda + \gamma \int_{\mathcal{I}} \left[\left(\frac{\partial \Omega(u)}{\partial x}\right)^2 + \left(\frac{\partial \Omega(u)}{\partial y}\right)^2\right] du,$$
(9.12)

where $\mathcal{I}$ is the image spatial domain and $\mathcal{W}$ is the wavelength range.

The arguments of the cost function C are the surface gradients $p(u)$ and $q(u)$, the index of refraction $\eta(u, \lambda)$ and the photogrammetric parameter set $\Omega(u)$. The weights α, β and γ control the contribution to the cost function of the integrability constraint, the spectral smoothness constraint on the refractive index and the spatial smoothness constraint on the surface scattering variables, respectively.

9.3.3 Variational Optimisation

With the cost function C at hand, we employ the calculus of variations to derive a set of Euler–Lagrange equations that yield the minimal solutions to the functional above. Subsequently, we arrive at an iterative update scheme as a result of discretising these equations on the spatial-spectral domain. The resulting Euler–Lagrange equations with respect to the argument functions are shown in Fig. 9.3. In the equations, the x, y subscripts imply partial derivatives with respect to the corresponding axis variable.

Moreover, we can employ the discrete approximation of the higher order derivatives in the equations shown in Fig. 9.3. To this end, let the spatial domain be discretised into a lattice with a spacing of ε_s between adjacent pixel sites and the wavelength domain in steps of ε_w. We index $p(u)$, $q(u)$ and $\Omega(u)$ according to the pixel coordinates (i, j) and $\eta(u, \lambda)$ according to the wavelength index k. With these ingredients, the partial derivatives can be approximated using finite differences as follows:

$$\overline{p}(u)|_{u=(i,j)} \triangleq \frac{1}{2}(p_{i,j-1} + p_{i,j+1}),$$

9.3 A Variational Approach to Shape Recovery

$$\sum_{l=1}^{M} \int_{\mathcal{W}} (R_l(u,\lambda) - f_l(u,\lambda)) \frac{\partial f_l(u,\lambda)}{\partial p(u)} \, d\lambda + \alpha(p_{yy}(u) - q_{xy}(u)) = 0$$

$$\sum_{l=1}^{M} \int_{\mathcal{W}} (R_l(u,\lambda) - f_l(u,\lambda)) \frac{\partial f_l(u,\lambda)}{\partial q(u)} \, d\lambda + \alpha(q_{xx}(u) - p_{yx}(u)) = 0$$

$$\sum_{l=1}^{M} \int_{\mathcal{I}} (R_l(u,\lambda) - f_l(u,\lambda)) \frac{\partial f_l(u,\lambda)}{\partial \eta(\lambda)} \, du + \beta \frac{\partial^2 \eta(\lambda)}{d\lambda^2} = 0$$

$$\sum_{l=1}^{M} \int_{\mathcal{W}} (R_l(u,\lambda) - f_l(u,\lambda)) \frac{\partial f_l(u,\lambda)}{\partial \Omega(u)} \, d\lambda + \gamma(\Omega_{xx}(u) + \Omega_{yy}(u)) = 0$$

Fig. 9.3 Euler–Lagrange equations for the cost function C

$$p^{(t+1)}(u) = \overline{p}^{(t)}(u) - \tfrac{1}{2}\hat{q}^{(t)}(u) + \frac{\varepsilon_s^2}{2\alpha} \sum_{l=1}^{M} \sum_{k=1}^{K} (R_l(u,\lambda_k)$$
$$- f_l^{(t)}(u,\lambda_k)) \frac{\partial f_l^{(t)}(u,\lambda_k)}{\partial p^{(t)}(u)}$$

$$q^{(t+1)}(u) = \overline{q}^{(t)}(u) - \tfrac{1}{2}\hat{p}^{(t)}(u) + \frac{\varepsilon_s^2}{2\alpha} \sum_{l=1}^{M} \sum_{k=1}^{K} (R_l(u,\lambda_k)$$
$$- f_l^{(t)}(u,\lambda_k)) \frac{\partial f_l^{(t)}(u,\lambda_k)}{\partial q^{(t)}(u)}$$

$$\eta^{(t+1)}(\lambda_k) = \overline{\eta}^{(t)}(\lambda_k) + \frac{\varepsilon_w^2}{2\beta} \sum_{l=1}^{M} \sum_{u \in \mathcal{I}} (R_l(u,\lambda_k) - f_l^{(t)}(u,\lambda_k)) \frac{\partial f_l^{(t)}(u,\lambda_k)}{\partial \eta^{(t)}(\lambda_k)}$$

$$\Omega^{(t+1)}(u) = \overline{\Omega}^{(t)}(u) + \frac{\varepsilon_s^2}{4\gamma} \sum_{l=1}^{M} \sum_{k=1}^{K} (R_l(u,\lambda_k) - f_l^{(t)}(u,\lambda_k)) \frac{\partial f_l^{(t)}(u,\lambda_k)}{\partial \Omega^{(t)}(u)}$$

Fig. 9.4 Line-search update equations for the optimisation with respect to the shape and photometric parameters

$$\overline{q}(u)|_{u=(i,j)} \triangleq \frac{1}{2}(q_{i-1,j} + q_{i+1,j}),$$

$$\hat{p}(u)|_{u=(i,j)} \triangleq \frac{\varepsilon_s^2}{4}(p_{i+1,j+1} + p_{i-1,j-1} - p_{i-1,j+1} - p_{i+1,j-1}),$$

$$\hat{q}(u)|_{u=(i,j)} \triangleq \frac{\varepsilon_s^2}{4}(q_{i+1,j+1} + q_{i-1,j-1} - q_{i-1,j+1} - q_{i+1,j-1}),$$

$$\overline{\eta}(\lambda_k) \triangleq \frac{1}{2}(\eta(\lambda_{k-1}) + \eta(\lambda_{k+1})),$$

$$\overline{\Omega}(u)|_{u=(i,j)} \triangleq \frac{1}{4}(\Omega_{i,j-1} + \Omega_{i,j+1} + \Omega_{i-1,j} + \Omega_{i+1,j}),$$

where $\hat{p}(u)$ and $\hat{q}(u)$ are the cross-derivatives of $p(u)$ and $q(u)$.

By substituting the finite differences into the Euler–Lagrange equations, we obtain a set of update equations for the model parameters with respect to the iteration number t. In Fig. 9.4, we show the set of resulting update equations, with the superscripts denoting the iteration number.

Note that the second right-hand terms of the update equations are the closed-form negative partial derivatives of the data closeness term with respect to each model parameter. These formulas are in fact instances of line-search optimisation, where a search from the current iterate is performed in the steepest gradient descent direction. This implies that the Euler–Lagrange equation in the function space is equivalent to gradient descent optimisation in the parameter space.

To enforce numerical stability on the update of parameter values over iterations, we can introduce a step length along the steepest descent direction. To this end, we employ the Wolfe condition (Nocedal and Wright 2000) to ensure that the step length delivers a sufficient decrease in the target function. For each update of the model parameters, we perform a backtracking line-search approach by starting with an initial long step length and contracting it upon insufficient decrease of the target function.

9.3.4 Relation to Shape from Shading

Recovering the shape of an object from its shading information is one of the classical problems in computer vision. The earliest approaches to shape from shading developed by Ikeuchi and Horn (1981), and by Horn and Brooks (1986), hinge on the compliance with the image irradiance equation and local surface smoothness. Zheng and Chellappa (1991) proposed a gradient consistency constraint that penalises differences between the image intensity gradient and the surface gradient for the recovered surface. Worthington and Hancock (1999) impose the Lambertian radiance constraint in a hard manner. Dupuis and Oliensis (1992) provided a means of recovering probably correct solutions with respect to the image irradiance equation. Prados and Faugeras (2003) have presented a perspective shape from shading approach applicable to the case where the light source and viewing directions are no longer coincident. Kimmel and Bruckstein (1995) have used the apparatus of level set methods to recover solutions to the eikonal equation.

The classical approaches to the original shape from shading problem (Ikeuchi and Horn 1981; Horn and Brooks 1986) make an implicit assumption that the radiance travelling from a small surface patch is dependent only on the orientation of the patch, and not on its position in space. This requires that the light sources and the viewer be located at infinity. The other assumptions are that the image is formed by orthographic image projection, and that the surface has homogeneous reflecting properties.

The classical shape from shading problem aims to recover the surface orientation from a single grey-scale image of Lambertian surfaces. This image is equivalent to a slice of a spectral image cube, which is the image of the scene in a single wavelength. The image irradiance equation for the shape from shading problem is

$$R(u) = \rho(\vec{L}.\overrightarrow{N(u)}) = \frac{-L_x p(u) - L_y q(u) + L_z}{\sqrt{p^2(u) + q^2(u) + 1}}, \quad (9.13)$$

where $R(u)$ is the value of the reflectance function at pixel u and ρ is the constant albedo across the surface. Here, we assume that $\|\vec{L}\| = 1$. We have omitted the wavelength index since the problem is posed in terms of a single wavelength.

9.3 A Variational Approach to Shape Recovery

We note that if the illuminant direction $\vec{L}$ and the surface albedo are known, the right-hand side is solely a function of the surface gradients $(p(u), q(u))$. This function is called the reflectance map, which can be denoted as $f(p(u), q(u))$. We also note that the function $f_I(u, \lambda)$ in Eq. (9.11) is a generalisation of the reflectance map since it involves photometric invariants in addition to the surface gradients. In other words, the problem of recovering both surface orientation and photometric invariant is a generalisation of the shape from shading problem.

Typically, the illuminant is assumed to be co-located with the viewer, i.e. $\vec{L} = [0, 0, 1]^T$. As a result, Eq. (9.13) becomes

$$R(u) = \frac{1}{\sqrt{p(u)^2 + q(u)^2 + 1}}. \tag{9.14}$$

Several approaches have aimed to tackle the problem in Eq. (9.14). Given a single irradiance equation per pixel, the problem of solving for both variables $p(u)$ and $q(u)$ is ill-posed. To this end, several authors have resorted to additional constraints such as smoothness and surface integrability. On one hand, the smoothness constraint enforces continuous, smooth changes of parameters across the spatial domain. On the other hand, the integrability constraint states that the surface depth is continuously twice differentiable.

Several formulations of the shape from shading problem have emerged in the form of functional optimisation. These target functionals are often expressed as a combination of the irradiance data error term and a regulariser representing the additional constraints. One of the popular variational formulations for shape from shading is that developed by Brooks and Horn (1985). Their approach enforces smoothness on the spatial variation of the surface normal $\vec{N}$, arriving at the following functional:

$$C_{\text{BH}} = \int_{\mathcal{I}} [(R(u) - \vec{L}.\vec{N(u)})^2 + \alpha(\|\vec{N_x(u)}\|^2 + \|\vec{N_y(u)}\|^2)$$
$$+ \beta(u)(\|\vec{N(u)}\|^2 - 1)] du, \tag{9.15}$$

where $\mathcal{I}$ is the image spatial domain.

In Eq. (9.15), the first term represents the error of the irradiance equation, while the second term is the smoothness error of the surface normal field, and the third term enforces a unit norm on the surface normals, i.e. $\|\vec{N(u)}\| = 1$. Here, the second and third terms play the role of regularising the cost functional, with their contributions weighted by α and $\beta(u)$.

Also aiming to develop a regulariser, Strat (1979) noted that the total change in depth along a closed curve on integrable surfaces should always be zero. Following this observation, the integral of the change in depth, i.e. $p(u) dx + q(u) dy$, is equal to zero. This is also called the loop integral, which is formally expressed as

$$\oint_C (p(u) dx + q(u) dy) = 0, \tag{9.16}$$

for all closed curves $\mathcal{C}$ in the spatial domain $\mathcal{I}$.

By enforcing the zero-loop integral along multiple closed curves in the image domain, the shape from shading problem becomes well-formed. This constraint can be rewritten as a least-square regulariser of the optimisation target function. Here, the left-hand side of Eq. (9.16) can be computed for each elementary square path that connects the eight surrounding neighbours of a given pixel. The resulting discrete approximation of the loop integral along this path is quantified based on the surface gradients in the neighbourhood. Let us denote this approximation at each pixel u as $e(u)$. Combining the loop integral with the irradiance equation, Strat's approach aims to minimise the following cost functional:

$$C_S = \int_{\mathcal{I}} \left(R(u) - (\vec{L}.\overrightarrow{N(u)})\right)^2 + \alpha e^2(u) \, du, \tag{9.17}$$

where the error in the loop integral is posed as a regulariser of the problem.

Later, Frankot and Chellappa (1988) presented an integrability constraint on smooth twice-differentiable surfaces for the shape from shading problem. The constraint implies that the cross-partial derivatives of $p(u)$ and $q(u)$ must be equal, i.e. $p_y(u) = q_x(u)$. With this constraint, the cost functional is formulated as

$$C_{FC} = \int_{\mathcal{I}} \left(R(u) - (\vec{L}.\overrightarrow{N(u)})\right)^2 + \alpha \left(\frac{\partial p(u)}{\partial y} - \frac{\partial q(u)}{\partial x}\right)^2 du. \tag{9.18}$$

Using an alternative coordinate system, Ikeuchi and Horn (1981) employed stereographic projection to parameterise the reflectance map. Instead of using the surface gradients (p, q), they project points on the Gaussian sphere onto a plane tangent to the sphere at the north pole. The projection results in the following gradient coordinates:

$$\begin{aligned} \tilde{p}(u) &= \frac{2p(u)}{1 + \sqrt{1 + p(u)^2 + q(u)^2}}, \\ \tilde{q}(u) &= \frac{2q(u)}{1 + \sqrt{1 + p(u)^2 + q(u)^2}}. \end{aligned} \tag{9.19}$$

In fact, Green's theorem (Riley et al. 2006) implies that the path integral of $p(u) dx + q(u) dy$ along a closed curve $\mathcal{C}$ equals the integrability over the interior region bounded by $\mathcal{C}$,

$$\oint_{\mathcal{C}} (p(u) dx + q(u) dy) = \int_{\mathcal{I}_\mathcal{C}} \left(\frac{\partial p(u)}{\partial y} - \frac{\partial q(u)}{\partial x}\right) du, \tag{9.20}$$

where $\mathcal{I}_\mathcal{C}$ is the region bounded by $\mathcal{C}$.

Therefore, if the total change of depth along every closed loop in the image is zero, the integrability term proposed by Frankot and Chellappa (1988) also equates zero for the respective interior region. This provides an equivalence between the

9.3 A Variational Approach to Shape Recovery

closed-loop integral constraint proposed by Strat (1979) and the surface integrability constraint proposed by Frankot and Chellappa (1988).

In another development, Ikeuchi and Horn (1981) re-parameterised the reflectance map $f(p,q)$ in the stereographic projection space, as $\tilde{f}(\tilde{p},\tilde{q})$. Further, they integrate the smoothness constraint on the resulting coordinates $(\tilde{p},\tilde{q})$ across the image pixels to arrive at the following target functional:

$$C_{\text{IH}} = \int_{\mathcal{I}} (R(u) - \tilde{f}(\tilde{p},\tilde{q}))^2 + \alpha(\tilde{p}_x^2 + \tilde{p}_y^2 + \tilde{q}_x^2 + \tilde{q}_y^2) \, du. \quad (9.21)$$

Similarly to Ikeuchi and Horn (1981), Smith (1982) formulated the shape from shading problem in an alternative space called the *lm* space, by transforming the gradients as follows:

$$\tilde{p}(u) = \frac{-p(u)}{\sqrt{1+p^2(u)+q^2(u)}},$$
$$\tilde{q}(u) = \frac{-q(u)}{\sqrt{1+p^2(u)+q^2(u)}}. \quad (9.22)$$

This transformation corresponds to an orthographic projection of the Gaussian sphere onto a tangent plane to the sphere at one of the poles. The change of space is followed by a re-parameterisation of the reflectance map. The resulting formulation of the problem is similar to that in Eq. (9.21) by integrating a smoothness constraint on the new coordinates across the image domain.

We note that all the cost functionals in Eqs. (9.15), (9.17), (9.18) and (9.21) fall under the general formulation in Eq. (9.12). Often, the authors derive Euler–Lagrange equations as a necessary condition for the optimal solutions to these problems. Subsequently, an iterative optimisation scheme emerges from the discretisation of the Euler–Lagrange equations on the image lattice. These variational approaches and the resulting discrete solutions, again, are special instances of the general optimisation framework we have discussed in Sect. 9.3.3.

9.3.5 Relation to Photometric Stereo

Photometric stereo methods employ multiple images to provide additional constraints for shape recovery. The intuition of this approach is that increasing the number of images provides additional information on the surface normal. This allows the surface orientation to be recovered without making smoothness assumptions.

In the early work by Woodham (1977, 1980), several grey-scale images were captured from the same view under various illumination directions. Therefore, each point in the scene corresponds to the same image location across the images. Since the variation in illumination direction affects the observed shading, the effect of each illumination direction is characterised by a separate reflectance map. We denote

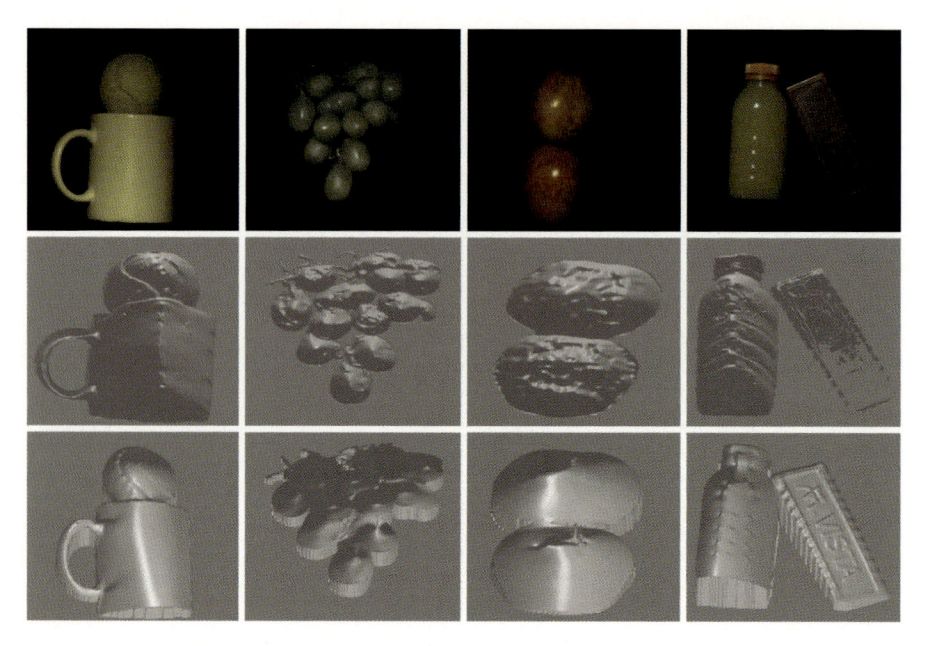

Fig. 9.5 Sample shapes recovered by shape from shading and photometric stereo. *First row*: a pseudo-colour rendering of spectral images illuminated by a frontal light source. *Second row*: shapes recovered by the shape from shading method in Worthington and Hancock (1999). *Third row*: shapes recovered by the photometric stereo method in Woodham (1980)

the reflectance map corresponding to the lth illuminant direction as $f_l(p(u), q(u))$, where $p(u)$ and $q(u)$ are the image gradients.

Suppose that the reflectance of the image captured under the lth illuminant direction (after normalising by the illuminant intensity) is R_l. In Woodham (1977, 1980), the author concentrated on Lambertian surfaces, i.e. $f_l = \rho(\vec{L}_l \cdot \overrightarrow{N(u)})$. Given a surface with albedo ρ, the image irradiance equation under the lth light direction represented by a unit vector $\vec{L}_l$ is

$$R_l(u) = \rho \vec{L}_l^T \overrightarrow{N(u)}. \tag{9.23}$$

Equation (9.23) has three unknowns on the right-hand side, provided that $\overrightarrow{N(u)}$ has unit length. Therefore, three images are sufficient to determine both the surface normal and the albedo.

Now we note that Eq. (9.23) is a special case of the general problem in Eq. (9.12) when the reflectance model is purely Lambertian and no additional constraints are required to over-determine the solution. Other photometric stereo methods (Tagare and deFigueiredo 1991) aim to solve the problem for non-Lambertian surfaces, where the reflectance map often includes a specular component. For these cases, the irradiance equation is no longer linear with respect to the surface normal direction. One of the solutions is to derive the related Euler–Lagrange equations, which lead to a non-linear iterative scheme over the parameters of the reflectance model.

9.4 Estimating the Illuminant Direction   193

In Fig. 9.5, we present sample shapes recovered by a shape from shading and a photometric stereo method. The top row shows the pseudo-colour rendering of the input spectral images illuminated by a frontal artificial light source matching the spectrum of the CIE standard illuminant D65. The images are synthesised for the spectral sensitivity of the human visual system, as described in Sect. 3.3.3. The second row shows the shapes recovered by a shape from shading method using robust kernels to regularise the smoothness of the surface normal (Worthington and Hancock 1999). The last row shows the shapes resulting from a photometric stereo setting proposed by Woodham (1980) when five illuminant directions are used, including the frontally lit images in the first row.

9.4 Estimating the Illuminant Direction

Estimating the illuminant direction is closely related to finding the surface orientation, because both of them jointly determine the image irradiance. This relationship is well known in shape from shading through the image irradiance equation (Horn and Brooks 1989). Hence, clues to finding the illuminant direction are often confounded with the uncertainty of the surface normal direction. In fact, without the prior knowledge of the illuminant direction, human perception is unable to disambiguate the convexity of the surface under observation from the shading in the input image. This phenomenon is called the convex-concave ambiguity.

As a result, knowing the illuminant direction is rather important to the determination of the surface orientation from image irradiance. Recall that the shape estimation framework presented in Sect. 9.3 requires known illuminant directions beforehand. However, estimating both the illuminant direction and surface orientation is often viewed as a "chicken-and-egg problem" due to their interaction.

In this section, we review methods that estimate a single light source direction from a single image. Methods for illuminant direction estimation can be classified into three broad categories. In Sect. 9.4.1, we provide an overview of methods that make statistical assumptions of the surface normal distribution. In Sect. 9.4.2, we describe the category that employs shading at object boundaries as cues for estimating the illuminant direction. In Sect. 9.4.3, we review methods that estimate the illuminant direction, the object shape and the material reflectance properties simultaneously in an optimisation approach. A number of the methods here aim at directional light sources, while others focus on point light sources.

9.4.1 Shading-Based Methods

In this section, we provide an overview of techniques for estimating the illuminant direction using shading statistics in a single image. One of the early works in this area was that of Pentland (1982). This work makes a simplifying assumption that change in surface orientation is distributed isotropically within each image region. The assumption holds true for a large class of objects, including convex objects with smoothly varying surface normals across their visible parts. The angle formed by the surface normal and the viewing direction gradually changes from 0 degree at singular points to 90 degrees at the occluding contour. In fact, we can establish a bijective map between these surface normals and points on the Gaussian hemisphere. Consequently, we can obtain statistics of the object surface orientation over this class via formulations on the Gaussian hemisphere.

In order to relate image intensity to the illuminant direction, Pentland (1982) assumed a Lambertian reflection model for the surface under study. Let us consider a global coordinate system, with the positive z-axis aligned with the viewing direction. Under an incident flux with radiance power L_0 travelling in the direction $\vec{L}$, a surface with albedo ρ and normal direction $\vec{N}$ yields the following image irradiance according to the Lambertian model:

$$I = L_0 \rho (\vec{L} \cdot \vec{N}), \qquad (9.24)$$

where $\cdot$ denotes the dot product operator. Here, for brevity, we have omitted the surface location and the wavelength of light from the parameterisation of the terms in Eq. (9.24).

Within a small homogeneous area on the surface, we can safely assume that there is little variation in the illumination and surface reflectance. Treating L_0, ρ and $\vec{L}$ as constants over the image spatial domain $\mathcal{I}$, we express the first derivative of the image irradiance I along an image direction $v = (v_x, v_y)$, denoted by I_v, as follows:

$$I_v = L_0 \rho \left(\vec{L} \cdot \frac{\partial \vec{N}}{\partial v} \right). \qquad (9.25)$$

With the above assumptions at hand, Pentland (1982) examined the influence of the illumination direction on the mean of the directional image gradients over the image spatial domain $\mathcal{I}$, denoted by $\mathbb{E}_{\mathcal{I}}(I_v)$, along various image directions v. Here, we employ the notation $\mathbb{E}_{\mathcal{I}}(\cdot)$ to indicate the expected value of the argument over the domain $\mathcal{I}$. Taking the expected value of both sides of Eq. (9.25), we arrive at

$$\mathbb{E}_{\mathcal{I}}(I_v) = L_0 \rho \left(\vec{L} \cdot \mathbb{E}_{\mathcal{I}} \left(\frac{\partial \vec{N}}{\partial v} \right) \right)$$
$$= L_0 \rho \left(L_x \mathbb{E}_{\mathcal{I}} \left(\frac{\partial N_x}{\partial v} \right) + L_y \mathbb{E}_{\mathcal{I}} \left(\frac{\partial N_y}{\partial v} \right) + L_z \mathbb{E}_{\mathcal{I}} \left(\frac{\partial N_z}{\partial v} \right) \right). \qquad (9.26)$$

Here, $\vec{L} = (L_x, L_y, L_z)^T$ is the illuminant direction and $\mathbb{E}_{\mathcal{I}}(\frac{\partial \vec{N}}{\partial v}) = (\mathbb{E}_{\mathcal{I}}(\frac{\partial N_x}{\partial v}), \mathbb{E}_{\mathcal{I}}(\frac{\partial N_y}{\partial v}), \mathbb{E}_{\mathcal{I}}(\frac{\partial N_z}{\partial v}))^T$ is the mean derivative of the surface normal in the direction v.

Under the assumption that the surface normal is isotropically distributed over the hemisphere facing the viewer, the mean derivative of the surface normal is assumed to be zero over the respective domain, i.e. $\mathbb{E}_{\mathcal{I}}(\frac{\partial \vec{N}}{\partial v}) = 0$. Thus, the z-component

9.4 Estimating the Illuminant Direction

of the surface normal derivative along any image direction is normally distributed with a zero mean, i.e. $\mathbb{E}_\mathcal{I}(\frac{\partial N_z}{\partial v}) = 0$. Furthermore, the x- and y-components of the mean surface normal derivative $\mathbb{E}_\mathcal{I}(\frac{\partial \vec{N}}{\partial v})$ are proportional to the differential step, i.e. $(\mathbb{E}_\mathcal{I}(\frac{\partial N_x}{\partial v}), \mathbb{E}_\mathcal{I}(\frac{\partial N_y}{\partial v})) = \kappa(v_x, v_y)$, where κ is a constant. Therefore, we can simplify Eq. (9.26) as

$$\mathbb{E}_\mathcal{I}(I_v) = (v_x, v_y) \begin{pmatrix} \hat{L}_x \\ \hat{L}_y \end{pmatrix}, \tag{9.27}$$

where $\hat{L}_x = \kappa L_0 \rho L_x$ and $\hat{L}_y = \kappa L_0 \rho L_y$.

Equation (9.27) provides a linear relationship between the directional derivative of the image irradiance to the illuminant direction. By considering the image irradiance derivatives along two or more image directions, we establish an overdetermined linear system to solve for $\hat{L}_x$ and $\hat{L}_y$. Subsequently, the tilt angle of the illuminant direction is computed by $\phi_L = \arctan(\frac{\hat{L}_y}{\hat{L}_x})$.

In Pentland (1982), the term $\kappa L_0 \rho$ was shown to be related to the variance of I_v as follows:

$$\kappa L_0 \rho = \sqrt{\mathrm{Var}_\mathcal{I}(I_v)} = \sqrt{\mathbb{E}_\mathcal{I}(I_v^2) - (\mathbb{E}_\mathcal{I}(I_v))^2}, \tag{9.28}$$

where $\mathrm{Var}_\mathcal{I}(I_v)$ is the variance of the directional image derivative I_v.

A detailed derivation of the formula 9.28 is given in Pentland (1982). From this equation, the three components of the normalised illuminant direction are recovered by

$$L_x = \frac{\hat{L}_x}{\kappa L_0 \rho}, \tag{9.29}$$

$$L_y = \frac{\hat{L}_y}{\kappa L_0 \rho}, \tag{9.30}$$

$$L_z = \left(1 - L_x^2 + L_y^2\right)^{\frac{1}{2}}. \tag{9.31}$$

Lee and Rosenfeld (1985) validated Pentland's method on synthetic images of spheres and confirmed the robustness of its illuminant tilt estimator. They arrived at an estimation of the tilt angle, which is equivalent to that proposed by Pentland (1982),

$$\phi_L = \arctan \frac{\mathbb{E}_\mathcal{I}(I_y)}{\mathbb{E}_\mathcal{I}(I_x)}.$$

On the other hand, Lee and Rosenfeld (1985) observed that the stability of the slant estimator is undermined when the slant angle is large, e.g. more than 40°. This is due to the small variance $\mathrm{Var}_\mathcal{I}(I_v)$ at large slant angles, which implies a small value of $\kappa L_0 \rho$ according to Eq. (9.28) and leads to unstable estimates of L_x and L_y by Eqs. (9.29) and (9.30). In addition, they commented that employing image derivatives does not necessarily reduce noise for the estimator. To this end, they proposed the following equations to estimate the illuminant slant angle θ_L, which are only based on expected values over the spatial domain $\mathcal{I}$:

$$\mathbb{E}_{\mathcal{I}}(I) = 4L_0\rho \frac{(\pi - \theta_L)\cos\theta_L + \sin\theta_L}{3\pi(1 + \cos\theta_L)}, \tag{9.32}$$

$$\mathbb{E}_{\mathcal{I}}(I^2) = \frac{L_0^2\rho^2}{4}(1 + \cos\theta_L). \tag{9.33}$$

To estimate the slant angle θ_L from the equations above, Zheng and Chellappa (1991) took the ratio of Eq. (9.32) and the square root of Eq. (9.33) to obtain

$$\frac{\mathbb{E}_{\mathcal{I}}(I)}{\mathbb{E}_{\mathcal{I}}(I^2)} = 8\frac{(\pi - \theta_L)\cos\theta_L + \sin\theta_L}{3\pi(1 + \cos\theta_L)^{3/2}}. \tag{9.34}$$

Since the right-hand side of Eq. (9.34) is a monotonically decreasing function of θ_L, there exists a unique solution for θ_L provided that $\frac{\mathbb{E}_{\mathcal{I}}(I)}{\mathbb{E}_{\mathcal{I}}(I^2)} \leq \frac{2\sqrt{2}}{3}$. In the case where $\frac{\mathbb{E}_{\mathcal{I}}(I)}{\mathbb{E}_{\mathcal{I}}(I^2)} > \frac{2\sqrt{2}}{3}$, the authors set $\theta_L = 0$. Nonetheless, Zheng and Chellappa (1991) commented on the formulation of Eqs. (9.32) and (9.33), saying that they should be normalised by the projection area of the sphere, including both the lit and shadowed parts, on the image plane. The proposed variants to these equations are expressed as

$$\mathbb{E}_{\mathcal{I}}(I) = \frac{2}{3\pi}L_0\rho\big((\pi - \theta_L)\cos\theta_L + \sin\theta_L\big), \tag{9.35}$$

$$\mathbb{E}_{\mathcal{I}}(I^2) = \frac{L_0^2\rho^2}{8}(1 + \cos\theta_L)^2. \tag{9.36}$$

Further, Zheng and Chellappa (1991) formulated a local voting method for estimating the illuminant tilt angle. The first method assumes that the surface can be locally approximated by spherical patches. With a similar derivation to that by Pentland's method (Pentland 1982), one can obtain a local estimate of the tilt angle at each image pixel. Finally, the estimates of the illuminant tilt angle are computed using the spatial average of values involving the local estimates over the spatial domain,

$$\phi_L = \arctan\frac{\mathbb{E}_{\mathcal{I}}\left(\frac{\tilde{y}_L}{\sqrt{\tilde{x}_L^2 + \tilde{y}_L^2}}\right)}{\mathbb{E}_{\mathcal{I}}\left(\frac{\tilde{y}_L}{\sqrt{\tilde{x}_L^2 + \tilde{y}_L^2}}\right)}, \tag{9.37}$$

where $\tilde{x}$ and $\tilde{y}$ are x- and y-coordinates of the illuminant direction estimated from the local neighbourhood of each pixel.

It has also been proven by Chojnacki and Brooks (Chojnacki et al. 1994) that the above method is only correct for estimating the tilt angle of the illuminant direction. In fact, the expression for estimating the slant angle has a limit at infinity

9.4 Estimating the Illuminant Direction

when it is integrated over the continuous spatial domain, even for spherical objects. The unbounded increase of this expression is realisable when squares of image irradiance derivatives are computed by means of finite differencing with increasing image resolutions.

Hence, Chojnacki and Brooks (Chojnacki et al. 1994) proposed the "disk method" to remedy this undesirable phenomenon. With this method, the authors focused on an estimator of the light source direction for a Lambertian sphere with a radius r. The new formulation disregards points close to the edge of the spatial domain, so that the expected values of squared image derivatives are guaranteed to be bounded. As a result, the domain of integration is confined to a closed set within the original spatial domain. Specifically, this domain was chosen to be a closed disk $\mathcal{D}$ centred at the origin of the image, with a radius τr, where $0 < \tau < 1$. With these ingredients, the tilt angle ϕ_L and the slant angle θ_L of the illuminant direction are given by

$$\phi_L = \arctan \frac{\mathbb{E}_\mathcal{D}(I_y)}{\mathbb{E}_\mathcal{D}(I_x)},$$

$$\theta_L = \arccos\left(\left(1 + h(\tau)\frac{(\mathbb{E}_\mathcal{D}(I_x))^2 + (\mathbb{E}_\mathcal{D}(I_y))^2}{(\mathrm{Var}_\mathcal{D}(I_v))^2}\right)^{-\frac{1}{2}}\right), \quad (9.38)$$

where $h(\tau) = -\frac{1}{2} - \frac{1}{2\tau^2}\log(1 - \tau^2)$. In Eq. (9.38), I_x, I_y and I_v are the image irradiance derivatives with respect to the x, y and v directions.

Note that the above methods of Pentland (1982) and Chojnacki and Brooks (Chojnacki et al. 1994) rely on the assumption that the distribution of image irradiance derivatives of the surface under study is close to the distribution of image irradiance derivatives over a sphere. To deal with general surfaces, usually composed of both convex and concave regions, Knill (1990), on the other hand, provided an estimator of both the illuminant direction and the surface relief by considering a stochastic process representing the surface depth. The author observed from a number of images that the illuminant direction has a biasing effect on the directional derivative of the image irradiance.

Specifically, if the surface under study is assumed to be Lambertian and is illuminated by a point source at infinity, the variance of the directional derivative of the image is maximised at the tilt direction of the illuminant. With v_L denoting the tilt direction of the illuminant $\vec{L} = (L_x, L_y, L_z)^T$, i.e. $v_L = (L_x, L_y)$, and $\mathcal{I}$ denoting the image spatial domain, the tilt angle ϕ_L of the illuminant satisfies

$$\phi_L = \mathrm{argmax}\, \mathbb{E}_\mathcal{I}\big[(I_{v_L})^2\big]. \quad (9.39)$$

In Eq. (9.39), we have made use of the assumption that the expected value of the directional derivative of the surface normal along an arbitrary direction is zero for surfaces that are not predominantly convex, i.e. $\mathbb{E}_\mathcal{I}(I_{v_L}) \doteq 0$. Under this assumption, the solution to the above problem is given by

$$\phi_L = \frac{1}{2}\arctan \frac{2\mathbb{E}_\mathcal{I}(I_x I_y)}{\mathbb{E}_\mathcal{I}(I_x^2) - \mathbb{E}_\mathcal{I}(I_y^2)}. \quad (9.40)$$

To estimate the slant angle of the illuminant direction, Knill (1990) resorted to high-order image statistics. The first of these is the luminance contrast, which can be obtained from the given image as

$$C = \frac{\mathbb{E}_{\mathcal{I}}(I^2) - (\mathbb{E}_{\mathcal{I}}(I))^2}{\mathbb{E}_{\mathcal{I}}(I^2)}. \tag{9.41}$$

The author then related the z-component of the illuminant direction, i.e. L_z, and the z-component of the surface normal, i.e. N_z, to the luminance contrast as

$$C = \frac{1 - L_z^2 + (3L_z^2 - 1)\mathbb{E}_{\mathcal{I}}(N_z^2)}{2L_z^2(\mathbb{E}_{\mathcal{I}}(N_z))^2} - 1. \tag{9.42}$$

As before, Knill (1990) observed that the variance of the derivative of luminance is greatest along a direction parallel to the tilt of the illuminant and is minimal along the orthogonal direction. The ratio $\mathcal{R}$ of these two variances, as we will show below, depends only on L_z and the moments of N_z. This ratio is defined as

$$\mathcal{R} = \frac{\mathbb{E}_{\mathcal{I}}(I_x^2)}{\mathbb{E}_{\mathcal{I}}(I_y^2)}. \tag{9.43}$$

The statistic above is related to L_z and high-order statistics of N_z as

$$\mathcal{R} = \frac{5\mathbb{E}_{\mathcal{I}}(N_z^2) + 2\mathbb{E}_{\mathcal{I}}(N_z^4) + 5\mathbb{E}_{\mathcal{I}}(N_z^6) - L_z^2[5\mathbb{E}_{\mathcal{I}}(N_z^2) - 6\mathbb{E}_{\mathcal{I}}(N_z^4) + 13\mathbb{E}_{\mathcal{I}}(N_z^6)]}{3\mathbb{E}_{\mathcal{I}}(N_z^2) - 2\mathbb{E}_{\mathcal{I}}(N_z^4) + 3\mathbb{E}_{\mathcal{I}}(N_z^6) - L_z^2[3\mathbb{E}_{\mathcal{I}}(N_z^2) - 10\mathbb{E}_{\mathcal{I}}(N_z^4) + 11\mathbb{E}_{\mathcal{I}}(N_z^6)]}. \tag{9.44}$$

If the surface gradients $p(u)$ and $q(u)$ are assumed to be normally distributed with equal standard deviations $\sigma_p = \sigma_q$, then the statistics of N_z in Eqs. (9.42) and (9.44) can be expressed as functions of σ_p. Substituting these functions into Eqs. (9.42) and (9.44) yields functions of the variables L_z and σ_p. These functions can be approximated with polynomials for the purpose of estimating the slant direction L_z of the illuminant direction and the standard deviation σ_p of the surface gradients.

9.4.2 Contour-Based Methods

So far, the methods above aim to estimate the illuminant direction from a single shading image by making statistical assumptions on the distribution of surface normals in the scene. Here, we review methods that rely on shading analysis along object boundaries. The key idea of the method is that the surface normals of occluding contours can be computed in a straightforward manner. Methods of this kind make use of the Lambertian reflectance model, by which image irradiance is expressed as a scalar multiple of the dot product of the illuminant direction and surface normal direction. In this way, the Lambertian image irradiance equations at multiple contour pixels allow the formation of an over-determined linear system with respect to

9.4 Estimating the Illuminant Direction

the tilt direction of the illuminant. Consequently, an analytical least-squares solution to the tilt direction is obtained. Note that, since the z-component of the surface normal is always zero at occluding contours, recovering the slant angle of the illuminant direction from only occluding contours is not a well-posed problem. Hence, we only focus on the estimation of the tilt direction in this section.

A classical approach in this category is perhaps the contour-based method for tilt angle estimation by Zheng and Chellappa (1991). Here, object boundaries are defined to consist of surface locations at which the zenith (slant) angle of the normal is constant. This is a general definition that includes occluding boundaries. Let the azimuth (tilt) angle of the surface normal at the jth boundary pixel be denoted as ϕ_j, $j = 1, 2, \ldots, N$, and the common zenith angle be θ. As a result, the surface normal can be expressed as $\vec{N}_j = (\cos\phi_j \sin\theta, \sin\phi_j \sin\theta, \cos\theta)^T$, and the light direction is $\vec{N}_j = (\cos\phi_L \sin\theta_L, \sin\phi_L \sin\theta_L, \cos\theta_L)^T$. With these ingredients, the Lambertian irradiance in Eq. (9.24) for the jth pixel can be rewritten as

$$I_j = L_0\rho(\cos\phi_j \sin\theta \cos\phi_L \sin\theta_L + \sin\phi_j \sin\theta \sin\phi_L \sin\theta_L + \cos\theta \cos\theta_L)$$

$$= (\cos\phi_j \sin\phi_j \cos\theta) \begin{pmatrix} L_0\rho \sin\theta \cos\phi_L \sin\theta_L \\ L_0\rho \sin\theta \sin\phi_L \sin\theta_L \\ L_0\rho \cos\theta_L \end{pmatrix} \quad \forall j = 1, 2, \ldots, N, \quad (9.45)$$

where I_j is the image irradiance at the jth pixel.

Assuming that the normal slant angle θ along the boundary is known, we can rewrite the system of Eqs. (9.45) in matrix form as follows:

$$\mathbf{I} = \mathbf{A}\mathbf{x}, \quad (9.46)$$

where

$$\mathbf{I} = \begin{pmatrix} I_1 \\ I_2 \\ \vdots \\ I_N \end{pmatrix}, \quad \mathbf{A} = \begin{pmatrix} \cos\phi_1 \sin\phi_1 \cos\theta \\ \cos\phi_2 \sin\phi_2 \cos\theta \\ \vdots \\ \cos\phi_N \sin\phi_N \cos\theta \end{pmatrix}, \quad \text{and}$$

$$\mathbf{x} = \begin{pmatrix} L_0\rho \sin\theta \cos\phi_L \sin\theta_L \\ L_0\rho \sin\theta \sin\phi_L \sin\theta_L \\ L_0\rho \cos\theta_L \end{pmatrix}.$$

If $N \geq 3$, Eq. (9.46) becomes an over-determined linear system with respect to $\mathbf{x}$. In this case, we can obtain the least-square solution as $\mathbf{x} = \mathbf{A}^*\mathbf{I}$, where $\mathbf{A}^* = (\mathbf{A}^T\mathbf{A})^{-1}\mathbf{A}^T$ is the pseudo-inverse of the $\mathbf{A}$ matrix. The tilt angle of the illuminant direction can be estimated from the first two components of the solution, denoted by x_1 and x_2, as follows:

$$\phi_L = \arctan\left(\frac{x_2}{x_1}\right). \quad (9.47)$$

While formulating Eq. (9.46), we have not discussed the issue of approximating the image irradiance I_j at boundary pixels not visible to the viewer. For further reading on this issue, the reader is referred to the relevant implementation details in Zheng and Chellappa's paper (Zheng and Chellappa 1991). In addition, when considering a closed boundary, the summations of cosines and sines of the tilt angle of the surface normal direction along the boundary are supposed to be zero. Due to sampling effects, the first two column sums of the **A** matrix may not be zero. To remedy this effect, the authors preprocessed the **I** vector and the first two columns of the **A** matrix by subtracting their means from the respective columns before solving Eq. (9.46).

More recently, the least-squares formulation above was re-examined by Nillius and Eklundh (2001) focusing on occluding contours as a cue to illuminant direction. In fact, this method is a special instance of Zheng and Chellappa's tilt estimator, in the sense that the algorithm focuses on occluding boundaries, for which the surface normal directions can be simply computed. Similar to the former method, Nillius and Eklundh dealt with missing irradiance information from occluding pixels by means of irradiance extrapolation. This paper also addresses how to correctly identified occluding boundaries among a variety of edges, which was not paid much attention in Zheng and Chellappa's work (Zheng and Chellappa 1991). Further, it includes a method for consolidating solutions obtained from various contours. Although the core of Nillius and Eklundh's work is reminiscent of the tilt estimator by Zheng and Chellappa (1991), it contains extensions including a preprocessing step for occluding boundary detection and a postprocessing step for fusing illuminant directions estimated from different contours.

The estimation method proceeds in three phases. In the first phase, edges are detected using an edge detector, such as the Canny edge detector (Canny 1986), and linked into chains. Subsequently, consecutive edges are grouped into occluding contours according to a number of criteria, including colour uniformity and continuity, edge saliency and smooth curvature of edges.

To commence the second phase, it is noted that, with respect to a coordinate system with the z-axis aligned to the viewing direction, surface normals at occluding contours always have a zero z-component, and the x- and y-components are equal to those of the local image irradiance gradients. To ensure that the illuminant direction estimation problem is well-posed, it is implicitly assumed that the irradiance at these locations is non-zero. This implies that the light source and the viewing positions are not co-located. Further, we need the irradiance at occluding contours as input for the tilt estimator. To this end, Nillius and Eklundh (2001) employed an extrapolation procedure to approximate the irradiance at the contours. This procedure relies on a shading model along a line in the image domain. After fitting this model, which is approximated with a power series, to the irradiance at the interior points of objects, one could obtain the extrapolated irradiance values at occluding contours.

Finally, the third phase detects the correct occluding contours based on the fitting error of the shading model above and the consensus between the illuminant direction estimated for the contours. The fusion of these estimates is posed as an inference process in a Bayesian network. The distributions of the estimates above and the true

9.4 Estimating the Illuminant Direction

value of the illuminant direction are computed using a standard message passing approach.

In a related development, Weinshall (1990) assumes a single point light source with constant intensity and constant albedo so as to estimate the tilt and slant angles of the illuminant direction. This approach can be easily extended to multiple light sources by examining the self-shadow edges cast by each light separately.

The tilt angle estimator is based on the proposition that the tilt angle is parallel to the angle of the tangent to the occluding contour at any point where the minimum of shading occurs, and perpendicular to that at any point where the maximum shading occurs. This applies to Lambertian surfaces and can be intuitively explained by the variation of the incident angle of the light at pixels along occluding boundaries. We note that these surface normals are parallel to the image plane. As a result, the incident angle reaches its minimum or maximum when the projection of the illuminant direction on the image plane is parallel or perpendicular to that of the surface normal along the occluding boundary, respectively. Hence, the shading at the respective locations is maximal or minimal given constant albedo and illuminant power.

Thus, the tilt angle of the illuminant can be estimated by first identifying points along an occluding contour where shading is minimal or maximal. Note that the robustness of the estimator increases with the number of points. If the (point or directional) illuminant is considered to be globally constant, one could improve the estimate by means of majority voting using contours with the greatest consensus.

In addition, the slant angle of the illuminant direction is given by

$$\theta_L = \arcsin \frac{\max_{\partial \mathcal{I}} I}{\max_{\mathcal{I}} I}, \tag{9.48}$$

which is the ratio of the maximal irradiance I along the occluding contour $\partial \mathcal{I}$ to the maximum value of I over the entire spatial domain $\mathcal{I}$.

Equation (9.48) can be briefly proved as follows. The irradiance of a Lambertian surface is $L_0 \rho \cos \theta_i$, where L_0, ρ and θ_i are the light irradiance power, the surface albedo and the incident angle. Further, if we assume that at least one location on the surface has its normal vectors aligned with the illuminant direction, then the maximal irradiance on the surface is $\max_{\mathcal{I}} I = L_0 \rho$.

From the previous proposition, $\max_{\partial \mathcal{I}} I$ is obtained when $\vec{N}$ is parallel to the projection of $\vec{L}$ on the image plane. In this case, the slant angle of the illuminant direction is $\theta_L = \frac{\pi}{2} - \theta_i$. As a result, $\max_{\partial \mathcal{I}} I = L_0 \rho \sin \theta_L$. Thus, the expressions of $\max_{\mathcal{I}} I$ and $\max_{\partial \mathcal{I}} I$ lead to $\frac{\max_{\partial \mathcal{I}} I}{\max_{\mathcal{I}} I} = \sin \theta_L$.

Note that this formula strictly assumes purely diffuse Lambertian reflectance. If the surface under study exhibits a fraction of specular reflection, the denominator $\max_{\mathcal{I}} I$ needs to be evaluated over the diffuse regions only. Otherwise the formula will yield an under-estimation of the slant angle.

Having presented the formulation of illuminant direction methods using occluding boundaries, we note that this problem is the dual to photometric stereo. Recall that, to estimate the single illuminant direction, we employ several known surface

9.4.3 Optimisation Methods

Optimisation methods often aim at recovering the lighting direction, the object shape and the reflectance simultaneously from a single image. One of the early works in this area is the variational framework introduced by Brooks and Horn (1985). Assuming Lambertian reflectance, the framework commences by posing a functional optimisation problem, as described in Eq. (9.15):

$$C_{\text{BH}} = \int_{\mathcal{I}} [(R(u) - \vec{L}.\overrightarrow{N(u)})^2 + \alpha(\|\overrightarrow{N_x(u)}\|^2 + \|\overrightarrow{N_y(u)}\|^2) + \beta(u)(\|\overrightarrow{N(u)}\|^2 - 1)] du. \tag{9.49}$$

Minimising the functional in Eq. (9.49) is a problem in the calculus of variations. Similarly to the variational approach presented in Sect. 9.3.3, the minimisers of this function are the solutions of the associated Euler–Lagrange equations. The resulting Euler–Lagrange equations, in turn, can also be discretised on the image lattice to arrive at iterative update equations with respect to the variables. To avoid the computational complexity resulting from a direct optimisation on all the variables involved, the authors estimated the light source direction and the surface normal field separately in alternating steps. In other words, each step only optimises either $\vec{L}$ or $\overrightarrow{N(u)}$ while fixing the value of the other variable.

When the light source direction $\vec{L}$ is fixed, the discrete update scheme for the surface normal field is given by

$$\overrightarrow{M(u)}^{(t+1)} = \overrightarrow{N(u)}^{(t)} + \frac{\varepsilon_s^2}{4\alpha}(R(u) - \vec{L}.\overrightarrow{N(u)}^{(t)})\vec{L} \tag{9.50}$$

$$\overrightarrow{N(u)}^{(t+1)} = \frac{\overrightarrow{M(u)}^{(t+1)}}{\|\overrightarrow{M(u)}^{(t+1)}\|}, \tag{9.51}$$

where t is the iteration index, and ε_s is the spatial distance between adjacent pixels. Here, Eq. (9.50) updates the surface normals in each iteration, whereas Eq. (9.51) constrains them to the unit length. We note that the update scheme in Eqs. (9.50) and (9.51) is readily applicable to a general reflectance model $f(\vec{L}, \overrightarrow{N(u)})$ other than the Lambertian one.

9.4 Estimating the Illuminant Direction

Once the surface normal $\vec{N(u)}$ has been obtained at every pixel, the global illuminant direction is the solution to the following linear equation:

$$\left(\int_{\mathcal{I}} \vec{N(u)} \vec{N(u)}^T \, du \right) \vec{L} = \int_{\mathcal{I}} R(u) \vec{N(u)} \, du, \tag{9.52}$$

where the matrix on the left-hand side has a size of 3×3, and the right-hand side is a 3-element column vector.

Equation (9.52) can then be discretised and translated to the following iterative update equation:

$$\vec{L}^{(t+1)} = \left(\sum_{\mathcal{I}} \vec{N(u)}^{(t+1)} \vec{N(u)}^{(t+1)^T} \, du \right)^{-1} \sum_{\mathcal{I}} R(u) \vec{N(u)}^{(t+1)} \, du. \tag{9.53}$$

The update scheme here converges slowly due to the alternation of the optimisation between the light source direction and the surface normal. Furthermore, the method may be prone to local minima, since the final solution depends on the initial values of the variables. Methods to accelerate the convergence speed include multilevel algorithms which initially solve the problem at a coarse image resolution and refine the results at finer resolutions (Terzopoulos 1983).

We note that the above method by Brooks and Horn (1985) is only applicable to diffuse Lambertian surfaces. To deal with specular surfaces, Hara et al. (2005) presented two methods that exploit both diffuse and specular reflection to estimate the light source position and the surface reflectance simultaneously from a single view. The first method separates the specular and diffuse components from an unpolarised image concurrently with an estimation of the diffuse and specular reflection parameters and the illuminant position through an iterative relaxation scheme. Estimating the diffuse reflection parameters is posed as a fitting problem to the Lambertian model, while estimating the specular reflection parameters is viewed as minimising the data error of the specular component with respect to a log-transformed Torrance–Sparrow model. To constrain the light source position, the authors utilised the law of reflection, that the surface normal at a specular peak is the bisector of the illuminant direction and the view direction. With the specular peaks detected, the light source position can be searched along the reflection direction about the surface normals of the peaks.

Further, the same authors also propose an alternative method to estimate the specular reflection parameters and the source position from only the specular component. Similar to the former approach, the optimisation scheme here aims to minimise the rendering error with respect to the log-transformed Torrance–Sparrow model. Unlike the previous approach, the latter method also aims to find the optimal position of the specular peaks in the image to provide a realistic constraint on the illuminant direction in case of off-specular peaks. Although adhering to the physical modelling of diffuse and specular reflection, these approaches are limited in practical settings, since they require camera geometric calibration and known object geometry as input.

Also making use of the Torrance–Sparrow model, Migita et al. (2008) presented a large-scale non-linear optimisation to recover the light source position, the shape and the reflectance simultaneously. The input to the algorithm is a set of images captured from a fixed viewpoint while the point light source moves around the scene, in a setting similar to that of photometric stereo. As before, the optimisation problem is formulated as a non-linear least-squares problem that minimises the difference between the measured image irradiance and that yielded by the Torrance–Sparrow model. The problem is then solved by a variant of the Levenberg–Marquardt method (Marquardt 1963) which includes a preconditioned conjugate gradient algorithm to find the search direction in each iteration. To ensure the convergence of the optimisation algorithm the method requires an appropriate initialisation. To this end, the authors initialised the light position and surface normals by a singular value decomposition of a matrix composed of the irradiance measured over the image pixels and light sources by assuming a Lambertian reflectance model on the scene. The algorithm is repeated with different initial values, and the optimal solution is the one that best fits the input images.

9.5 Notes

The problem of recovering the shape and photometric parameters for a real-world scene has found applications in inserting virtual objects in a real scene (Nishino and Nayar 2004), creating new synthetic images with novel illumination conditions (Zhang and Yang 2001) and tracing digital forgery (Johnson and Farid 2005).

For shape and light source direction recovery, most of the existing methods can be categorised into three broad classes: shading-, shadow- and specularity-based methods. Shading-based methods (Zhang and Yang 2001; Wang and Samaras 2002) generally rely on critical points and image intensity features to detect the light source direction. The main drawback of these methods stems from the fact that the detection of critical points and intensity features is often prone to noise corruption. Shadow-based methods (Sato et al. 1999, 2001), on the other hand, employ changes in the image radiance distribution induced by cast shadows in the imagery. Although they are effective, these methods often fail under single illuminants since, in these cast shadows, the image irradiance tend to be too low to provide reliable information.

Specular highlights provide a strong cue for estimating the light source direction. However, most of the existing methods require a calibration step. Powell et al. (2001) use three spheres with known relative positions as calibration objects in order to triangulate light source positions from specularity. Zhou and Kambhamettu (2002) proposed an iterative method that utilises a stereo image pair of a sphere where the highlights are used to estimate the directions of the light sources and the shading is employed to estimate their intensities. Wong et al. (2008) utilised the specular highlights on a calibration sphere to solve the problem in a closed-form manner. Extending the work in Zhou and Kambhamettu (2002, 2008) proposed an iterative approach for estimating the area light sources from specularity observed on two spheres. Schnieders et al. (2010) also dealt with area light sources but, unlike the approach in Zhou and Kambhamettu (2008), they proposed a closed-form solution using a single view of two identical spheres.

Other methods require full or partial knowledge of the geometry to be available (Wang and Samaras 2002; Hara et al. 2005; Zhang and Yang 2001; Sato et al. 2001) or assume distant illuminant and viewpoint (orthographic projection) (Pentland 1982; Sato et al. 1999). To avoid imposing strong assumptions on the scene geometry, one could view the total surface radiance as a sum of the specular and diffuse reflection components. This is consistent with a number of references elsewhere in the literature and would permit the modelling of the diffuse and specular terms as a function of reflection angle variables, photogrammetric variables and a wavelength-dependent index of refraction. This can also provide a link between photometric parameter estimation and reflectance geometry. Further, these photometric parameters can be recovered by making use of the techniques presented earlier in the book.

References

Beckmann, P., & Spizzichino, A. (1963). *The scattering of electromagnetic waves from rough surfaces*. New York: Pergamon.

Brooks, M. J., & Horn, B. K. P. (1985). Shape and source from shading. In *International joint conference on artificial intelligence* (pp. 932–936).

Canny, J. (1986). A computational approach to edge detection. *IEEE Transactions on Pattern Analysis and Machine Intelligence*, *8*(6), 679–698.

Chojnacki, W., Brooks, M. J., & Gibbins, D. (1994). Revisiting Pentland's estimator of light source direction. *Journal of the Optical Society America A*, *11*(1), 118–124.

Dupuis, P., & Oliensis, J. (1992). Direct method for reconstructing shape from shading. In *Proceedings of the IEEE conference on computer vision and pattern recognition* (pp. 453–458).

Frankot, R. T., & Chellappa, R. (1988). A method of enforcing integrability in shape from shading algorithms. *IEEE Transactions on Pattern Analysis and Machine Intelligence*, *4*(10), 439–451.

Hara, K., Nishino, K., & Ikeuchi, K. (2005). Light source position and reflectance estimation from a single view without the distant illumination assumption. *IEEE Transactions on Pattern Analysis and Machine Intelligence*, *27*(4), 493–505.

Hecht, E. (2002). *Optics* (4th ed.). Reading: Addison-Wesley.

Horn, B. K. P., & Brooks, M. J. (1986). The variational approach to shape from shading. *CVGIP*, *33*(2), 174–208.

Horn, B. K. P., & Brooks, M. J. (1989). *Shape from shading*. Cambridge: MIT Press.

Huynh, C. P., & Robles-Kelly, A. (2010). A solution of the dichromatic model for multispectral photometric invariance. *International Journal of Computer Vision*, *90*(1), 1–27.

Ikeuchi, K., & Horn, B. K. P. (1981). Numerical shape from shading and occluding boundaries. *Artificial Intelligence*, *17*(1–3), 141–184.

Johnson, M. K., & Farid, H. (2005). Exposing digital forgeries by detecting inconsistencies in lighting. In *Proceedings of the 7th workshop on multimedia and security* (pp. 1–10).

Kimmel, R., & Bruckstein, A. M. (1995). Tracking level sets by level sets: a method for solving the shape from shading problem. *Computer Vision and Image Understanding*, *62*(2), 47–48.

Knill, D. C. (1990). Estimating illuminant direction and degree of surface relief. *Journal of the Optical Society America A*, *7*(4), 759–775.

Lee, C.-H., & Rosenfeld, A. (1985). Improved methods of estimating shape from shading using the light source coordinate system. *Artificial Intelligence, 26*(2), 125–143.

Marquardt, D. (1963). An algorithm for least-squares estimation of nonlinear parameters. *SIAM Journal on Applied Mathematics, 11*, 431–441.

Migita, T., Ogino, S., & Shakunaga, T. (2008). Direct bundle estimation for recovery of shape, reflectance property and light position. In *Proceedings of the 10th European conference on computer vision: part III* (pp. 412–425).

Nillius, P., & Eklundh, J. (2001). *Automatic estimation of the projected light source direction.*

Nishino, K., & Nayar, S. (2004). Eyes for relighting. *ACM Transactions on Graphics, 23*(3), 704–711.

Nocedal, J., & Wright, S. (2000). *Numerical optimization*. Berlin: Springer.

Pentland, A. P. (1982). Finding the illuminant direction. *Journal of the Optical Society of America, 72*(4), 448–455.

Powell, M., Sarkar, S., & Goldgof, D. (2001). A simple strategy for calibrating the geometry of light sources. *IEEE Transactions on Pattern Analysis and Machine Intelligence, 23*(9), 1022–1027.

Prados, E., & Faugeras, O. (2003). Perspective shape from shading and viscosity solutions. In *IEEE international conference on computer vision* (Vol. II, pp. 826–831).

Riley, K., Hobson, M., & Bence, S. (2006). *Mathematical methods for physics and engineering*. Cambridge: Cambridge University Press.

Sato, I., Sato, Y., & Ikeuchi, K. (1999). Illumination distribution from shadows. In *Computer vision and pattern recognition* (pp. 1306–1312).

Sato, I., Sato, Y., & Ikeuchi, K. (2001). Stability issues in recovering illumination distribution from brightness in shadows. In *Computer vision and pattern recognition* (pp. 400–407).

Schnieders, D., Wong, K.-Y. K., & Dai, Z. (2010). Polygonal light source estimation. In *Asian conference on computer vision* (pp. 96–107).

Shafer, S. A. (1985). Using color to separate reflection components. *Color Research and Application, 10*(4), 210–218.

Smith, G. B. (1982). The recovery of surface orientation from image irradiance. In *Proceedings of the DARPA image understanding workshop*, Palo Alto, CA, USA (pp. 132–141).

Strat, T. (1979). *A numerical method for shape-from-shading from a single image*. Master's thesis, Department of Electrical Engineering and Computer Science, Massachusetts Institute of Technology, Cambridge, MA, USA.

Tagare, H. D., & deFigueiredo, R. J. P. (1991). A theory of photometric stereo for a class of diffuse non-Lambertian surfaces. *IEEE Transactions on Pattern Analysis and Machine Intelligence, 13*(2), 133–152.

Terzopoulos, D. (1983). Multilevel computational processes for visual surface reconstruction. *Computer Vision, Graphics and Image Understanding, 24*, 52–96.

Torrance, K., & Sparrow, E. (1967). Theory for off-specular reflection from roughened surfaces. *Journal of the Optical Society of America, 57*(9), 1105–1112.

Vernold, C. L., & Harvey, J. E. (1998). A modified Beckmann–Kirchoff scattering theory for non-paraxial angles. In *Proceedings of the SPIE: Vol. 3426. Scattering and surface roughness* (pp. 51–56).

Wang, Y., & Samaras, D. (2002). Estimation of multiple illuminants from a single image of arbitrary known geometry. In *European conference on computer vision* (pp. 272–288).

Weinshall, D. (1990). *The shape of shading and the direction of illumination from shading on occluding contours*. Artificial intelligence memo 1264; center for biological information processing memo 60.

Wolff, L. B. (1994). Diffuse-reflectance model for smooth dielectric surfaces. *Numbers, 11*, 2956–2968.

Wong, K.-Y. K., Schnieders, D., & Li, S. (2008). Recovering light directions and camera poses from a single sphere. In *European conference on computer vision* (pp. 631–642).

Woodham, R. (1980). Photometric method for determining surface orientation from multiple images. *Optical Engineering, 19*(1), 139–144.

References

Woodham, R. J. (1977). *A cooperative algorithm for determining surface orientation from a single view* (pp. 635–641).

Worthington, P. L., & Hancock, E. R. (1999). New constraints on data-closeness and needle map consistency for shape-from-shading. *IEEE Transactions on Pattern Analysis and Machine Intelligence*, *21*(12), 1250–1267.

Zhang, Y., & Yang, Y.-H. (2001). Multiple illuminant direction detection with application to image synthesis. *IEEE Transactions on Pattern Analysis and Machine Intelligence*, *23*(8), 915–920.

Zheng, Q., & Chellappa, R. (1991). Estimation of illuminant direction, albedo, and shape from shading. *IEEE Transactions on Pattern Analysis and Machine Intelligence*, *13*(7), 680–702.

Zhou, W., & Kambhamettu, C. (2002). Estimation of illuminant direction and intensity of multiple light sources. In *European conference on computer vision* (pp. 206–220).

Zhou, W., & Kambhamettu, C. (2008). A unified framework for scene illuminant estimation. *Image and Vision Computing*, *26*(3), 415–429.

Chapter 10
Polarisation of Light

The polarisation of light is a concept describing the distribution of its electromagnetic field at different oscillation orientations in the plane perpendicular to the propagation direction. It has long been a well-studied subject in astronomy (Hall 1951), optics (Born and Wolf 1999; Mandel and Wolf 1995) and crystallography. However, polarimetric imaging is a somewhat recent development, with the early work done by Wolff (1989).

Although the human vision system is oblivious to polarisation, a number of animals, such as mantis shrimps, naturally possess a polarisation vision system (Marshall et al. 1991). Biology researchers have also noticed evidence for biophysical mechanisms of polarisation coding in fish (Hawryshyn 2000). With recent advances in camera technologies, these polarisation effects can be captured by devices such as polarimeters and more recently, polarisation cameras (Wolff 1997; Wolff and Andreou 1995; Wolff et al. 1997). In (Wolff 1997; Wolff and Andreou 1995; Wolff et al. 1997), Wolff et al. developed a liquid crystal polarisation video camera with twisted nematic liquid crystals that were electro-optically controlled to replace the need for a mechanical rotation of linear polarisers. The development of these portable, low-cost and fast polarisation camera sensors potentially extends the applications of polarisation imaging to areas such as target detection and segmentation (Goudail et al. 2004; Sadjadi and Chun 2004) and material property recovery (Wolff and Boult 1989).

This chapter explores computational models of polarimetric imaging and its applications to the recovery of shape and material properties. We commence the chapter by providing the physical background on polarisation of light as an electromagnetic wave in Sect. 10.1. In particular, we elaborate on the polarisation effects caused by surface reflection, including both diffuse and specular reflection. We describe the related physical processes in detail in Sects. 10.2.1 and 10.2.2. We also draw on the Fresnel reflection theory to formulate a computational model of polarimetric imaging in Sect. 10.3.1. Subsequently, we review techniques concerning the use of polarimetric imaging for the problem of shape and material refractive index recovery in Chap. 11.

Notation	Description
$I_\vartheta(u, \lambda)$	Transmitted irradiance at a pixel u and a wavelength λ corresponding to a polariser angle ϑ.
$I_{max}(u, \lambda)$ and $I_{min}(u, \lambda)$	Maximum and minimum transmitted values on the transmitted radiance sinusoid (TRS).
$I_{un}(u, \lambda)$	Unpolarised irradiance.
$I_{i\parallel}$ and $I_{i\perp}$	Parallel and perpendicular components of the incident irradiance.
$I_{R\parallel}$ and $I_{R\perp}$	Parallel and perpendicular components of the reflected radiance.
$I_{T\parallel}$ and $I_{T\perp}$	Parallel and perpendicular components of the transmitted radiance.
$\vec{E}$	Electric field vector with components $\vec{E}_x$ and $\vec{E}_y$, wave number k and phase difference ε.
$\vec{E}_{un}$ and $\vec{E}_{pol}$	Unpolarised and completely polarised components.
$F_\parallel$ and $F_\perp$	Parallel and perpendicular Fresnel reflection coefficients.
$\eta(u, \lambda)$	Refractive index at a pixel u and a wavelength λ.
θ_i	Incident angle between the illuminant direction and the surface normal.
ϕ	Phase of polarisation.
ρ	Degree of linear polarisation.
$L(u, \lambda)$	The illuminant power at the pixel u and the wavelength λ.
$S(u, \lambda)$	Spectral reflectance at the pixel u and the wavelength λ.
$g(u), k(u)$	Shading factor and specular coefficient at pixel u.

Fig. 10.1 Notation used in Chap. 10

10.1 Polarisation of Electromagnetic Waves

In this section, we provide a brief background on the theory of the polarisation of electromagnetic waves. Based on the wave theory, light propagation results in harmonic vibrations in its transmission medium. As an electromagnetic wave, light is associated with a magnetic field and an electric field that are mutually orthogonal and vibrate perpendicular to the propagation direction. Such electric and magnetic fields can be represented as field vectors parallel to a plane perpendicular to the propagation direction. According to Born and Wolf (1999) and Hecht (2002), the polarisation of light characterises the pattern of vibration of the electric and magnetic field vectors as light propagates across space and time.

Throughout the section, we introduce a number of concepts relating polarisation, diffuse and specular reflection and the Fresnel transmission theory. For mnemonic purposes, in Fig. 10.1, we provide a list of notation frequently used in the chapter. In addition, we consider a right-handed reference coordinate system where the origin is located at the viewpoint, and the positive z-axis coincides with the propagation direction of the reflected light as observed in the line of sight. The positive x-axis points towards the right-hand side of the field of view. Within this right-handed

10.1 Polarisation of Electromagnetic Waves

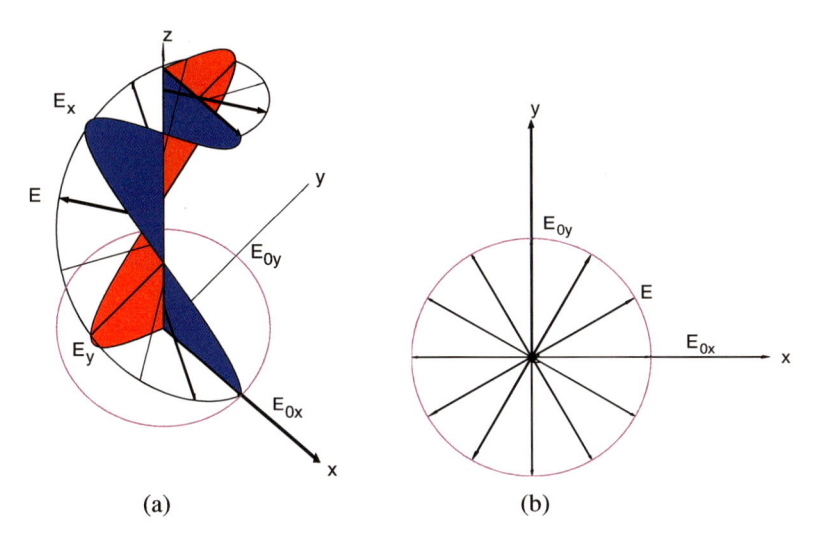

Fig. 10.2 The polarisation of an electromagnetic wave. (**a**) The polarisation of light is caused by the rotation of its electric field ($\vec{E}$) along its propagation direction z. The electric field can be decomposed into the two orthogonal harmonic components $\vec{E}_x$ and $\vec{E}_y$ with magnitudes E_{0x} and E_{0y}. (**b**) The projection of the trajectory for the front of $\vec{E}$ as projected onto the x–y plane perpendicular to the propagation direction

reference system, surface normals can be specified by their azimuth (tilt) and zenith (slant) angles with respect to the positive x- and z-axes.

The electric field of an electromagnetic wave can be viewed as a superposition of two harmonic plane waves with orthogonal planes of vibration (Born and Wolf 1999). Figure 10.2(a) depicts a circularly polarised illumination flux with its electric field vector $\vec{E}$ rotating in a helicoidal pattern about the propagation direction of the reflected light as observed from the viewpoint. In the figure, the propagation direction is along the z-axis. As shown in Fig. 10.2(a), the electric field vector can be decomposed into two sinusoidal components $\vec{E}_x$ and $\vec{E}_y$ vibrating in the x–z and y–z planes. In the figure, the two components have equal magnitudes and $\vec{E}_y$ is shifted by 1/4-wavelength ($\frac{\pi}{2}$) with respect to $\vec{E}_x$. As a result, the front of the electric field follows a trajectory with a circular projection in the x–y plane as it varies with respect to time and space. Figure 10.2(b) illustrates the circular rotation of the electric field, while its x and y components vibrate in their respective planes.

In general, the polarisation of a light wave can range from unpolarised to completely polarised. When the phase difference between its orthogonal components varies in a random manner with respect to time, the resultant electric field is incoherent or unpolarised, producing an electric field vector rotating isotropically and randomly in the plane perpendicular to the light propagation direction. On the other hand, completely polarised light is the result of the coherent superposition of two orthogonal harmonic components with a fixed phase difference with respect to time. In this case, the light wave is elliptically polarised since the front of the electric

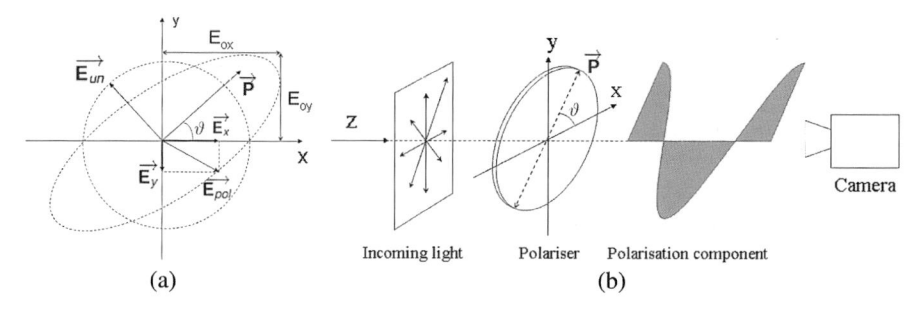

Fig. 10.3 (a) The superposition of an unpolarised wave $\vec{E}_{\text{un}}$ with an elliptically polarised one $\vec{E}_{\text{pol}}$. The elliptically polarised wave can be further decomposed into two component vectors $\vec{E}_x$ and $\vec{E}_y$ whose lengths oscillate sinusoidally with magnitudes E_{0x} and E_{0y}. The shape and orientation of the major and minor axes of the ellipse are determined by the ratio $\frac{E_{0x}}{E_{0y}}$ and the phase difference between $\vec{E}_x$ and $\vec{E}_y$. (b) The transmitted wave through a polariser with the transmission axis direction $\vec{P}$ oriented at an angle ϑ with respect to the x-axis

field traces out an ellipse whose major and minor axes are determined by the phase difference of the harmonic components.

A general polarisation state can be represented as a mixture of an unpolarised state with a completely polarised state. Consider a light wave with wavelength λ propagating in the z direction. Let the unpolarised and completely polarised components be $\vec{E}_{\text{un}}$ and $\vec{E}_{\text{pol}}$, respectively. The fronts of $\vec{E}_{\text{un}}$ and $\vec{E}_{\text{pol}}$ can be projected onto the x–y plane so as to trace a circle and an ellipse as shown in Fig. 10.3(a). The total polarisation state $\vec{E}_{\text{pol}}$ can be further decomposed into the x and y components $\vec{E}_x$ and $\vec{E}_y$. The latter polarisation components are sinusoidal in nature, oscillating in the x–z and y–z planes with magnitudes E_{0x} and E_{0y}, respectively. Mathematically, the vectorial form of $\vec{E}_x$ and $\vec{E}_y$ at a point with coordinates $[0, 0, z_0]^T$ at time t is expressed as

$$\vec{E}_x = E_{0x} \cos(kz_0 - \omega t), \tag{10.1}$$

$$\vec{E}_y = E_{0y} \cos(kz_0 - \omega t + \varepsilon), \tag{10.2}$$

where the wave number k is given by $k = \frac{2\pi}{\lambda}$, ω is the frequency of the wave and ε is the phase difference between the two components. The vectors $\vec{E}_x$ and $\vec{E}_y$ along the x- and y-axes have magnitudes E_{0x} and E_{0y}, respectively.

Thus, the total electric field vector $\vec{E}$ can be obtained as a linear combination of the unpolarised and the completely polarised field vectors (Hecht 2002), which is given by

$$\vec{E} = \vec{E}_{\text{un}} + \left[E_{0x} \cos \psi, E_{0y} \cos(\psi + \varepsilon) \right]^T, \tag{10.3}$$

where $[\cdot]^T$ denotes the transpose of a vector and $\psi = kz_0 - \omega t$.

10.1.1 Transmitted Radiance Sinusoid (TRS)

Equation (10.3) provides a general analytical expression of the electric field with respect to time and space. Now we consider the decomposition of the above light wave into linear polarisation components oriented at different directions in the plane perpendicular to the propagation direction (i.e. the x–y plane). The power of these polarisation components can be measured by placing a linear polariser in front of the camera, as shown in Fig. 10.3(b). In effect, the polariser only transmits the polarisation component of the incoming light that vibrates along its transmission axis. By rotating the polariser, one can capture linear polarisation components of different orientations in the plane orthogonal to the light propagation direction.

Let the transmission axis $\vec{P}$ lie on the x–y plane at an angle ϑ with respect to the x-axis, i.e. $\vec{P} = [\cos\vartheta, \sin\vartheta]^T$. For the randomly polarised component $\vec{E}_{un}$, the radiance transmitted along any direction $\vec{P}$ is half of its original energy I_{un}, which is given by the time-averaged square of its amplitude $I_{un} = \langle \vec{E}_{un}^2 \rangle_t$, where $\langle . \rangle_t$ denotes average over time. Similarly, the transmitted radiance for the completely polarised component $\vec{E}_{pol}$ is the time-averaged energy along the orientation $\vec{P}$ per unit area. Therefore, the total transmitted radiance at the pixel u and wavelength λ of the incoming light through the polariser is related to the electric field as

$$I_\vartheta(u,\lambda) = \langle |\vec{E} \cdot \vec{P}|^2 \rangle_t$$

$$= \frac{1}{2} I_{un} + \langle \left((E_{0x} \cos\psi \cos\vartheta + E_{0y} \cos(\psi+\varepsilon) \sin\vartheta \right)^2 \rangle_t$$

$$= \frac{1}{2} I_{un} + E_{0x}^2 \cos^2\vartheta \langle \cos^2\psi \rangle_t + E_{0y}^2 \sin^2\vartheta \langle \cos^2(\psi+\varepsilon) \rangle_t$$

$$+ E_{0x} E_{0y} \sin\vartheta \cos\vartheta \left(\langle \cos(2\psi+\varepsilon) \rangle_t + \cos\varepsilon \right)$$

$$= \frac{1}{2} \left(I_{un} + E_{0x}^2 \cos^2\vartheta + E_{0y}^2 \sin^2\vartheta + E_{0x} E_{0y} \cos\varepsilon \sin 2\vartheta \right), \quad (10.4)$$

where, as before, $\langle . \rangle_t$ denotes the average value over time.

In Eq. (10.4), we use the equalities $\langle \cos^2\psi \rangle_t = \langle \cos^2(kz_0 - \omega t) \rangle_t = \frac{1}{2}$, $\langle \cos^2(\psi+\varepsilon) \rangle_t = \langle \cos^2(kz_0 - \omega t + \varepsilon) \rangle_t = \frac{1}{2}$ and $\langle \cos(2\psi+\varepsilon) \rangle_t = 0$. The right-hand side of this equation can be rewritten as a sinusoid function,

$$I_\vartheta(u,\lambda) = \frac{1}{2} \left(I_{un} + \frac{1}{2}(E_{0x}^2 + E_{0y}^2) + \frac{1}{2}(E_{0x}^2 - E_{0y}^2) \cos 2\vartheta + E_{0x} E_{0y} \cos\varepsilon \sin 2\vartheta \right)$$

$$= \frac{1}{2} \left(I_{un} + \frac{1}{2}(E_{0x}^2 + E_{0y}^2) + I_s \cos 2(\vartheta - \phi) \right), \quad (10.5)$$

where $I_s = (\frac{1}{4}(E_{0x}^2 + E_{0y}^2)^2 - E_{0x}^2 E_{0y}^2 \sin^2\varepsilon)^{1/2}$ and $\phi = \frac{1}{2} \arctan(\frac{2 E_{0x} E_{0y} \cos\varepsilon}{E_{0x}^2 - E_{0y}^2})$.

Equation (10.5) shows that the transmitted radiance of a polarisation component is a sinusoidal function of the transmission angle ϑ with an amplitude $\frac{1}{2} I_s$ and

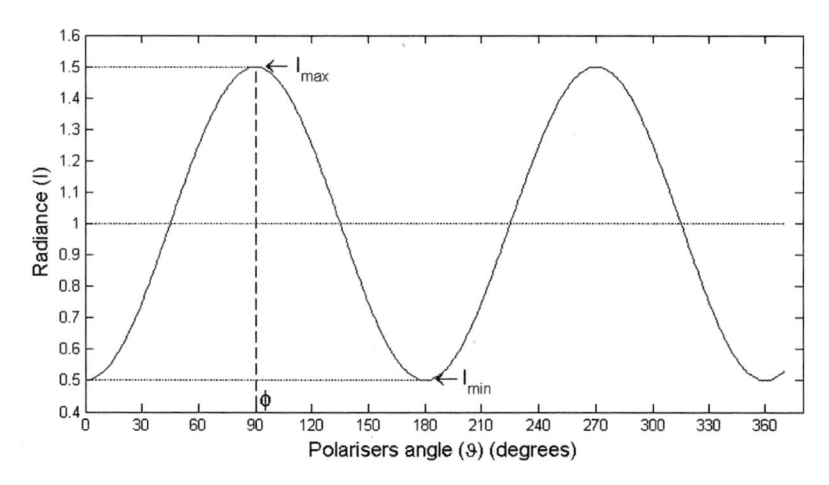

Fig. 10.4 The transmitted radiance through a linear polariser varies sinusoidally with respect to the polariser orientation angle

a phase ϕ. This sinusoid is termed the transmitted radiance sinusoid (TRS). The variation along the TRS is depicted in Fig. 10.4, where the radiance I fluctuates along a sinusoid curve bounded between $I_{\min}$ and $I_{\max}$.

We note that the formulation in Eq. (10.5) applies to monochromatic light travelling through a particular point in space. To generalise the formula for spectropolarimetric imaging, we index an image with respect to the wavelength λ and the image location u, in addition to the polariser angle ϑ. As an alternative, we consider a spectral polarimetric image as consisting of N spectral image cubes $\mathcal{I}_{\vartheta_1}, \mathcal{I}_{\vartheta_2}, \ldots, \mathcal{I}_{\vartheta_N}$. Each cube is captured at N discrete values of the polariser orientation angle ϑ, which we denote as ϑ_j, $j = \{1, 2, \ldots, N\}$.

The variation of the image intensity at pixel u, wavelength λ and polariser orientation ϑ_j is then given by

$$I_{\vartheta_j}(u, \lambda) = \frac{I_{\max} + I_{\min}}{2} + \frac{I_{\max} - I_{\min}}{2} \cos(2\vartheta_j - 2\phi), \quad (10.6)$$

where $I_{\min}$ and $I_{\max}$ are the minimum and maximum radiance, respectively, at the pixel u and ϕ is the corresponding polarisation angle. For brevity, we have dropped the wavelength, pixel and polarisation angle variables from the notation for $I_{\min}$ and $I_{\max}$.

With the polarisation components $I_{\min}$, $I_{\max}$ and ϕ above, we derive the unpolarised intensity I_{un} as the average intensity of the TRS:

$$I_{\text{un}} = \frac{I_{\max} + I_{\min}}{2}. \quad (10.7)$$

This intensity is depicted as a flat line with respect to wavelength in Fig. 10.4.

10.1 Polarisation of Electromagnetic Waves

Further we define the degree of linear polarisation ρ to be

$$\rho = \frac{I_{max} - I_{min}}{I_{max} + I_{min}}. \tag{10.8}$$

The degree of partial polarisation varies between 0 and 1. Unpolarised light has a zero degree of polarisation. At the other extreme, completely polarised light has a degree polarisation of 1, where the minimal intensity I_{min} on the TRS is zero. This case occurs when the polariser angle coincides with Brewster's angle (Hecht 2002).

10.1.2 Decomposing Polarimetric Images

The recovery of I_{min}, I_{max} and ϕ in Eq. (10.6) from a captured image sequence $\mathcal{I}_{\vartheta_1}, \mathcal{I}_{\vartheta_2}, \ldots, \mathcal{I}_{\vartheta_M}$ may be effected in a number of ways. By acquiring three images I_0, I_{45} and I_{90} at polariser orientations of $0°$, $45°$ and $90°$ respectively, Wolff (1997) computed the phase ϕ, unpolarised intensity I_{un} and degree of polarisation ρ using the following equations:

$$\phi = \begin{cases} \frac{1}{2}\arctan(\frac{I_0 + I_{90} - 2I_{45}}{I_{90} - I_0}) + \frac{\pi}{2}, & \text{if } I_{90} \geq I_0, \\ \frac{1}{2}\arctan(\frac{I_0 + I_{90} - 2I_{45}}{I_{90} - I_0}) + \pi, & \text{if } I_{90} < I_0 \text{ and } I_{45} < I_0, \\ \frac{1}{2}\arctan(\frac{I_0 + I_{90} - 2I_{45}}{I_{90} - I_0}), & \text{if } I_{90} < I_0 \leq I_{45}. \end{cases} \tag{10.9}$$

$$I_{un} = I_0 + I_{90}, \tag{10.10}$$

$$\rho = \frac{I_{90} - I_0}{(I_{90} + I_0 \cos 2\phi)}. \tag{10.11}$$

After computing the unpolarised intensity and degree of polarisation, we can recover I_{max} and I_{min} using the relations in Eqs. (10.7) and (10.8). However, the method in Wolff (1997) is susceptible to noise corruption since it employs only three images.

Alternatively, capturing polarisation images of a scene at three or more polariser orientation angles, Atkinson and Hancock (2006) fitted a sinusoidal curve to the points whose coordinates are pairs of measured intensities and polariser angles. Their fitting method is based on the Levenberg–Marquardt non-linear least-squares algorithm (Marquardt 1963). This method is, however, not efficient since the optimisation has to be performed per pixel.

Several other authors (Nayar et al. 1997; Huynh et al. 2010) have obtained a more stable estimation of the intensity, phase and degree of polarisation by solving an over-determined linear system of equations. To do this, Eq. (10.6) can be rewritten for each pixel site u and wavelength λ in the following vector form:

$$I_{\vartheta_j}(u, \lambda) = \begin{bmatrix} 1 \\ \cos(2\vartheta_j) \\ \sin(2\vartheta_j) \end{bmatrix}^T \begin{bmatrix} \frac{I_{max} + I_{min}}{2} \\ \frac{I_{max} - I_{min}}{2}\cos(2\phi) \\ \frac{I_{max} - I_{min}}{2}\sin(2\phi) \end{bmatrix} = \mathbf{a}_j^T \mathbf{x}, \tag{10.12}$$

where

$$\mathbf{a}_j = \begin{bmatrix} 1 & \cos(2\vartheta_j) & \sin(2\vartheta_j) \end{bmatrix}^T \quad \text{and} \quad \mathbf{x} = \begin{bmatrix} \frac{I_{\max}+I_{\min}}{2} \\ \frac{I_{\max}-I_{\min}}{2}\cos(2\phi) \\ \frac{I_{\max}-I_{\min}}{2}\sin(2\phi) \end{bmatrix}.$$

After collecting $M \geq 3$ measurements at three or more polariser orientations, we arrive at the following over-determined linear system:

$$\mathbf{I} = \mathbf{A}\mathbf{x}, \tag{10.13}$$

where

$$\mathbf{I} = \begin{bmatrix} I_{\vartheta_1}(u,\lambda) \\ I_{\vartheta_2}(u,\lambda) \\ \vdots \\ I_{\vartheta_M}(u,\lambda) \end{bmatrix} \quad \text{and} \quad \mathbf{A} = \begin{bmatrix} \mathbf{a}_1^T \\ \mathbf{a}_2^T \\ \vdots \\ \mathbf{a}_M^T \end{bmatrix}.$$

Equation (10.13) is well-constrained when the number of polariser angles is $M \geq 3$. Moreover, the coefficient matrix $\mathbf{A}$ depends solely on the polariser angles and, therefore, allows for an efficient solution of Eq. (10.13) over all the image pixels and wavelengths simultaneously.

Having obtained the solution $\mathbf{x}^* = [\mathbf{x}_1, \mathbf{x}_2, \mathbf{x}_3]^T$, the maximal and minimal radiance on the sinusoid and the phase of polarisation at each pixel u and wavelength λ can be recovered in a straightforward manner by making use of the following relations:

$$I_{\max} = \mathbf{x}_1 + \sqrt{\mathbf{x}_2^2 + \mathbf{x}_3^2},$$

$$I_{\min} = \mathbf{x}_1 - \sqrt{\mathbf{x}_2^2 + \mathbf{x}_3^2},$$

$$\phi = \frac{1}{2}\arctan\frac{\mathbf{x}_3}{\mathbf{x}_2}.$$

10.2 Polarisation upon Reflection

In Sect. 10.1, we have formulated the transmitted irradiance of light through a linear polariser. This formulation is concerned with the polarisation of light in general, whether it is caused by reflection upon a surface, refraction through the boundary between media, or scattering through the atmosphere, haze or fog. In this section, we focus our analysis on polarisation caused by the reflection of unpolarised incident light upon a dielectric surface. Since diffuse and specular reflection differ significantly in physical process, we analyse each of these cases separately.

10.2 Polarisation upon Reflection

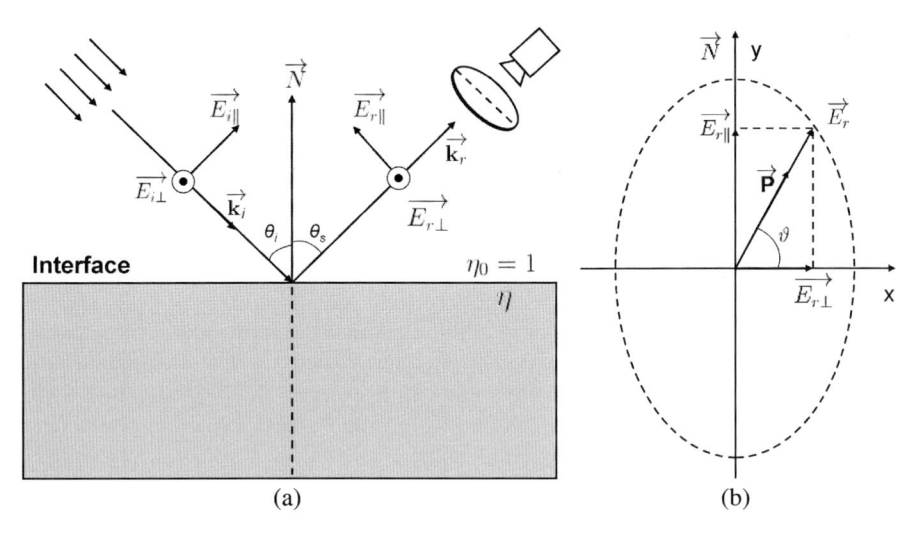

Fig. 10.5 (a) Polarisation upon specular reflection from a dielectric surface with a normal vector $\vec{N}$ and refractive index η. The electric fields of the incident and reflected waves are represented by vectors rotating in a plane perpendicular to the propagation direction. These electric field vectors can be decomposed into two orthogonal components which are parallel and perpendicular to the plane of incidence/reflection. The vectors $\vec{E}_{i\parallel}$ and $\vec{E}_{i\perp}$ are the components of the electric field incident on the material-air surface boundary. The vectors $\vec{E}_{r\parallel}$ and $\vec{E}_{r\perp}$ are the wave components reflected from the surface. (b) The reflected electric field, where the surface normal $\vec{N}$ is aligned with the y-axis. $\vec{E}_{r\parallel}$ vibrates along the y-axis because it is in the plane of reflection. The perpendicular component $\vec{E}_{r\perp}$ lies in the x-axis, and $\vec{P}$ is the orientation of the polariser's transmission axis

10.2.1 Polarisation upon Specular Reflection

Specular reflection is a result of direct reflection on the media-material interface. Figure 10.5(a) shows a diagram of the specular reflection process. Here, an incident light ray $\vec{k}_i$ impinges on the surface at an angle θ_i with the surface normal $\vec{N}$ and is reflected in the direction of the ray $\vec{k}_r$. According to the law of reflection, the reflection angle θ_s between the surface normal $\vec{N}$ and the viewing direction is equal to the incident angle θ_i. In addition, we denote the relative refractive index of the material to the air as η.

We quantify the radiance of the incident and reflected light by examining the electric fields associated with the rays $\vec{k}_i$ and $\vec{k}_r$. The vectors representing these fields are perpendicular to the light propagation direction and can be decomposed into two orthogonal harmonic components, i.e. one in the plane of reflection and the other one perpendicular to it. In Fig. 10.5(a), we have denoted $\vec{E}_{i\parallel}$ and $\vec{E}_{i\perp}$ as the parallel and perpendicular components of the incident electric field. Similarly, $\vec{E}_{r\parallel}$ and $\vec{E}_{r\perp}$ are the parallel and perpendicular components of the light waves reflected in the direction $\vec{k}_r$.

In Fig. 10.5(b), we present a cross section of the reflected electric field as viewed from the camera's position, where the surface normal $\vec{N}$ is aligned with the y-axis. In this view, $\vec{E}_{r\|}$ vibrates along the y-axis because it is co-planar with vector $\vec{N}$. The perpendicular component $\vec{E}_{r\perp}$ lies in the x-axis as it is orthogonal to $\vec{E}_{r\|}$. In the figure, $\vec{P}$ is the orientation of the polariser's transmission axis.

Fresnel Reflection Ratio

According to Fresnel reflection theory, the radiance of the reflected light is attenuated due to the change of the wave velocity after reflection. The attenuation ratio can be decomposed into components in the planes parallel and perpendicular to the incident plane. We refer to these components as the Fresnel reflection coefficients $F_\perp$ and $F_\|$, respectively, and they are given by

$$F_\perp(\theta_i, \eta) = \left(\frac{a - \cos\theta_i}{a + \cos\theta_i}\right)^2, \tag{10.14}$$

$$F_\|(\theta_i, \eta) = F_\perp(\theta_i, \eta) \times \left(1 + \left(\frac{a - \sin\theta_i \tan\theta_i}{a + \sin\theta_i \tan\theta_i}\right)^2\right), \tag{10.15}$$

where $a = (\eta^2 - \sin^2\theta_i)^{\frac{1}{2}}$.

From these definitions, we note that $F_\perp \geq F_\|$, i.e. the radiance of the perpendicular component is always greater than or equal to that of the parallel component.

Let us denote the radiance of the parallel and perpendicular reflection components as $I_{r\|}$ and $I_{r\perp}$, respectively. They are related to the respective Fresnel reflection coefficients and incoming irradiance as follows:

$$I_{r\|} = F_\| I_{i\|}, \tag{10.16}$$

$$I_{r\perp} = F_\perp I_{i\perp}, \tag{10.17}$$

where $I_{i\|}$ and $I_{i\perp}$ are the parallel and perpendicular components of the incident irradiance.

Since we assume that the incident light is unpolarised, its electric field vector oscillates in all directions perpendicular to the propagation direction with equal energy and probability. This leads to $E_{i\|} = E_{i\perp}$, i.e. the perpendicular and parallel components of the incident light are equal in amplitude. In other words, $I_{i\|} = I_{i\perp}$ as the radiance of these components is proportional to the square of the electric field amplitude. With this equality, we take the ratio of the left-hand and right-hand sides of Eqs. (10.17) and (10.16) to obtain

$$\frac{I_{r\|}}{I_{r\perp}} = \frac{F_\|}{F_\perp}. \tag{10.18}$$

We refer to the right-hand side of Eq. (10.18) as the Fresnel reflection ratio, which is a function of the reflection angle and the material refractive index.

Phase of Specular Polarisation

Here we aim to relate the phase of specular polarisation ϕ to the surface geometry. Let us consider the diagram in Fig. 10.5(b). Since the incident light ray $\vec{k}_i$ is randomly polarised, its orthogonal components $\vec{E}_{i\|}$ and $\vec{E}_{i\perp}$ are incoherent, i.e. their phase difference varies randomly. Since we focus our study on dielectric materials, there is no phase shift between the perpendicular and parallel components upon reflection (Born and Wolf 1999). Therefore, the reflection components $\vec{E}_{r\|}$ and $\vec{E}_{r\perp}$ are also incoherent, i.e. their phase difference ε is $\langle\cos\varepsilon\rangle_t = 0$, where $\langle.\rangle_t$ is a notation for an average quantity over time.

Thus, substituting the condition $\langle\cos\varepsilon\rangle_t = 0$ into Eq. (10.4), the reflected radiance becomes

$$I_\vartheta = \frac{1}{2}\left(I_{\text{un}} + E_{r\perp}^2 \cos^2\vartheta + E_{r\|}^2 \sin^2\vartheta\right)$$

$$= \frac{1}{2}\left(I_{\text{un}} + I_{r\perp}\cos^2\vartheta + I_{r\|}\sin^2\vartheta\right). \tag{10.19}$$

The formula above effectively means that the radiance of a reflected polarisation component is a convex combination of the parallel and perpendicular reflection components. In fact, Eq. (10.19) is consistent with the Fresnel reflectance model introduced by Wolff and Boult (1991).

Since $F_\perp \geq F_\|$, from Eq. (10.18), we deduce that $I_{r\perp} \geq I_{r\|}$. As a result, the maximal reflected radiance occurs when $|\cos\vartheta| = 1$, i.e. $\vartheta = 0(\bmod \pi)$. With reference to the diagram in Fig. 10.5(b), this is the case when the polariser's transmission axis is perpendicular to the plane of incidence/reflection. In contrast, the minimal reflection is observed when the polariser transmission axis lies on the plane of reflection. In this case, the transmission axis coincides with the tilt direction of the surface normal when observed from the viewpoint. From Fig. 10.4, we also note that the TRS reaches its peak when the polariser angle coincides with the phase angle. Therefore, the azimuth (tilt) direction of the surface normal is perpendicular to the phase of specular polarisation.

10.2.2 Polarisation upon Diffuse Reflection

Recall that, in several cases, diffuse reflection is the result of multiple scattering from micro-facets (Oren and Nayar 1995) on an object's surface. On the other hand, for translucent or multi-layered materials, diffuse reflection is attributed to incident light penetrating the surface, scattering within the dielectric body and refracting back into the transmission medium. Here, we may assume that, after penetration, light is largely depolarised by the internal scattering process within the dielectric body. Later, the refraction of scattered light from the dielectric body through the material-air boundary induces polarisation. Therefore, the following theory of polarisation, which is subject to Snell's law of refraction and Fresnel reflection theory, applies to translucent or multi-layered materials with subsurface scattering.

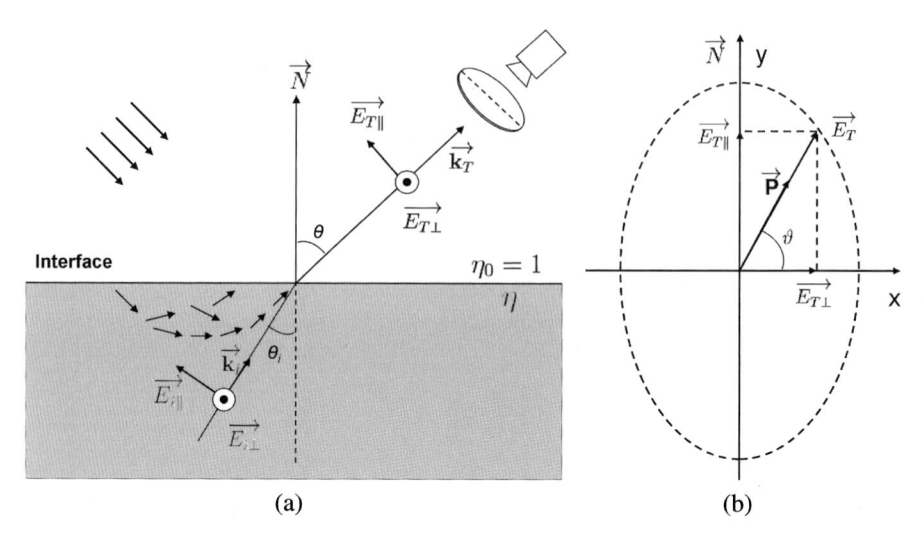

Fig. 10.6 (a) Polarisation upon diffuse reflection from a dielectric surface with a normal vector $\vec{N}$ and refractive index η. The incident light waves penetrate the surface, scatter inside the dielectric body and finally refract through the material-air boundary. The electric fields of the waves incident on and transmitted through the material-air interface are represented by vectors rotating in a plane perpendicular to the propagation direction. These electric field vectors, in turn, can be decomposed into two orthogonal components which are parallel and perpendicular to the plane of emittance, which contains both the surface normal and the emittance direction. The vectors $\vec{E}_{i\|}$ and $\vec{E}_{i\perp}$ are the components of the internal incident field on the material-air surface boundary before refraction. The vectors $\vec{E}_{T\|}$ and $\vec{E}_{T\perp}$ are the components of the wave emerging from the surface after refraction. (b) The emitted electric field, where the surface normal $\vec{N}$ is aligned with the y-axis. $\vec{E}_{T\|}$ vibrates along the y-axis because it is in the plane of reflection. The perpendicular component $\vec{E}_{T\perp}$ lies in the x-axis, and $\vec{P}$ is the orientation of the polariser's transmission axis

Figure 10.6(a) shows a diagram depicting the diffuse reflection process described above. Here, refraction is the result of the change of velocity as the internally scattered ray $\vec{k}_i$ travels from the material body with refractive index η relative to the air whose index is unity. Assume that this ray is incident on the surface boundary at an angle θ_i. After emerging from the surface, the transmitted ray $\vec{k}_T$ is bent at an emittance (reflection) angle θ with respect to the surface normal $\vec{N}$. The propagation direction of the transmitted ray and the surface normal both lie in the plane of reflection, as shown in Fig. 10.6(a). Note that θ is also the zenith angle of the surface normal with respect to the viewing direction.

Note that the electric field vectors associated with the rays $\vec{k}_i$ and $\vec{k}_T$ are always perpendicular to their propagation directions. Furthermore, these electric field vectors can be decomposed into two orthogonal harmonic components, i.e. one in the plane of reflection and the other one perpendicular to it. In Fig. 10.6(a), we have denoted $\vec{E}_{i\|}$ and $\vec{E}_{i\perp}$ as the parallel and perpendicular components of the electric field associated with $\vec{k}_i$. Similarly, $\vec{E}_{T\|}$ and $\vec{E}_{T\perp}$ are the parallel and perpendicular components of the transmitted waves travelling in the direction $\vec{k}_T$.

10.2 Polarisation upon Reflection

Furthermore, Fig. 10.6(b) presents a cross section of the emitted electric field as viewed from the camera's position, where the surface normal $\vec{N}$ is aligned with the y-axis. In this view, $\vec{E}_{T\parallel}$ vibrates along the y-axis because it is co-planar with $\vec{N}$. The perpendicular component $\vec{E}_{T\perp}$ lies in the x-axis as it is orthogonal to $\vec{E}_{T\parallel}$. In the figure, $\vec{P}$ is the orientation of the polariser's transmission axis.

Fresnel Transmission Ratio

We now formulate the radiance of the emitted polarisation components as a result of the diffuse reflection process above. To commence, let us denote the radiance of the parallel and perpendicular components emitted at the material surface as $I_{T\parallel}$ and $I_{T\perp}$ respectively. By the law of energy conservation, the light power transmitted through the material-air boundary is complementary to the reflected components. Therefore, the transmission coefficients of light through this boundary are $1 - F_\perp$ and $1 - F_\parallel$, according to the Fresnel reflection theory. As a result, the parallel and perpendicular components of the emitted light are derived as follows:

$$I_{T\perp} = (1 - F_\perp)I_{i\perp}, \qquad (10.20)$$

$$I_{T\parallel} = (1 - F_\parallel)I_{i\parallel}, \qquad (10.21)$$

where $I_{i\parallel}$ and $I_{i\perp}$ are, respectively, the irradiance of the parallel and perpendicular components of the internally scattered light when it arrives at the material-air boundary.

Here we assume that the light scattered inside the dielectric body is unpolarised. Therefore, its electric field vector oscillates in all directions perpendicular to the propagation direction with equal energy and probability. This means that the magnitudes $E_{i\parallel}$ and $E_{i\perp}$ of the perpendicular and parallel components of the internally scattered ray $\mathbf{k}_i$ are equal, i.e. $E_{i\parallel} = E_{i\perp}$. This, in turns, leads to $I_{i\parallel} = I_{i\perp}$ as the radiance is proportional to the square of the electric field amplitude. With this in mind, we take the ratio of the left-hand and right-hand sides of Eqs. (10.20) and (10.21) to obtain

$$\frac{I_{T\perp}}{I_{T\parallel}} = \frac{1 - F_\perp}{1 - F_\parallel}, \qquad (10.22)$$

where we refer to the right-hand side of Eq. (10.22) as the Fresnel transmission ratio, which is a function of the normal zenith angle and the refractive index.

Phase of Diffuse Polarisation

We further explore the relationship between the phase of polarisation ϕ resulting from subsurface scattering and the surface geometry. To do this, we consider the sketch in Fig. 10.6(b). Note that, since the internal light ray $\vec{\mathbf{k}}_i$ in Fig. 10.6(a) is scattered randomly (depolarised), its orthogonal components $\vec{E}_{i\|}$ and $\vec{E}_{i\perp}$ are incoherent, i.e. their phase difference varies randomly. For dielectric materials, the phase of polarised light is preserved upon refraction (Born and Wolf 1999). Therefore, the components $\vec{E}_{T\|}$ and $\vec{E}_{T\perp}$ of the transmitted light are also incoherent after refraction through the material-air interface. This imposes a condition on their phase difference ε that $\langle\cos\varepsilon\rangle_t = 0$.

By substituting the equality $\langle\cos\varepsilon\rangle_t = 0$ into Eq. (10.4), we obtain the transmitted radiance in a similar form to Eq. (10.19) as follows:

$$I_\vartheta = \frac{1}{2}\left(I_{\text{un}} + E_{T\perp}^2 \cos^2\vartheta + E_{T\|}^2 \sin^2\vartheta\right)$$
$$= \frac{1}{2}\left(I_{\text{un}} + I_{T\perp} \cos^2\vartheta + I_{T\|} \sin^2\vartheta\right). \qquad (10.23)$$

The formula above effectively means that the radiance of a polarisation component oriented at an arbitrary angle is a convex combination of the parallel and perpendicular components of reflection. In fact, Eq. (10.23) is consistent with the Fresnel reflectance model introduced by Wolff and Boult (1991). In addition, since $F_\perp \geq F_\|$, we can deduce from Eq. (10.22) that $I_{T\perp} \leq I_{T\|}$.

As a result, the maximal transmitted radiance occurs when $\vartheta = \pm\frac{\pi}{2}(\bmod\pi)$, i.e. when the transmission axis lies in the plane of reflection. Moreover, since the plane of reflection contains the surface normal vector, this orientation direction indicates the tilt direction of the surface normal as observed from the viewpoint. Also note that, according to Eq. (10.5), the maximal value of the transmitted radiance is attained when the polariser's angle coincides with the phase angle. Therefore, the phase angle coincides with the azimuth (tilt) angle of the surface normal or differs from it by π.

10.3 Polarimetric Reflection Models

Little work has been done on the modelling of polarisation upon surface reflection. Wolff and Boult derived a polarisation reflectance model (Wolff and Boult 1991) based on the Fresnel reflection theory and demonstrated its application to the classification of dielectrics and metals. In addition, although the Wolff reflectance model (Wolff 1994) was originally designed for unpolarised reflection, it can be decomposed into polarisation components and is applicable to polarised light sources.

In contrast, a vast body of literature has been dedicated to the modelling of unpolarised reflectance for rough surfaces. Oren and Nayar (1995) generalised the Lambertian model to account for variation of reflected light with respect to the viewing direction. In an earlier work, Torrance and Sparrow (1967) formulated a reflectance model based on the Fresnel reflection theory to explain the off-specular spike phenomenon, in which the specular spike occurred at a viewing angle larger than the specular reflection angle. Beckmann (Beckmann and Spizzichino 1963) modelled the surface reflectance as a wave scattering process. However, none of the latter models takes into account polarisation effects.

10.3.1 A Polarimetric Reflection Model for Rough Surfaces

In this section, we relate the components of specular polarisation to surface geometry and several photometric invariants of the material. In particular, we derive expressions of the parallel and perpendicular polarisation components, which are also the maximal and minimal components, i.e. $I_{\min}$ and $I_{\max}$, described in Sect. 10.1.1. This derivation establishes a link between polarimetric components and the diffuse and specular components of the corresponding unpolarised image. The modelling is general and applies to spatially varying illumination sources. Here we depart from the conventional form of the Torrance-Sparrow reflectance model (Torrance and Sparrow 1967) described in Chap. 4. Making use of this model, we derive a formulation of the $I_{\min}$ and $I_{\max}$ polarisation components in a similar manner to that in Torrance et al. (1966). Through this formulation, we relate the amplitude of the TRS curve to the power spectrum of the illumination.

According to the Torrance and Sparrow model, the reflection of light from rough surfaces at moderate and large angles of incidence does not adhere to a perfect diffuse distribution (Torrance and Sparrow 1967). Specifically, the maximal reflected radiance occurs at an angle larger than the specular angle. To account for this phenomenon, Torrance and Sparrow model rough surfaces as being composed of mirror-like micro-facets whose slope is randomly distributed. Note that this model also relates specular reflection to the source radiance spectrum for both direct illumination and inter-reflection.

Let us consider an incoming radiant flux incident on the surface under study with a spectral power $L(u, \lambda)$ at wavelength λ and pixel u. To preserve generality, we consider source radiance originating from either illuminants or reflecting surfaces, i.e. inter-reflection, with spatially varying power spectrum. Further, given an incident angle θ_i with the mean local surface normal $\vec{N}$ and the diffuse surface reflectance $S(u, \lambda)$, the total unpolarised radiance $I_{\text{un}}(u, \lambda)$ reflected from the surface is given by

$$I_{\text{un}}(u, \lambda) = L(u, \lambda) \cos \theta_i \left(w_d S(u, \lambda) + w_s R_s(u, \lambda) \right). \quad (10.24)$$

In Eq. (10.24), the first term is the diffuse reflectance component, which obeys Lambert's cosine law with a weight w_d. In fact, the original model further includes an ambient component (Torrance and Sparrow 1967). Here, we incorporate ambient light into the diffuse component since ambient reflection is widely regarded as unpolarised light and only causes a global additive shift in the reflectance spectrum of the scene (Wyszecki and Stiles 2000). The second term in Eq. (10.24) accounts for the specular components of direct reflection and inter-reflection making use of a weight w_s.

Further, under the assumption of a perfect polariser with constant transmission over its surface, the unpolarised radiance $I_{\text{un}}(u, \lambda)$ reflected from the scene is equal to twice the average radiance of the polarimetric components over all possible polariser orientations. Therefore, $I_{\text{un}}(u, \lambda) = I_{\max}(u, \lambda) + I_{\min}(u, \lambda)$. This is consistent with Fig. 10.4, where $\frac{1}{2} I_{\text{un}}(u, \lambda)$ is a flat line across all the polariser angles.

As mentioned earlier, the specular term in Eq. (10.24) is contributed by mirror-like micro-facets oriented in the bisector between the sources and viewing direction. Hence, following Torrance and Sparrow (1967), we can write the off-specular reflection term as a function of the micro-facet distribution, making use of the Fresnel reflection theory. In summary, the specular reflectance R_s is given by

$$R_s(u, \lambda) = \frac{F(\psi, n(\lambda)) P(\alpha) G}{\pi \cos\theta_i \cos\theta_s}. \tag{10.25}$$

In Eq. (10.25), θ_s denotes the scattering angle between the mean normal $\vec{N}$ and the viewing direction $\vec{V}$. Treating these micro-facets as perfect mirrors, Torrance and Sparrow related specular reflection to the Fresnel reflection coefficient $F(\psi, n(\lambda))$, which depends on the half-angle ψ between the light direction $\vec{L}$ and viewing direction $\vec{V}$, and the material refractive index $n(\lambda)$.

In addition, the specular component depends on the probability of micro-facet normals oriented halfway between the light source and the viewing directions, which we denote $P(\alpha)$, where α is the angle between the micro-facet normal and the mean surface normal. The geometric factor G accounts for the masking and shadowing of one facet by another. Thus, by combining Eqs. (10.24) and (10.25), the total unpolarised surface radiance is expressed as

$$I_{\text{un}}(u, \lambda) = I_d(u, \lambda) + \frac{w_s P(\alpha) G}{\pi \cos\theta_s} F(\psi, n(\lambda)) L(u, \lambda). \tag{10.26}$$

Note that, in Eq. (10.26), the ambient and diffuse components have been combined into a single term $I_d(u, \lambda) = w_d \cos\theta_i S(u, \lambda) L(u, \lambda)$, which is generally unpolarised. In the second term, the source radiance $L(u, \lambda)$ can originate from either an illuminant or a reflecting surface.

Moreover, since the diffuse reflection component is generally unpolarised, half of its power, i.e. I_d, is attenuated after being transmitted through a linear polariser, irrespective of the orientation of its transmission axis. On the other hand, specularly reflected and specularly inter-reflected light is polarised after leaving the reflecting surface. This is why the transmitted radiance of these components observed through a linear polariser varies as a sinusoidal function of the polariser angular orientation as shown in Fig. 10.4. For specular polarisation, the maximal and minimal intensities, i.e. $I_{\max}$ and $I_{\min}$, are observed at the polarisation orientations perpendicular and parallel to the plane of reflection, respectively. Denoting $I_\perp$ and $I_\parallel$ as the transmitted radiance of the polarisation components at these orientations, we have $I_{\max} = I_\perp$ and $I_{\min} = I_\parallel$.

We further note that the unpolarised Fresnel reflection coefficient can be decomposed into a perpendicular and a parallel component as follows:

$$F(\psi, n(\lambda)) = \frac{1}{2} \left(F_\perp(\psi, n(\lambda)) + F_\parallel(\psi, n(\lambda)) \right).$$

10.3 Polarimetric Reflection Models

Thus, by decomposing Eq. (10.26) into a perpendicular component $I_\perp$ and a parallel component $I_\parallel$, we arrive at the following expressions:

$$I_{\max}(u,\lambda) = I_\perp(u,\lambda) = \frac{1}{2}\left(I_d(u,\lambda) + \frac{w_s P(\alpha) G}{\pi \cos\theta_s} F_\perp(\psi, n(\lambda)) L(u,\lambda)\right), \quad (10.27)$$

$$I_{\min}(u,\lambda) = I_\parallel(u,\lambda) = \frac{1}{2}\left(I_d(u,\lambda) + \frac{w_s P(\alpha) G}{\pi \cos\theta_s} F_\parallel(\psi, n(\lambda)) L(u,\lambda)\right). \quad (10.28)$$

On the right-hand sides of Eqs. (10.27) and (10.28), the diffuse components I_d are attenuated by half when going through the linear polariser, whereas the specular reflection is governed by the parallel and perpendicular Fresnel coefficients. It is worth noting that our derivation here is consistent with the polarisation reflectance model formulated by Wolff and Boult (1991) based on the Fresnel reflection theory.

Due to the equality $I_d(u,\lambda) = w_d \cos\theta_i S(u,\lambda) L(u,\lambda)$ in Eq. (10.26), we can rewrite Eqs. (10.27) and (10.28) as

$$I_{\max}(u,\lambda) = g^*(u) D(u,\lambda) + k^*(u) \mathcal{K}_\perp(u,\lambda), \quad (10.29)$$

$$I_{\min}(u,\lambda) = g^*(u) D(u,\lambda) + k^*(u) \mathcal{K}_\parallel(u,\lambda), \quad (10.30)$$

where the terms in Eqs. (10.29) and (10.30) are defined as

$$g^*(u) = \frac{1}{2} w_d \cos\theta_i,$$

$$k^*(u) = \frac{w_s P(\alpha) G}{2\pi \cos\theta_s},$$

$$D(u,\lambda) = S(u,\lambda) L(u,\lambda),$$

$$\mathcal{K}_\perp(u,\lambda) = F_\perp(\psi, n(\lambda)) L(u,\lambda),$$

$$\mathcal{K}_\parallel(u,\lambda) = F_\parallel(\psi, n(\lambda)) L(u,\lambda).$$

Further, the model can be rewritten in the following vectorial form:

$$\mathbf{I}_{\max}(u) = g^*(u) \mathbf{D}(u) + k^*(u) \boldsymbol{\mathcal{K}}_\perp(u), \quad (10.31)$$

$$\mathbf{I}_{\min}(u) = g^*(u) \mathbf{D}(u) + k^*(u) \boldsymbol{\mathcal{K}}_\parallel(u), \quad (10.32)$$

with the following vectorial notation:

$$\mathbf{D}(u) \triangleq [D(u,\lambda_1), \ldots D(u,\lambda_l)]^T,$$

$$\mathbf{I}_{\max}(u) \triangleq [I_{\max}(u,\lambda_1), \ldots, I_{\max}(u,\lambda_l)]^T,$$

$$\mathbf{I}_{\min}(u) \triangleq [I_{\min}(u,\lambda_1), \ldots, I_{\min}(u,\lambda_l)]^T,$$

$$\boldsymbol{\mathcal{K}}_\perp(u) \triangleq [\mathcal{K}_\perp(u,\lambda_1), \ldots \mathcal{K}_\perp(u,\lambda_l)]^T,$$

Fig. 10.7 The maximal and minimal polarisation components $\mathbf{I}_{\max}(u)$ and $\mathbf{I}_{\min}(u)$ belong to two-dimensional subspaces $P_\perp$ and $P_\parallel$. $P_\perp$ is spanned by the diffuse radiance spectrum $\mathbf{D}(u)$ and $\mathcal{K}_\perp(u)$, whereas $P_\parallel$ is spanned by $\mathbf{D}(u)$ and $\mathcal{K}_\parallel(u)$

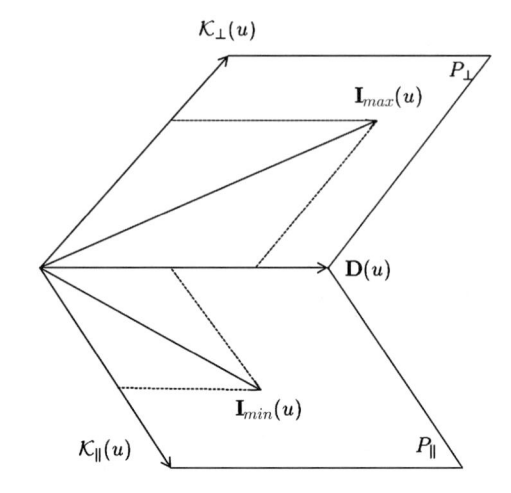

$$\mathcal{K}_\parallel(u) \triangleq \left[\mathcal{K}_\parallel(u, \lambda_1), \ldots \mathcal{K}_\parallel(u, \lambda_l)\right]^T,$$

where l is the number of sampled wavelengths of the captured images.

Equations (10.31) and (10.32) give rise to a geometrical interpretation of our polarimetric reflection model. Let us consider the space of l-dimensional discrete radiance spectra. In this space, the maximal and minimal polarimetric components $\mathbf{I}_{\max}(u)$ and $\mathbf{I}_{\min}(u)$ lie in two two-dimensional subspaces, each of which is spanned by the diffuse radiance spectrum $\mathbf{D}(u)$ and the component-wise products of the illuminant spectrum and the corresponding wavelength-dependent Fresnel reflection coefficients ($\mathcal{K}_\perp(u)$ and $\mathcal{K}_\parallel(u)$). Figure 10.7 depicts these linear subspaces as the two planes $P_\perp$ and $P_\parallel$. We note that these two planes intersect at the diffuse radiance vector $\mathbf{D}(u)$. In the following section, we will relate the geometric interpretation in Fig. 10.7 to the dichromatic reflection model introduced by Shafer (1985).

10.3.2 Relation to the Dichromatic Reflection Model

Due to the fact that $I_{\text{un}}(u, \lambda) = I_{\max}(u, \lambda) + I_{\min}(u, \lambda)$, we can combine Eqs. (10.29) and (10.30) to arrive at

$$I_{\text{un}}(u, \lambda) = \bigl(g(u)D(u, \lambda) + k(u)\bigr)L(u, \lambda), \qquad (10.33)$$

where

$$\begin{aligned} g(u) &= 2g^*(u), \\ k(u) &= 2k^*(u)\langle F(\psi, n(\lambda))\rangle_\lambda, \end{aligned} \qquad (10.34)$$

and $\langle . \rangle_\lambda$ denotes the average across the wavelength domain. Here, the specular coefficient in Eq. (10.34) can be estimated as the spectral average of the Fresnel reflection coefficient since the latter is assumed to vary slowly with respect to the wavelength.

10.3 Polarimetric Reflection Models

The formulation in Eq. (10.33) is reminiscent of the dichromatic reflection model (Shafer 1985), where the diffuse reflection component consists of a wavelength-independent geometric factor $g(u)$ and the normalised diffuse radiance of the object material $D(u, \lambda)$. On the other hand, the specular coefficient $k(u)$ can be considered wavelength independent in most cases, because the Fresnel term $F(\psi, n(\lambda))$ only varies slightly with respect to the wavelength and the other factors are wavelength independent. This means that the dichromatic reflection model in Shafer (1985) is indeed a particular case of the Torrance and Sparrow model under the assumption that the Fresnel reflection coefficients are almost constant across the wavelength spectrum. Furthermore, under the same assumption, $\mathcal{K}_\perp(u)$ and $\mathcal{K}_\parallel(u)$ are both collinear to the light spectrum $\mathbf{L}(u) \triangleq [L(u, \lambda_1), \ldots L(u, \lambda_l)]^T$. In this case, the two polarimetric planes $P_\perp$ and $P_\parallel$ in Fig. 10.7 degenerate into the dichromatic plane spanned by the light spectrum and the unpolarised diffuse component.

10.3.3 Reflection Component Separation

Among the large number of methods aiming to recover local estimates of the colour of specular reflection (Nayar et al. 1997; Sato and Ikeuchi 1994; Novak and Shafer 1992), several methods employ polarisation methods for reflection component separation by modelling wavelength-dependent inter-reflection on surfaces. Specific to the use of polarimetric data, Nayar et al. (1997) used colour and polarisation information to obtain constraints on the reflection components at each image point. Umeyama and Godin (2004) separated the diffuse from the specular component by maximising the probabilistic independence between these two components via independent component analysis (ICA). In this section, we utilise the formulation of the polarimetric components $I_{\max}$ and $I_{\min}$ given in Eqs. (10.27) and (10.28) for the purpose of separating off-specular reflection. Furthermore, we explore the methods in Nayar et al. (1997) and Umeyama and Godin (2004).

A Propagation and Optimisation Method

In this section, we employ the formulation of polarimetric components $I_{\max}$ and $I_{\min}$ in Eqs. (10.27) and (10.28) to separate off-specular reflection using polarisation images. To commence, we estimate the illumination spectrum per pixel. With the estimated illuminant spectra, we propagate the separation process from purely diffuse pixels to those having a specular component.

By subtracting Eq. (10.28) from Eq. (10.27), the illuminant spectrum is related to the polarisation components $I_{\max}$ and $I_{\min}$ as

$$I_{\max}(u, \lambda) - I_{\min}(u, \lambda) = \frac{w_s P(\alpha) G}{\pi \cos \theta_s} L(u, \lambda) \times \left(F_\perp(\psi, n(\lambda)) - F_\parallel(\psi, n(\lambda)) \right). \tag{10.35}$$

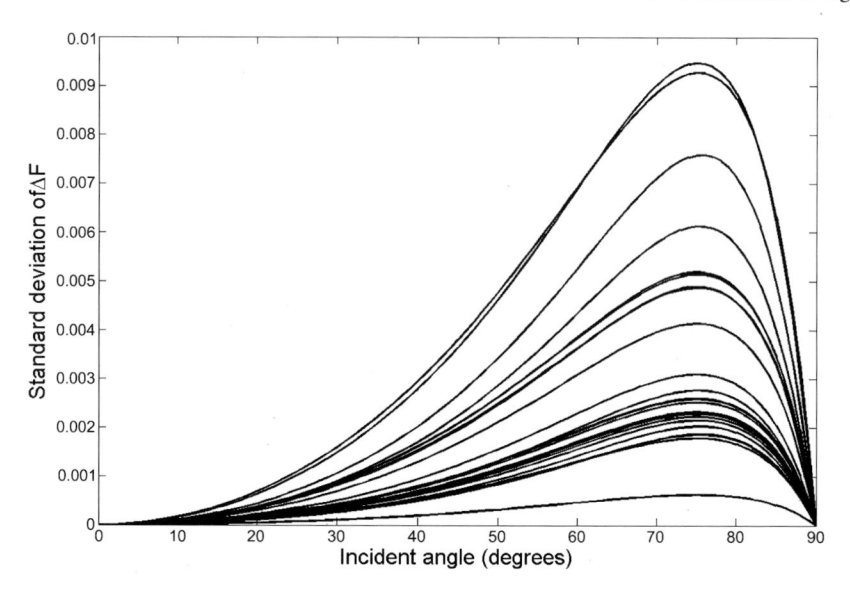

Fig. 10.8 Variation of the difference between the Fresnel reflection components $\Delta F \triangleq F_\perp - F_\parallel$ across the visible spectrum for 24 different materials

It is often the case that the Fresnel coefficients in Eq. (10.35) vary slowly with respect to the wavelength. As a consequence, so does their difference. To confirm this hypothesis, we compute the quantity $F_\perp(\psi, n(\lambda)) - F_\parallel(\psi, n(\lambda))$ at various angles of incidence between 0 and 90 degrees across the visible range for 24 plastic and liquid materials reported by Kasarova et al. (2007). The standard deviation of this quantity over the wavelengths is plotted in Fig. 10.8. In the figure, each material is represented by a trace that shows the standard deviation at various incident angles. We observe that the Fresnel component difference $F_\perp(\psi, n(\lambda)) - F_\parallel(\psi, n(\lambda))$ varies slowly across wavelengths, with a negligible standard deviation of no more than 0.01. This observation justifies the applicability of our component reflection separation problem for both narrowband spectral images and trichromatic images.

By the observation above, we can assume that the term $F_\perp(\psi, n(\lambda)) - F_\parallel(\psi, n(\lambda))$ in Eq. (10.35) is constant with respect to the wavelength. This implies that, if partial polarisation occurs, i.e. $F_\perp(\psi, n(\lambda)) > F_\parallel(\psi, n(\lambda))$, then the power spectrum of the light $L(u, \lambda)$ is proportional to the polarisation amplitude spectrum $\hat{L}(u, \lambda) \triangleq I_{\max}(u, \lambda) - I_{\min}(u, \lambda)$, i.e. $L(u, \lambda) = h(u)\hat{L}(u, \lambda)$, where $h(u)$ is a pixel-indexed scaling factor. More importantly, since $\hat{L}(u, \lambda)$ can be computed per pixel, this quantity serves as an estimate of the spatially varying source spectrum.

We now aim to recover the terms $g(u)$, $D(u, \lambda)$ and $k(u)$ in Eq. (10.33) from the unpolarised image radiance $I_{un}(u, \lambda) = I_{\max}(u, \lambda) + I_{\min}(u, \lambda)$. Here, we rely on the assumption that the diffuse radiance $D(u, \lambda)$ varies much more slowly within a spatial neighbourhood than the specular and the inter-reflection components. This stems from the fact that a local neighbourhood of the scene is generally composed of the same material. Under this assumption, we can estimate the diffuse radiance

10.3 Polarimetric Reflection Models

Algorithm 10.1 *ComponentSeparation*($I_{\max}, I_{\min}, t_{\mathrm{amp}}$)

Require: $\mathbf{I}_{\max}(u), \mathbf{I}_{\min}(u)$: The maximum and minimum polarisation components for all pixels u

Ensure: $\mathbf{I}_d(u), \mathbf{I}_s(u)$: the diffuse and specular components for all pixels u

1: $\mathbf{I}_{\mathrm{un}}(u) \leftarrow \frac{1}{2}(\mathbf{I}_{\max}(u) + \mathbf{I}_{\min}(u)) \forall u$
2: $\mathbf{L}(u) \leftarrow (\mathbf{I}_{\max}(u) - \mathbf{I}_{\min}(u)) \forall u$
  //Initialisation of the map of diffuse pixels
3: **for all** pixel u **do**
4:   **if** $\|\mathbf{L}(u)\| \leq t_{\mathrm{amp}}$ **then**
5:     $explore(u) \leftarrow$ **true**
6:     $\mathbf{I}_d(u) \leftarrow \mathbf{I}_{\mathrm{un}}(u)$
7:     $\mathbf{I}_s(u) \leftarrow 0$
8:   **end if**
9: **end for**
10: **while** unexplored pixels remain **do**
11:   **for all** unexplored pixel u **do**
12:     **if** a number of neighbouring pixels $v_1, \ldots v_m$ of u have been explored **then**
13:       Estimate diffuse radiance $\mathbf{D}(u)$ as in Eq. (10.36)
14:       Solve problem (10.38) for $g(u)$ and $k(u)$
15:       $\mathbf{I}_d(u) \leftarrow g(u)\mathbf{D}(u)$ //Diffuse component
16:       $\mathbf{I}_s(u) \leftarrow k(u)\mathbf{L}(u)$ //Specular reflection & inter-reflection component
17:       $explore(u) \leftarrow$ **true**
18:     **end if**
19:   **end for**
20: **end while**
21: **return** $\mathbf{I}_d(u), \mathbf{I}_s(u) \forall$ pixel u

---

at a specular pixel as a weighted average of its neighbouring diffuse pixels. This idea lends itself to the development of an algorithm that propagates diffuse radiance information from diffuse to specular pixels.

Algorithm 10.1 proceeds as follows. The initial map of diffuse pixels is established from the polarisation amplitude at each pixel. Initially, a pixel u is considered diffuse if the norm of the polarisation amplitude vector $\mathbf{L}(u)$ does not exceed a preset threshold t_{amp}. At pixels satisfying this condition, we recover the diffuse radiance directly as $I_{\mathrm{un}}(u, \lambda) = \frac{1}{2}(I_{\max}(u, \lambda) + I_{\min}(u, \lambda))$, setting the specular component to zero. Specular pixels, i.e. those whose polarisation amplitude vector $\mathbf{L}(u)$ is large in magnitude, are marked as candidates for processing.

With the initial diffuse pixels at hand, we propagate diffuse radiance information throughout the image. Suppose that we aim to recover the specular component at a pixel u, with some of its neighbourhood pixels $v_1, \ldots, v_m$ already processed or considered diffuse from the start. Let us denote the diffuse components at these neighbouring pixels as $D(v_1, \lambda), \ldots, D(v_m, \lambda)$. The diffuse radiance $D(u, \lambda)$ at the pixel u is estimated as a weighted average of the known diffuse radiance in the neighbourhood as follows:

$$D(u, \lambda) = \frac{\sum_{j=1}^{m} \omega_j D(v_j, \lambda)}{\sum_{j=1}^{m} \omega_j}. \quad (10.36)$$

In Eq. (10.36), the weight ω_j is a similarity measure between the diffuse radiance spectrum at the pixel u and v_j. In the case where u and v_j share the same diffuse radiance, $\mathbf{D}(v_j) \triangleq [D(v_j, \lambda_1), \ldots, D(v_j, \lambda_l)]^T$ lies on the plane spanned by the vectors $\mathbf{I}_{\max}(u)$ and $\mathbf{I}_{\min}(u)$. However, this may not be the case due to noise corruption and variations of albedo in the image. Hence, the weight ω_j can be quantified as the cosine of the angle between $\mathbf{D}(v_j)$ and its projection onto the plane spanned by $\mathbf{I}_{\max}(u)$ and $\mathbf{I}_{\min}(u)$. Since the linear projection matrix $\mathbf{Q}$ onto this plane is given by $\mathbf{Q} = \mathbf{A}(\mathbf{A}^T \mathbf{A})^{-1} \mathbf{A}^T$, where $\mathbf{A} = [\mathbf{I}_{\max}(u), \mathbf{I}_{\min}(u)]$, the orthogonal projection of $\mathbf{D}(v_j)$ onto this plane is $\mathbf{QD}(v_j)$. Following this, the weight ω_j is computed as follows:

$$\omega_j = \frac{\mathbf{D}(v_j) \cdot \mathbf{QD}(v_j)}{\|\mathbf{D}(v_j)\| \|\mathbf{QD}(v_j)\|}, \quad (10.37)$$

where $\cdot$ denotes the dot product and $\|\cdot\|$ denotes the L^2-norm of a vector.

With the radiance spectra of the diffuse component $D(u, .)$ and the illumination $\hat{L}(u, .)$ at hand, the separation problem can be reduced to recovering the diffuse and specular coefficients $g(u)$ and $k(u)$. This is effected by solving the following non-negative least-squares optimisation problem:

$$\text{minimise } \left\| \mathbf{I}_{\text{un}}(u) - [\mathbf{D}(u) | \mathbf{L}(u)] \begin{bmatrix} g(u) \\ k(u) \end{bmatrix} \right\|$$

$$\text{subject to } [g(u) k(u)]^T \geq \mathbf{0}, \quad (10.38)$$

where the vectors $\mathbf{D}(u) = [D(u, \lambda_1), \ldots D(u, \lambda_l)]^T$, $\mathbf{L}(u)$ and $\mathbf{I}_{\text{un}}(u) = [I_{\text{un}}(u, \lambda_1), \ldots I_{\text{un}}(u, \lambda_l)]^T$ are known. Once the solutions for $g(u)$ and $k(u)$ are obtained, the diffuse and specular reflection components can be reconstructed by making use of the vectors $\mathbf{D}(u)$ and $\mathbf{L}(u)$. Note that the optimisation problem (10.38) is well-constrained if the number of bands is $l \geq 3$ and $\mathbf{D}(u)$ and $\mathbf{L}(u)$ are non-collinear.

The entire component separation process is summarised in the pseudo-code shown in Algorithm 10.1. The algorithm takes at input the maximum and the minimum polarisation components, and the threshold t_{amp}. It delivers at output the diffuse and specular reflection images. In the pseudo-code, the initialisation of the diffuse pixel map is done in Lines 3–9, while the diffuse radiance information propagation is performed at Lines 10–20. Note that our method only employs a single threshold for the polarisation magnitude.

A Geometric Method

In Nayar et al. (1997), the authors considered polarisation upon specular reflection from dielectric surfaces; i.e. only the specular component causes the polarisation of

10.3 Polarimetric Reflection Models

the reflected light. Under this assumption, they expressed the transmitted radiance as a linear combination of the form

$$I_\vartheta(u,\lambda) = I_d(u,\lambda) + I_s(u,\lambda) + I_{sv}(u,\lambda) \cos\left(2(\vartheta - \phi(u,\lambda))\right), \tag{10.39}$$

where $I_d(u,\lambda)$ and $I_s(u,\lambda)$ are the diffuse and specular components of the corresponding unpolarised image, $I_{sv}(u,\lambda)$ is the amplitude of the TRS curves and $\phi(u,\lambda)$ is the phase of linear polarisation at pixel u and wavelength λ. We note that this formula is equivalent to that of the TRS in Eq. (10.6).

Let $\mathbf{I}_d(u)$, $\mathbf{I}_s(u)$, $\mathbf{I}_{sv}(u)$, $\mathbf{I}_{max}(u)$ and $\mathbf{I}_{min}(u)$ denote vectors whose elements are the wavelength indexed values of $I_d(u)$, $I_s(u,\lambda)$, $I_{sv}(u,\lambda)$, $I_{max}(u,\lambda)$ and $I_{min}(u,\lambda)$, respectively. These vectors are related as

$$\mathbf{I}_{max}(u) = \mathbf{I}_d(u) + \mathbf{I}_s(u) + \mathbf{I}_{sv}(u), \tag{10.40}$$

$$\mathbf{I}_{min}(u) = \mathbf{I}_d(u) + \mathbf{I}_s(u) - \mathbf{I}_{sv}(u). \tag{10.41}$$

With $\mathbf{I}_{max}(u)$ and $\mathbf{I}_{min}(u)$ recovered from polarisation images using one of the procedures described in Sect. 10.1.2, we obtain from Eqs. (10.40) and (10.41) the TRS amplitude as

$$\mathbf{I}_{sv}(u) = \frac{1}{2}\left(\mathbf{I}_{max}(u) - \mathbf{I}_{min}(u)\right). \tag{10.42}$$

and the unpolarised irradiance as

$$\mathbf{I}_{un}(u) \triangleq \mathbf{I}_d(u) + \mathbf{I}_s(u) = \frac{1}{2}\left(\mathbf{I}_{max}(u) + \mathbf{I}_{min}(u)\right). \tag{10.43}$$

Let us consider the problem of estimating the diffuse component at the surface point corresponding to a pixel u. The point is deemed to be polarised if the maximal colour channel value of the unpolarised triplet $\mathbf{I}_{un}$ exceeds a threshold and the angle between $\mathbf{I}_{max}(u)$ and $\mathbf{I}_{min}(u)$ is sufficiently large. Otherwise, the point is not sufficiently polarised and is assumed to be purely diffuse.

If the point is considered to be polarised, we relate the Fresnel reflection ratio to $I_s(u,\lambda)$ and $I_{sv}(u,\lambda)$ as

$$R(u,\lambda) = \frac{I_s(u,\lambda) - I_{sv}(u,\lambda)}{I_s(u,\lambda) + I_{sv}(u,\lambda)}, \tag{10.44}$$

since $I_s(u,\lambda) - I_{sv}(u,\lambda)$ and $I_s(u,\lambda) + I_{sv}(u,\lambda)$ are the minimal and maximal intensities on the TRS due to the polarisation of the specular component.

Similar to the previous method, Nayar et al. (1997) also assumed that the Fresnel coefficients $F_\perp$ and $F_\parallel$ are constant with respect to the wavelength. This assumption is reasonable within the visible spectrum given the small variation of the material refractive index within this range. As a result, the Fresnel reflection ratio is considered constant with respect to the wavelength within this range, and we can denote $R(u,\lambda) \approx \tilde{R}(u)$.

Fig. 10.9 Estimating the diffuse reflection at a point u using a diffuse neighbouring point v by the method of Nayar et al. The diffuse component $\mathbf{I}_d(u)$ is constrained to lie on the specular line defined by $\mathbf{I}_{\max}(u) - \mathbf{I}_{\min}(u)$

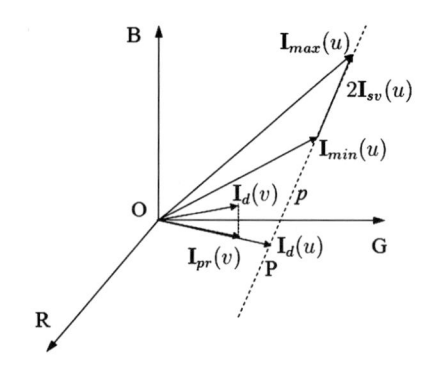

From Eq. (10.44), we obtain

$$\mathbf{I}_s(u) = \frac{1 + \tilde{R}(u)}{1 - \tilde{R}(u)} \mathbf{I}_{sv}(u). \tag{10.45}$$

Combining Eqs. (10.42) and (10.45), we have

$$\mathbf{I}_s(u) = \frac{1 + \tilde{R}(u)}{2(1 - \tilde{R}(u))} \left(\mathbf{I}_{\max}(u) - \mathbf{I}_{\min}(u) \right). \tag{10.46}$$

Substituting the expression of $\mathbf{I}_s(u)$ in Eq. (10.46) into Eq. (10.43), we relate the diffuse component $\mathbf{I}_d(u)$ to $\mathbf{I}_{\min}(u)$ and $\mathbf{I}_{\max}(u)$ as

$$\mathbf{I}_d(u) = \frac{\mathbf{I}_{\min}(u) - \tilde{R}(u)\mathbf{I}_{\max}(u)}{1 - \tilde{R}(u)}. \tag{10.47}$$

Equation (10.47) shows that for a polarised point u, $\mathbf{I}_d(u)$, $\mathbf{I}_{\max}(u)$ and $\mathbf{I}_{\min}(u)$ are collinear. In Fig. 10.9, we visually illustrate the relationship between these vectors using a three-dimensional coordinate system whose axes represent the R, G and B colour channels. In this coordinate system, an RGB triplet is represented by a vector emanating from the origin O. Due to Eq. (10.42), $\mathbf{I}_{sv}(u)$ is parallel to the specular line connecting the end points of $\mathbf{I}_{\max}(u)$ and $\mathbf{I}_{\min}(u)$.

Let P be the terminal point of vector $\mathbf{I}_d(u)$. By Eq. (10.47), P lies on the specular line, i.e. $\mathbf{I}_{sv}(u)$. Hence, the location of P can be parameterised as

$$\mathbf{I}_d(u) = \mathbf{I}_{\min}(u) - p \frac{\mathbf{I}_{sv}(u)}{\|\mathbf{I}_{sv}(u)\|}, \tag{10.48}$$

where p is the distance of P from $\mathbf{I}_{\min}$.

The diffuse component of the polarised point u is estimated using those of the neighbours. A neighbouring point v is considered for the estimation of $\mathbf{I}_d(u)$ if it has a low degree of polarisation or if its diffuse component has already been computed. Furthermore, the diffuse vector $\mathbf{I}_d(v)$ is required to lie close to the plane $\mathcal{P}$ that spans $\mathbf{I}_{\max}(u)$ and $\mathbf{I}_{\min}(u)$.

10.3 Polarimetric Reflection Models

If a neighbouring point v satisfies all the above constraints, it is assumed to have the same diffuse chromaticity as the point u. In Fig. 10.9, we annotate the diffuse reflection component at v as the vector $\mathbf{I}_d(v)$ originating from O. Ideally, $\mathbf{I}_d(v)$ lies in the plane $\mathcal{P}$, and its intersection with the specular line $\mathbf{I}_{max}(u) - \mathbf{I}_{min}(u)$ gives the location of P.

However, due to noise corruption, this may not be the case. Therefore, the location of P can be estimated as the intersection of the projection of $\mathbf{I}_d(v)$ on the plane $\mathcal{P}$, which we denote $\mathbf{I}_{pr}(v)$, with the specular line. Using the geometric relationship illustrated in Fig. 10.9, the position of P is estimated from the diffuse colour of pixel v as

$$p_v = \|\mathbf{I}_{sv}(u)\| \frac{(\mathbf{I}_{max}(u) \times \mathbf{I}_{min}(u)) \cdot (\mathbf{I}_d(v) \times \mathbf{I}_{min}(u))}{(\mathbf{I}_{max}(u) \times \mathbf{I}_{min}(u)) \cdot (\mathbf{I}_d(v) \times \mathbf{I}_{sv}(u))}. \tag{10.49}$$

A stable estimation of the diffuse component at pixel u can be obtained from the neighbouring diffuse pixels of u. We denote the estimate of p for each v as p_v. Then the final estimate of p is given as a weighted average,

$$\bar{p} = \frac{\sum_v w_v p_v}{\sum_v w_v}, \tag{10.50}$$

where the weight w_v is set as $\|\mathbf{I}_d(v)\|$ based on the fact that neighbouring points with a longer diffuse vector are more robust for the estimation in the presence of a point. With the estimate $\bar{p}$, the diffuse component at the point u is estimated by Eq. (10.48).

A Blind Source Separation Method

Umeyama and Godin (2004) posed the task of separating reflection components as an independent source separation problem and tackled it making use of ICA. Departing from the TRS, the authors rewrote Eq. (10.39) in the following form:

$$I_\vartheta(u, \lambda) = I_d(u, \lambda) + f(\vartheta) I_{sv}(u, \lambda), \tag{10.51}$$

where $f(\vartheta) = \cos(2(\vartheta - \phi(u, \lambda)))$.

The formulation above results in an implicit assumption that $I_s(u, \lambda) = 0$, i.e. the specular component is completely polarised. First, we vectorise the polarisation image acquired at a wavelength λ and the polariser angle ϑ_j, $j = 1, \ldots, M$, into a row vector $\mathbf{I}_{\vartheta_j}$ so as to express Eq. (10.51) over the image in a matrix form given by

$$\begin{bmatrix} \mathbf{I}_{\vartheta_1} \\ \vdots \\ \mathbf{I}_{\vartheta_M} \end{bmatrix} = \begin{bmatrix} 1 & f(\vartheta_1) \\ \vdots & \vdots \\ 1 & f(\vartheta_M) \end{bmatrix} \begin{bmatrix} \mathbf{I}_d \\ \mathbf{I}_{sv} \end{bmatrix}, \tag{10.52}$$

where $\mathbf{I}_d$ and $\mathbf{I}_{sv}$ are the vectorised form of the diffuse and specular images.

We can rewrite Eq. (10.52) in a compact form as follows:

$$\mathbf{I} = \mathbf{B}\mathbf{x}, \tag{10.53}$$

where

$$\mathbf{I} = \begin{bmatrix} \mathbf{I}_{\vartheta_1} \\ \vdots \\ \mathbf{I}_{\vartheta_M} \end{bmatrix}, \quad \mathbf{B} = \begin{bmatrix} 1 & f(\vartheta_1) \\ \vdots & \vdots \\ 1 & f(\vartheta_M) \end{bmatrix}, \quad \text{and} \quad \mathbf{x} = \begin{bmatrix} \mathbf{I}_d \\ \mathbf{I}_{sv} \end{bmatrix}.$$

Here, the observed transmitted radiance $\mathbf{I}$ is written as a product of two matrices. Hence, the component separation problem is viewed as a matrix decomposition of the observed matrix $\mathbf{I}$ to recover $\mathbf{I}_d, \mathbf{I}_{sv}$ and $f(\vartheta_j), j = 1, \ldots, M$. Since $\mathbf{B}, \mathbf{I}$ and $\mathbf{x}$ are of rank 2, we can employ singular value decomposition (SVD) to approximate $\mathbf{I}$. To this end, let $\mathbf{I} = \mathbf{U}\mathbf{D}\mathbf{V}^T$. With this notation, the rank-2 approximation of $\mathbf{I}$ is then given by

$$\mathbf{I} = \hat{\mathbf{U}}\mathbf{W}\mathbf{W}^{-1}\hat{\mathbf{D}}\hat{\mathbf{V}}^T, \tag{10.54}$$

where $\hat{\mathbf{D}} = \text{diag}(\sigma_1, \sigma_2)$, $\sigma_1 \geq \sigma_2$ with σ_1 and σ_2 the two greatest singular values of $\mathbf{I}$, i.e. the top two diagonal elements of $\mathbf{D}$, the matrices $\hat{\mathbf{U}}$ and $\hat{\mathbf{V}}$ are composed of the columns of $\mathbf{U}$ and $\mathbf{V}$ corresponding to $\hat{\mathbf{D}}$ and $\mathbf{W}$ is a 2×2 non-singular matrix.

With this approximation, the decomposition of $\mathbf{I}$ into $\mathbf{B}$ and $\mathbf{x}$ is given by

$$\mathbf{B} = \hat{\mathbf{U}}\mathbf{W}, \tag{10.55}$$

$$\mathbf{x} = \mathbf{W}^{-1}\hat{\mathbf{D}}\hat{\mathbf{V}}^T. \tag{10.56}$$

The component separation problem is now reduced to determining $\mathbf{W}$. Let us represent $\mathbf{W}$ as

$$\mathbf{W} = \begin{bmatrix} b_{11} & b_{12} \\ b_{21} & b_{22} \end{bmatrix}. \tag{10.57}$$

Subsequently, we express the first column of both sides of Eq. (10.55) as

$$\begin{bmatrix} 1 \\ \vdots \\ 1 \end{bmatrix} = \hat{\mathbf{U}} \begin{bmatrix} b_{11} \\ b_{21} \end{bmatrix}. \tag{10.58}$$

Since $\hat{\mathbf{U}}$ is known and orthogonal, the above equality allows a solution to the first column of $\mathbf{W}$ given by

$$\begin{bmatrix} b_{11} \\ b_{21} \end{bmatrix} = \hat{\mathbf{U}}^T \begin{bmatrix} 1 \\ \vdots \\ 1 \end{bmatrix}. \tag{10.59}$$

10.3 Polarimetric Reflection Models

To determine the second column of $\mathbf{W}$, we rewrite it in the polar form:

$$\mathbf{W} = \begin{bmatrix} r_1 \cos\gamma & r_2 \cos\beta \\ r_1 \sin\gamma & r_2 \sin\beta \end{bmatrix}, \quad (10.60)$$

where $r_1 > 0, r_2 > 0$ and r_1 and γ are already known.

Since the specular component $I_{sv}(u, \lambda)$ in Eq. (10.51) has a scaling ambiguity, we constrain the determinant of $\mathbf{W}$ to one without loss of generality, i.e.

$$\det(\mathbf{W}) = r_1 r_2 \sin(\beta - \gamma) = 1. \quad (10.61)$$

At each fixed value of β, we can determine r_2, $\mathbf{W}$ and the diffuse and specular components by making use of Eq. (10.56). To determine the optimal value of β, Umeyama and Godin assume probabilistic independence between the diffuse and the specular component. Following this assumption, the authors used the histograms of intensity values in the diffuse and specular images to compute the mutual information between these two components. The computation is repeated by changing β gradually, where the optimal value of β is chosen to be the one that minimises the mutual information.

Reflection Component Separation on Polarimetric Trichromatic Images

In this section, we demonstrate the utility of the above methods for the purpose of reflection component separation on real images. Although the propagation and optimisation method is formulated for narrowband images, it can be applied in a straightforward manner to trichromatic images. On the other hand, the method of Nayar et al. (1997) was designed for trichromatic images, while Umeyama and Godin's method (Umeyama and Godin 2004) can be applied to the colour channels of trichromatic images separately.

Polarimetric imagery may be acquired by attaching a linear polariser to the front optics of a camera. With this set-up, polarisation images are captured while the polariser transmission axis is rotated in angular increments. To minimise internal reflection and glare within the optics, one can use a multi-coated polariser with the transmission assumed to be uniform over its surface. In Fig. 10.10, we show results for polarisation images captured by a Nikon D80, with a polariser rotated between 0 and 180 degrees with respect to the horizontal axis.

The top row of Fig. 10.10 shows the unpolarised images reconstructed by averaging the input images over the polariser orientations in the first three columns. In addition, the last three columns of this row show the lower bound $I_{\min}$ of the TRS curve. In the first three columns of the second row in Fig. 10.10, we show the diffuse component recovered by the optimisation method presented earlier. The specular and inter-reflection components are shown in the right-hand columns. We repeat this pattern in the third and fourth rows of the figure, where we show the results yielded by the methods of Nayar et al. (1997) and Umeyama and Godin (2004), respectively.

| Unpolarised images | Minimum polarisation component (I_{min}) |

Diffuse component
by propagation & optimisation method

Specular component
by propagation & optimisation method

Diffuse component
by method in (Nayar et al., 1997)

Specular component
by method in (Nayar et al., 1997)

Diffuse component
by method in (Umeyama and Godin, 2004)

Specular component
by method in (Umeyama and Godin, 2004)

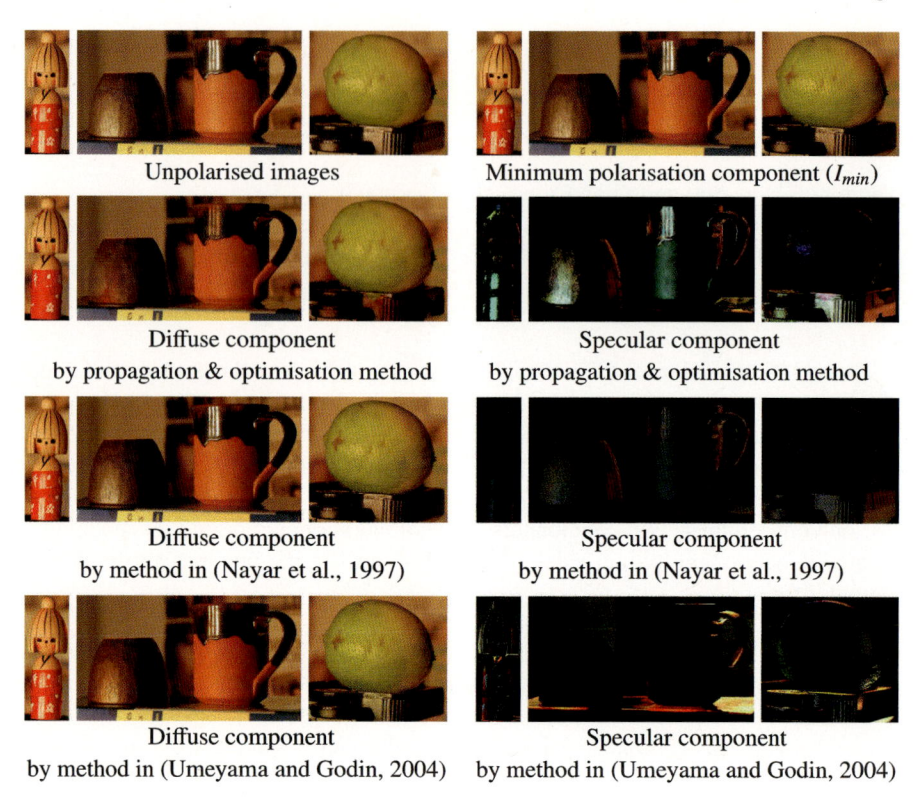

Fig. 10.10 Reflection component separation on real images using the algorithms described in Sect. 10.3.3

We note that the I_{min} images in the first row show the radiance values transmitted through a polariser when its orientation is aligned to the surface normal direction. Note that this lower bound is equivalent to the result produced by the use of cross-polarisers in front of the illuminant and the camera optics. Such a cross-polariser set-up has been widely used to remove glare and specularities from object surfaces. However, cross-polarisers are unable to remove limb-brightening effects. For example, strong specularities and specular inter-reflection near the object limbs still persist in the I_{min} images. Although it is counter-intuitive, this phenomenon is theoretically plausible as I_{min} still consists of a specular component, as can be seen in Eq. (10.28).

Moreover, by close inspection of these images, one can observe that the propagation and optimisation method reduces the limb brightening on the objects, consistent with Lambert's cosine law. With this method, specularity has also been removed from most parts of the object, except for places in the smooth, highly specular upper part due to pixel value saturation. Interestingly, the resulting specularity/inter-reflection images are indeed spatially varying in colour and clearly show the rough patterns on the surface of the dark-coloured cup and the lime.

10.4 Notes

In this chapter, we have elaborated on the influence of the polarisation properties of emitted and reflected light on the image irradiance. To achieve reliable scene understanding, we are not only required to focus on higher-level tasks such as recognition or classification, but we also have to recover the object shape, the illuminant power spectrum, the material photometric properties and the position of the light with respect to the camera.

Polarimetric imaging provides an alternative modality of image acquisition for purposes of shape recovery and surface reconstruction from multiple viewpoints. A proof of concept of this is the work in (Rahmann and Canterakis, 2001; Atkinson and Hancock 2005a, 2005b), where polarisation imaging is used for shape recovery of specular surfaces making use of the correspondences between phase images. The inverse is also true. Recall that Drbohlav and Sára (2001) have shown how to disambiguate surface orientations from uncalibrated photometric stereo using images corresponding to different polarisation angles of the incident and emitted light. This is possible since polarisation is inherently related to the object surface normal. Further, since the phase of polarisation is related to the plane of incidence, it can provide information on topographical artifacts on the surface and can be used in conjunction with the index of refraction so as to indicate changes in the object's composition.

Thus, polarimetric analysis can tackle tasks related to recognition and classification through the estimation of surface orientation and photometric invariants, including the index of refraction. We dedicate Chap. 11 solely to the problem of shape and refractive index recovery. This is an important theoretical development, since it relates polarisation to the shape, material index of refraction and other photometric variables. Further, polarisation permits the recovery of phase maps, i.e. the distribution of polarisation phase over the image lattice. This opens up the use of statistics (Atkinson and Hancock 2007) or the derivative of the degree of polarisation (Miyazaki et al. 2004) to refine the surface normal estimates.

References

Atkinson, G., & Hancock, E. R. (2005a). Multi-view surface reconstruction using polarization. In *International conference on computer vision* (pp. 309–316).
Atkinson, G., & Hancock, E. R. (2005b). Recovery of surface height using polarization from two views. In *CAIP* (pp. 162–170).
Atkinson, G. A., & Hancock, E. R. (2006). Recovery of surface orientation from diffuse polarization. *IEEE Transactions on Image Processing*, 15(6), 1653–1664.
Atkinson, G. A., & Hancock, E. R. (2007). Shape estimation using polarization and shading from two views. *IEEE Transactions on Pattern Analysis and Machine Intelligence*, 29(11), 2001–2017.
Beckmann, P., & Spizzichino, A. (1963). *The scattering of electromagnetic waves from rough surfaces*. New York: Pergamon.
Born, M., & Wolf, E. (1999). *Principles of optics: electromagnetic theory of propagation, interference and diffraction of light* (7th ed.). Cambridge: Cambridge University Press.

Drbohlav, O., & Sára, R. (2001). Unambigous determination of shape from photometric stereo with unknown light sources. In *International conference on computer vision* (pp. 581–586).

Goudail, F., Terrier, P., Takakura, Y., Bigué, L., Galland, F., & DeVlaminck, V. (2004). Target detection with a liquid-crystal-based passive stokes polarimeter. *Applied Optics, 43*(2), 274–282.

Hall, J. S. (1951). Some polarization measurements in astronomy. *Journal of the Optical Society of America, 41*(12), 963–966.

Hawryshyn, C. W. (2000). Ultraviolet polarization vision in fishes: possible mechanisms for coding e-vector. *Philosophical Transactions: Biological Sciences, 355*(1401), 1187–1190.

Hecht, E. (2002). *Optics* (4th ed.). Reading: Addison-Wesley.

Huynh, C. P., Robles-Kelly, A., & Hancock, E. R. (2010). Shape and refractive index recovery from single-view polarisation images. In *IEEE conference on computer vision and pattern recognition*.

Kasarova, S. N., Sultanova, N. G., Ivanov, C. D., & Nikolo, I. D. (2007). Analysis of the dispersion of optical plastic materials. *Optical Materials, 29*, 1481–1490.

Mandel, L., & Wolf, E. (1995). *Optical coherence and quantum optics*. Cambridge: Cambridge University Press.

Marquardt, D. (1963). An algorithm for least-squares estimation of nonlinear parameters. *SIAM Journal on Applied Mathematics, 11*, 431–441.

Marshall, N. J., Land, M. F., King, C. A., & Cronin, T. W. (1991). The compound eyes of Mantis Shrimps (Crustacea, Hoplocarida, Stomatopoda). I. Compound eye structure: the detection of polarized light. *Philosophical Transactions: Biological Sciences, 334*(1269), 33–56.

Miyazaki, D., Kagesawa, M., & Ikeuchi, K. (2004). Transparent surface modeling from a pair of polarization images. *IEEE Transactions on Pattern Analysis and Machine Intelligence, 26*(1), 73–82.

Nayar, S. K., Fang, X. q. S., & Boult, T. (1997). Separation of reflection components using color and polarization. *International Journal of Computer Vision, 21*(3), 163–186.

Novak, C., & Shafer, S. (1992). Anatomy of a color histogram. In *Proceedings of the IEEE conference on computer vision and pattern recognition* (pp. 599–605).

Oren, M., & Nayar, S. K. (1995). Generalization of the Lambertian model and implications for machine vision. *International Journal of Computer Vision, 14*(3), 227–251.

Rahmann, S., & Canterakis, N. (2001). Reconstruction of specular surfaces using polarization imaging. In *IEEE conference on computer vision and pattern recognition* (Vol. 1, pp. 149–155).

Sadjadi, F. A., & Chun, C. S. L. (2004). Remote sensing using passive infrared stokes parameters. *Optical Engineering, 43*, 2283–2291.

Sato, Y., & Ikeuchi, K. (1994). Temporal-color space analysis of reflection. *Journal of the Optical Society of America A, 11*(11), 2990–3002.

Shafer, S. A. (1985). Using color to separate reflection components. *Color Research and Application, 10*(4), 210–218.

Torrance, K., & Sparrow, E. (1967). Theory for off-specular reflection from roughened surfaces. *Journal of the Optical Society of America, 57*(9), 1105–1112.

Torrance, K. E., Sparrow, E. M., & Birkebak, R. C. (1966). Polarization, directional distribution, and off-specular peak phenomena in light reflected from roughened surfaces. *Journal of the Optical Society of America, 56*, 916–924.

Umeyama, S., & Godin, G. (2004). Separation of diffuse and specular components of surface reflection by use of polarization and statistical analysis of images. *IEEE Transactions on Pattern Analysis and Machine Intelligence, 26*(5), 639–647.

Wolff, L. B. (1989). Using polarization to separate reflection components. In *Computer vision and pattern recognition* (pp. 363–369).

Wolff, L. B. (1994). *Diffuse-reflectance model for smooth dielectric surfaces* (pp. 2956–2968).

Wolff, L. B. (1997). Polarization vision: a new sensory approach to image understanding. *Image and Vision Computing, 15*(2), 81–93.

Wolff, L. B., & Andreou, A. G. (1995). Polarization camera sensors. *Image and Vision Computing, 13*(6), 497–510.

References

Wolff, L. B., & Boult, T. E. (1989). Polarization/radiometric based material classification. In *Computer vision and pattern recognition* (pp. 387–395).

Wolff, L. B., & Boult, T. E. (1991). Constraining object features using a polarization reflectance model. *IEEE Transactions on Pattern Analysis and Machine Intelligence*, *13*(7), 635–657.

Wolff, L. B., Mancini, T. A., Pouliquen, P., & Andreou, A. G. (1997). Liquid crystal polarization camera. *IEEE Transactions on Robotics and Automation*, *13*(2), 195–203.

Wyszecki, G., & Stiles, W. S. (2000). *Color science: concepts and methods, quantitative data and formulae*. New York: Wiley.

Chapter 11
Shape and Refractive Index from Polarisation

We now discuss the recovery of the surface normals and the refractive index from spectro-polarimetric images captured from a single viewpoint. Shape and material properties such as refractive index are usually coexisting factors that influence the appearance of an object to an observer and the polarisation properties of the emitted radiation. In an early work (Torrance et al. 1966) the authors measured the specular reflectance distribution of rough surfaces for different polarisation orientations. The reflectance model attributes polarisation to specular reflection from a collection of small, randomly disposed, mirror-like facets that constitute the surface area. The model includes a specular reflection component based on the Fresnel reflection theory and a micro-facet distribution function. Other reflectance models such as the Torrance–Sparrow model (Torrance and Sparrow 1967) and the Wolff model (Wolff 1994) are motivated by the Fresnel reflection theory. As a result, these reflectance models consider the reflected light as a combination of polarisation components parallel and perpendicular to the plane of reflection, and are applicable to polarised light sources. In these models, the material properties and the geometry of the reflection process are expressed in a single equation with multiple degrees of freedom. As a result, the simultaneous recovery of the photometric and shape parameters becomes an under-constrained problem.

In this chapter, we exploit the combination of the Fresnel reflection theory, material dispersion equations and the surface integrability constraint to estimate the shape and refractive index simultaneously. We draw constraints from the latter two to render the recovery problem well-posed. These two constraints reduce the number of parameters in the spectral dimension to the number of dispersion coefficients. As a result, the formulation permits the use of an iterative procedure to find an approximately optimal solution. Further, the iterative optimisation approach is computationally efficient due to the use of closed-form solutions for the recovery of the zenith angle of surface normals and the refractive index.

The techniques discussed here are applicable to convex and continuously twice-differentiable surfaces with material refractive index following a dispersion equation. Here, we focus on dielectric surfaces that undergo diffuse polarisation due to subsurface scattering and transmission from the object surface into the air. The diffuse polarisation of the reflection process is modelled by the Fresnel reflection theory and Snell's law. We depart from the phase angle and the maximal and minimal radiance recovered from the input polarimetric imagery as described in Sect. 10.1.2 to present, in Sect. 11.1, a method for disambiguating and estimating the azimuth angle of the surface normals from the wavelength indexed spectrum of phase angles. In addition, we compute the Fresnel transmission ratio from the maximal and minimal radiance, from which we estimate the zenith angle of the surface normal and the refractive index. In Sect. 11.2, we formulate the estimation as an optimisation problem that takes into account the data error for the Fresnel transmission ratio, the surface integrability and the dispersion of material refractive index across the spectrum. Having formulated the objective function, we present the optimisation process in Sect. 11.3.

The process of shape and refractive index recovery can be summarised as follows.

1. Decomposition of the spectral polarimetric imagery into polarisation components, including the phase, maximal and minimal radiance of the transmitted radiance sinusoid (TRS), as described in Sect. 10.1.2.
2. Rough estimation of the azimuth angle from the phase angle for each pixel and wavelength, allowing a 180-degree ambiguity. The relationship between these two angles is described in Sect. 11.1.
3. Joint estimation of the zenith angle and refractive index from the Fresnel transmission ratio, as presented in Sect. 11.2.
4. Disambiguation between the two candidates of the azimuth angle using shading information indicated by the estimated zenith angle, as presented in Sect. 11.1.1.
5. Re-estimation of the azimuth angle as presented in Sect. 11.1.2.
6. Integration of the surface normal field to reconstruct the surface depth.

It is noted that the 180-degree ambiguity between the candidates of the azimuth angle resulting from step 2 does not affect the joint estimation of the zenith angle and refractive index in step 3 because the objective functions involved in this estimation are invariant to the 180-degree shift in the azimuth angle. As we shall show in Sect. 11.2.3, the relevant objective function consists of the square of the cosine and sine of the azimuth angle. Once the zenith angle has been obtained, we perform disambiguation of the azimuth angle based on the shading information suggested by the estimated zenith angle.

Before further formalism, in Fig. 11.1, we provide the list of commonly used notation in this chapter for mnemonic purposes. To relate the formulation in this chapter to the concepts introduced in Chap. 10, we maintain a consistent nomenclature and notation between the two chapters.

11.1 Azimuth Angle and Phase of Diffuse Polarisation

To remedy the ill-posed nature of the problem, Miyazaki et al. (2003) assumed that the histogram of zenith angles for a given object was similar to that of a sphere

11.1 Azimuth Angle and Phase of Diffuse Polarisation

Notation	Description
$I_\vartheta(u, \lambda)$	Transmitted irradiance at a pixel u and a wavelength λ corresponding to a polariser angle ϑ.
$\epsilon(u, \lambda)$	Fitting error of the transmitted radiance sinusoid (TRS).
$\phi(u, \lambda)$	Phase of polarisation.
$\alpha(u)$ and $\theta(u)$	Azimuth and zenith angles at pixel u.
$I_{\max}(u, \lambda)$	Maximum transmitted radiance on the TRS.
$I_{\min}(u, \lambda)$	Minimum transmitted radiance on the TRS.
$I_{T\parallel}$ and $I_{T\perp}$	Parallel and perpendicular components of the transmitted radiance.
$F_\parallel$ and $F_\perp$	Parallel and perpendicular Fresnel reflection coefficients.
$R(u, \lambda)$	Fresnel transmission ratio at a pixel u and a wavelength λ.
$\eta(u, \lambda)$	Refractive index at a pixel u and a wavelength λ.
$I_{\text{un}}(u, \lambda)$	Unpolarised radiance at a pixel u and a wavelength λ.
$\vec{N}(u)$	Surface normal at pixel u.
$\mathcal{Z}(u)$	Surface depth at pixel u.
$C_m(u), m = 1, \ldots, M$	Dispersion coefficients of Cauchy's equation.
$B_m(u)$ and $D_m(u), m = 1, \ldots, M$	Dispersion coefficients of Sellmeier's equation.
$\mathcal{S}, \mathcal{W}$	Image spatial and spectral domains.
$\beta(u)$	The local weight of the surface integrability constraint.
K	The number of wavelengths in the spectral image.

Fig. 11.1 Notation used throughout Chap. 11

and used this property to recover a mapping from the degree of polarisation to the zenith angle. Despite being effective, their approach is limited to only surfaces with a uniform distribution of surface normal directions. Moreover, the mapping is not necessarily consistent across different material refractive indices. The related work in Miyazaki et al. (2002, 2003) employed the degree of polarisation in the visible and far-infrared regions to resolve the surface orientation of transparent objects. The main drawback of the method in Miyazaki et al. (2002) is the need for an omnidirectional diffuse illumination source, which limits its applicability in real-world settings. Moreover, the method requires measurements of the energy emitted in the far-infrared spectrum, which is susceptible to thermal noise due to the heating of the object under study.

Using a similar experimental set-up involving a spherical optical diffuser, Saito et al. (1999) were able to recover the shape of transparent objects with known refractive index. Rahmann (1999) presented a method for computing the light source position and the orientation of rough surfaces through the polarisation of specular highlights. However, because the method in Rahmann (1999) relies on the existence of highlights reflected from flat surfaces, it cannot be applied to objects which exhibit purely diffuse reflectance. In a subsequent development, Rahmann (2000) employed level sets to reconstruct the surface of an object from a single polarisation image. This was done under the assumption of weak perspective camera projection. Atkinson and Hancock (2006) recovered the surface orientation from the measured diffuse polarisation of smooth dielectric surfaces. However, in their work, they assumed a known refractive index in order to estimate the zenith angle of surface normals from the degree of polarisation. Overall, methods for inferring surface orientation from polarisation images captured from a single view either assume a known refractive index or a known surface orientation distribution, or require complex instrument set-up.

11.1.1 Disambiguation of the Azimuth Angle

Recall that, in Sect. 10.2.2, we observed that the maximum transmitted radiance, i.e. $I_{\max}$, occurs when the polariser's angle is parallel to the plane of reflection. Moreover, note that the TRS in Fig. 10.4 reaches its maximum at the phase angle. Consequently, the phase angle ϕ must either coincide with the azimuth (tilt) angle α of the surface normal or differ from it by π radians. In other words, at each pixel u, either of the two cases given by $\alpha(u) = \phi$ and $\alpha(u) = \phi \pm \pi$ must hold. This reduces the problem of recovering the azimuth of the surface normal to that of disambiguating the two possible solutions for $\alpha(u)$.

To this end, we assume that the surface under study is convex; i.e. the surface normals point in the opposite direction to the gradient of surface shading. This assumption allows us to select the candidate azimuth angle with the closer orientation to the negative gradient direction. The surface shading at a pixel u is given by $\cos\theta(u)$, where $\theta(u)$ is the zenith angle of the surface normal. Suppose that the zenith angle has been obtained through the process described in Sect. 11.3. By sorting the zenith angles available, we are able to detect both the boundary and singular points on the surface and use them to disambiguate the azimuth angles.

The disambiguation process commences by first rotating the azimuth angles by 180° where necessary to satisfy the convexity requirement. In addition, for convex objects, the surface normals at the occluding boundary always point outward from the body of the object. We make use of this observation to initialise the azimuth angles at the occluding boundary. Also, since diffuse polarisation occurs most strongly at the occlusion boundary, the disambiguation of azimuth angle based on polarisation performs most reliably at these positions. On the other hand, weak polarisation is often observed at singular points, i.e. where the surface normal is perpendicular to the image plane. Therefore, abrupt changes in azimuth angle are permitted near these locations. As a result, we start smoothing the azimuth angle at the boundary pixels and propagate the operation towards the singular points of the surface. The smoothing operation proceeds iteratively, aiming to preserve the directional similarity of the azimuth angle within a neighbourhood.

We note that the disambiguation method above only copes with convex surfaces. To fully recover the height of a complex surface with both convex and concave parts, one could consider the method proposed by Zhu and Shi (2006). The algorithm in Zhu and Shi (2006) flips surface patches and adjusts their heights so as to satisfy a global integrability constraint. It resolves this global disambiguation problem by computing the optimal height of singular points. To compute the height at the singular points, the authors construct a graph with singular points as vertices, on which the global integrability constraint can be stated as a max-cut problem. The solution to this problem can then be found by making use of semi-definite programming. Having obtained the height at singular points, the height of the whole surface is propagated from these points via the application of fast marching and patch stitching.

11.1.2 Estimation of the Azimuth Angle

In general, the disambiguation process above provides an estimation of the azimuth angle $\phi^*(u, \lambda)$ at each pixel u and wavelength λ of the spectral imagery. However, the estimated phase angle for each wavelength may vary widely due to weak polarisation and noise corruption. To overcome this drawback, the computer vision community has turned its attention to the use of multiple images. Rahmann and Canterakis (2001) proposed a polarisation imaging method to recover the shape of specular surfaces. Their approach made use of the correspondences between the polarisation phases recovered from multiple views. They showed that three views were sufficient for surface reconstruction. Atkinson and Hancock (2005) also made use of the link between the phase and the degree of polarisation for shape recovery. Using two views, they proposed a method to disambiguate the two candidates for the azimuth angle of surface normals. They also resolved the mapping from the degree of specular polarisation to the zenith angle of surface normals. However, the method in Atkinson and Hancock (2005) assumes the homography between the views is known in order to match points across the imagery. This work was later extended in Atkinson and Hancock (2007), where robust statistics were used to refine the correspondence estimates between the two views of an object.

Here, we make use of the weighted mean of the spectral phase of polarisation across the spectrum as an alternative to the phase angle at any given wavelength. The weights are chosen to reflect the fitting error of the TRS curve to the intensity-polariser angle pairs at each wavelength. We quantify the fitting error $\epsilon(u, \lambda)$ using the ℓ^2-norm of the residual $\epsilon(u, \lambda)$ of Eq. (10.13), i.e. $\epsilon(u, \lambda) = \|\mathbf{I} - \mathbf{Ax}\|_2$, where we have solved Eq. (10.13) for $\mathbf{x}$ given the known input $\mathbf{I}$ and $\mathbf{A}$.

The weight $w(u, \lambda)$ associated with the phase of polarisation at each wavelength is defined via a kernel weighting function. To this end, we employ the Epanechnikov kernel, which is a popular choice in the parameter estimation literature. The kernel is given by

$$w(u, \lambda) = \begin{cases} 1 - \frac{\epsilon^2(u,\lambda)}{h}, & \text{if } \frac{\epsilon^2(u,\lambda)}{h} < 1, \\ 0, & \text{otherwise,} \end{cases} \qquad (11.1)$$

where h is a bandwidth parameter.

Since the azimuth angle is a directional quantity, instead of averaging the disambiguated azimuth angles $\phi^*(u, \lambda)$ directly, we estimate the mean of the sines and cosines of these angles for each pixel site as follows:

$$\langle \sin(\phi^*(u)) \rangle_\lambda = \frac{\sum_\lambda \sin(\phi^*(u, \lambda)) w(u, \lambda)}{\sum_\lambda w(u, \lambda)},$$

$$\langle \cos(\phi^*(u)) \rangle_\lambda = \frac{\sum_\lambda \cos(\phi^*(u, \lambda)) w(u, \lambda)}{\sum_\lambda w(u, \lambda)},$$

where $\langle \cdot \rangle_\lambda$ denotes the mean value across wavelengths. Thus, the estimated azimuth angle at pixel u becomes

$$\alpha^*(u) = \begin{cases} \arctan(\frac{\langle \sin(\phi^*(u)) \rangle_\lambda}{\langle \cos(\phi^*(u)) \rangle_\lambda}) & \text{if } \langle \cos(\phi^*(u)) \rangle_\lambda > 0, \\ \arctan(\frac{\langle \sin(\phi^*(u)) \rangle_\lambda}{\langle \cos(\phi^*(u)) \rangle_\lambda}) + \pi & \text{if } \langle \cos(\phi^*(u)) \rangle_\lambda < 0, \\ \frac{\pi}{2} & \text{if } \langle \cos(\phi^*(u)) \rangle_\lambda = 0. \end{cases} \quad (11.2)$$

11.2 Zenith Angle and Refractive Index

To fully determine the surface shape, one needs to estimate the zenith angle of the surface normals with respect to the viewing direction. Following the previous section, where we showed how the azimuth angle of surface normals can be estimated from the phase of polarisation, we now provide an account of the joint estimation of the zenith angle of surface normals and material refractive index from the polarimetric spectral imagery under consideration.

Miyazaki et al. (2004) disambiguated the two possible zenith angles which yield the same degree of specular polarisation by physically tilting the observed object by a small angle. Thilak et al. (2007) presented a non-linear least-squares estimation algorithm to extract the complex index of refraction and the zenith angle of the surface normals from multiple images illuminated by unpolarised light sources. However, their method requires prior knowledge of the light source positions relative to the observer. Furthermore, it employs a polarimetric bidirectional reflectance distribution function (BRDF) of light scattering that is limited to the case where the light source direction, the surface normal direction and the view direction are co-planar. Here, following Sect. 10.2.2, we conclude that the TRS curve reaches its minimal value when the transmission axis is perpendicular to the plane of reflection. This is, in fact, the complement of the observation made in Sect. 10.2.2 regarding $I_{\max}$. As a result, we have the equalities $I_{\max} = I_{T\|}$ and $I_{\min} = I_{T\perp}$ and thus obtain the ratio of the recovered minimal and maximal radiance on the TRS curve as

$$\frac{I_{\min}}{I_{\max}} = \frac{I_{T\perp}}{I_{T\|}} = \frac{1 - F_\perp}{1 - F_\|}. \quad (11.3)$$

11.2 Zenith Angle and Refractive Index

Following the expressions of the Fresnel reflection coefficients in Wolff (1994) and Snell's law of refraction, we can relate the ratio on the right-hand side of Eq. (11.3) to the zenith angle and refractive index through the equation

$$\frac{I_{\min}}{I_{\max}} = \left(\frac{\cos\theta(u)\sqrt{\eta^2(u,\lambda) - \sin^2\theta(u)} + \sin^2\theta(u)}{\eta(u,\lambda)} \right)^2. \qquad (11.4)$$

The right-hand side of Eq. (11.4) is the Fresnel transmission ratio. To further simplify computation, we take its square root to obtain the following function with respect to the zenith angle and the refractive index:

$$R(u,\lambda) \triangleq \frac{\cos\theta(u)\sqrt{\eta^2(u,\lambda) - \sin^2\theta(u)} + \sin^2\theta(u)}{\eta(u,\lambda)}. \qquad (11.5)$$

Furthermore, from Eq. (11.4) we note that the above function can be related to the components $I_{\max}$ and $I_{\min}$ recovered in Sect. 10.1.2 as follows:

$$R(u,\lambda) = \sqrt{\frac{I_{\min}}{I_{\max}}}. \qquad (11.6)$$

We note in passing that prior literature has mainly focused on the recovery of the zenith angle of the surface normals and the index of refraction from the degree of polarisation (Atkinson and Hancock 2005, 2006, 2007; Miyazaki et al. 2002, 2004; Thilak et al. 2007). Indeed, all of these methods depart from the Fresnel reflection theory to arrive at an equation similar to Eq. (11.4). However, the main limitation to their practical application resides on their reliance upon either known index of refraction (Atkinson and Hancock 2005, 2006, 2007; Miyazaki et al. 2002, 2004), or imagery captured from multiple viewpoints (Miyazaki et al. 2004) or under several known light source directions (Thilak et al. 2007). The need for multiple measurements and instrumental set-ups makes them impractical for shape and material analysis on real-world data.

Moreover, the estimation of the zenith angle and index of refraction cannot be performed without further constraints. The reason is that the Fresnel transmission ratio only provides a single equation per wavelength relating the zenith angle and the index of refraction; i.e. the number of variables to be estimated is one more than the number of equations. Therefore, the problem is ill-posed in general.

To deal with these limitations, we present two additional constraints in order to recover the zenith angle and the refractive index in a well-formed manner. While the integrability constraint enforces spatial consistency between neighbouring surface locations, the dispersion constraint aims to resolve the ill-posedness of the joint estimation of the zenith angle and refractive index. In Sect. 11.2.1, we formulate the integrability constraint over the spatial domain, i.e. the object surface. Later, in Sect. 11.2.2, we employ the material dispersion equations and utilise them to constrain the space of solutions for the refractive index across the wavelength domain.

11.2.1 Integrability Constraint

We commence by formulating the integrability constraint with respect to the zenith and azimuth angles of the surface normal. Note that the surface under study can be represented by a two-dimensional twice-differentiable function with a continuous second derivative (Frankot and Chellappa 1988). As a result, its cross-derivatives are the same irrespective of the order of the differentiated variable.

For the reference coordinate system presented before, let the surface height function at the pixel u be $\mathcal{Z}(u)$. By definition, the normalised surface normal at u is given by

$$\overrightarrow{N(u)} = \frac{1}{\sqrt{\mathcal{Z}_x^2 + \mathcal{Z}_y^2 + 1}}[-\mathcal{Z}_x, -\mathcal{Z}_y, 1]^T, \tag{11.7}$$

where $\mathcal{Z}_x$ and $\mathcal{Z}_y$ are the surface gradients on the x- and y-axes on our coordinate system.

The normalised surface normal direction can also be represented with respect to the azimuth and zenith angles as follows:

$$\overrightarrow{N(u)} = \begin{bmatrix} \cos\alpha(u)\sin\theta(u) \\ \sin\alpha(u)\sin\theta(u) \\ \cos\theta(u) \end{bmatrix}. \tag{11.8}$$

From Eqs. (11.7) and (11.8), we have

$$\mathcal{Z}_x = -\cos\alpha(u)\tan\theta(u), \tag{11.9}$$

$$\mathcal{Z}_y = -\sin\alpha(u)\tan\theta(u). \tag{11.10}$$

Recall that, in Sect. 11.1.2, we obtained the normal azimuth angles $\alpha(u)$ and we treat them as constants in Eqs. (11.9) and (11.10). As a consequence, the cross-derivatives can be rewritten as $\mathcal{Z}_{xy} = -\cos\alpha(u)\frac{\partial \tan\theta(u)}{\partial y}$ and $\mathcal{Z}_{yx} = -\sin\alpha(u)\frac{\partial \tan\theta(u)}{\partial x}$. Since the integrability constraint states that $\mathcal{Z}_{xy} = \mathcal{Z}_{yx}$, it can be parameterised with respect to the azimuth and zenith angles as

$$\cos\alpha(u)\frac{\partial \tan\theta(u)}{\partial y} = \sin\alpha(u)\frac{\partial \tan\theta(u)}{\partial x}. \tag{11.11}$$

According to the chain rule, $\partial \tan\theta(u) = \frac{\partial\theta(u)}{\cos^2\theta(u)}$. Therefore, the integrability constraint in Eq. (11.11) can be rewritten as

$$\cos\alpha(u)\theta_y(u) = \sin\alpha(u)\theta_x(u), \tag{11.12}$$

where $\theta_x(u)$ and $\theta_y(u)$ are the derivatives of $\theta(u)$ with respect to x and y.

We note in passing that polarisation imaging can also be combined with photometric stereo. Drbohlav and Šára (2001) showed how to disambiguate surface

11.2 Zenith Angle and Refractive Index

orientations from uncalibrated photometric stereo images by altering the polarisation angles of the incident and emitted light. Their method uses two projections of the object normals onto two planes perpendicular to the viewing direction and the illumination direction. When combined with the integrability constraint, this yields a method that can cope with both the bas-relief (Belhumeur et al. 1997) and convex-concave ambiguities. Atkinson and Hancock (2007) disambiguated the surface normals by combining polarisation data with photometric stereo using three known light source positions.

11.2.2 Material Dispersion

In order to impose further constraints on the wavelength domain, we note that the index of refraction for a wide range of materials is governed by the material dispersion equation (Born and Wolf 1999). Specifically, the dispersion equation constrains the refractive index spectrum to a low-dimensional space with dimension equal to the number of dispersion coefficients. The number of these coefficients is generally lower than the number of spectral bands. The implication is hence that the number of variables to be estimated in the spectral domain can be significantly reduced, especially for spectral images with a high spectral resolution.

Thus, we introduce dispersion equations as a constraint on the refractive index for our optimisation scheme. Among several approximating functions of material dispersion in the physics literature, perhaps Cauchy's and Sellmeier's (Born and Wolf 1999) are the most popular. In the former, Cauchy modelled the empirical relationship between the refractive index of a material and the wavelength of light as follows:

$$\eta(u, \lambda) = \sum_{m=1}^{M} C_m(u) \lambda^{-2(m-1)}, \quad (11.13)$$

where the index of refraction depends solely on the wavelength and the material-specific dispersion coefficients $C_m(u)$, $m \in \{1, \ldots, M\}$.

In addition, Sellmeier's dispersion equation (Sellmeier 1871) can handle anomalous dispersive regions by including additional coefficients to represent vacuum wavelengths, i.e. where the wave front moves across a vacuum, and holds for a wide range of wavelengths, including the ultraviolet, visible and infrared spectra. Sellmeier's dispersion equation is given by

$$\eta^2(u, \lambda) = 1 + \sum_{m=1}^{M} \frac{B_m(u) \lambda^2}{\lambda^2 - D_m(u)}, \quad (11.14)$$

where $B_m(u)$ and $D_m(u)$ are the material-specific dispersion coefficients.

The dispersion equations above allow a representation of the index of refraction as a linear combination of M rational functions of wavelength. With these representations, the estimation of refractive index can be treated as that of computing the dispersion coefficients. In practice, an expansion containing up to the sixth term is sufficient to represent a wide range of materials including crystals, liquids, glasses, gases and plastics (Kasarova et al. 2007). For spectral imagery comprising more than seven wavelength indexed bands, the number of equations relating the Fresnel transmission ratio to the zenith angle and refractive index exceeds the number of dispersion coefficients, rendering the problem solvable. As a result, the problem becomes a well-constrained one that can be formulated in a minimisation setting.

11.2.3 Objective Function

Having introduced the integrability and material dispersion constraints in Sects. 11.2.1 and 11.2.2, we now focus on the formulation of an objective function for the estimation of the zenith angle and refractive index. The rationale behind the cost function presented here lies in the use of the additional constraints, including integrability and dispersion equations to allow the recovery of the shape and the index of refraction to be performed without prior knowledge or predetermined illumination conditions.

The cost function aims at satisfying Eq. (11.6), which equates the square root of the Fresnel transmission ratio defined in Eq. (11.5) to the quantity $\sqrt{\frac{I_{min}}{I_{max}}}$ while taking into account the integrability of the surface and the material dispersion equation. Our objective function is given by two terms. The first of these accounts for the error of the Fresnel transmission ratio $R(u, \lambda)$ in Eq. (11.5) with respect to the ratio $r(u, \lambda) \triangleq \sqrt{\frac{I_{min}}{I_{max}}}$ as computed from the image radiance. The second term measures the error of the integrability constraint described in Eq. (11.12). Thus, the cost function is given by

$$\mathcal{E} = \int_S \int_W (R(u, \lambda) - r(u, \lambda))^2 d\lambda \, du$$

$$+ \beta(u) \int_S \left(\cos\alpha(u) \frac{\partial \tan\theta(u)}{\partial y} - \sin\alpha(u) \frac{\partial \tan\theta(u)}{\partial x} \right)^2 du, \quad (11.15)$$

subject to the chosen dispersion equation, i.e. Eq. (11.13) or (11.14), where S is the image spatial domain and W is the wavelength range.

In Eq. (11.15), we assume an estimate of the azimuth angle $\alpha(u)$, up to a 180-degree ambiguity, has been obtained, as described in Sect. 11.1, and treat it as a constant in the cost function. We note that this cost function is invariant to the 180-degree shift in the azimuth angle, i.e.

$$(\cos\alpha(u)\theta_y(u) - \sin\alpha(u)\theta_x(u))^2 = (\cos(\alpha(u) + \pi)\theta_y(u) - \sin(\alpha(u) + \pi)\theta_x(u))^2.$$

Therefore, we can utilise the rough estimate of the azimuth angle obtained in Sect. 11.1 without an adverse effect on the estimation of the zenith angle and the refractive index.

In addition, the Fresnel transmission ratio $R(u, \lambda)$ is related to the zenith angle and refractive index through Eq. (11.5). The regularisation parameter $\beta(u)$ is spatially varying and weighs the relative importance between the data closeness and surface smoothness imposed by the integrability constraint. Here, we use the spatial dependence of $\beta(u)$ on the surface location so as to reflect the reliability of the azimuth angle $\alpha(u)$ estimated from polarisation information. To quantify the reliability of the estimate $\alpha(u)$, we employ the degree of polarisation as follows:

$$\beta(u) = \gamma \left\langle \frac{I_{\max}(u, \lambda) - I_{\min}(u, \lambda)}{I_{\max}(u, \lambda) + I_{\min}(u, \lambda)} \right\rangle_\lambda, \qquad (11.16)$$

where γ is a scaling constant for the whole image and $\langle . \rangle_\lambda$ denotes the mean value across wavelengths.

11.3 Optimal Zenith Angle and Refractive Index

In this section, we adopt an iterative approach to the recovery of both the zenith angle and the index of refraction. The algorithm proceeds in a coordinate descent manner (Boyd and Vandenberghe 2004) to minimise the cost functional in Eq. (11.15). The step sequence of the minimisation strategy is summarised in Algorithm 11.1. At each iteration, the first step of the coordinate descent approach minimises the objective function with respect to the zenith angle $\{\theta(u) \mid u \in \mathcal{S}\}$ while fixing the index of refraction. The second step recovers the index of refraction $\{\eta(u, \lambda) \mid u \in \mathcal{S}, \lambda \in \mathcal{W}\}$ for the current estimate of the zenith. We iterate between these two steps until convergence is reached.

The algorithm is initialised with a uniform material refractive index η_0 across both the spatial and spectral domains, as indicated in Line 1. It terminates once the estimation of the parameters stabilises, i.e. the change between the estimates in successive iterations falls below a predetermined threshold τ_θ for the zenith angle and a threshold τ_η for the refractive index. For both the zenith angle and the refractive index, the change is measured as the L^1-norm of the difference between successive estimates. In the following two subsections, we elaborate further on the details of the optimisation steps above.

11.3.1 Zenith Angle Computation

Here we derive a solution for the zenith angle while keeping the refractive index fixed at each iteration. We note that the original cost functional in Eq. (11.15) involves a non-linear Fresnel transmission ratio function $R(u, \lambda)$ with respect to the zenith angle $\theta(u)$. To minimise this cost functional, we need to solve a highly non-linear Euler–Lagrange equation whose analytical solution cannot be derived in a

Algorithm 11.1 Estimation of the zenith angle and refractive index from a polarimetric spectral image with K wavelength indexed bands

---
Require: Fresnel transmission ratio $r(u, \lambda)$ for each pixel u and band $\lambda \in \{\lambda_1, \ldots, \lambda_K\}$.
Ensure: $\{\theta(u), \eta(u, \lambda) \mid u \in \mathcal{S}, \lambda \in \mathcal{W}\}$, where
  $\theta(u)$: the zenith angle at the pixel u.
  $\eta(u, \lambda)$: the refractive index at pixel u and wavelength λ.
1: $\eta(u, \lambda) \leftarrow \eta_0 \forall u \in \mathcal{S}, \lambda \in \mathcal{W}$
2: **while true do**
3: $\quad \theta_{\text{old}}(u, \lambda) \leftarrow \theta(u, \lambda)$
4: $\quad \eta_{\text{old}}(u, \lambda) \leftarrow \eta(u, \lambda)$
5: $\quad \theta(u) \leftarrow \text{argmin}_{\theta(u)} \mathcal{E}$
6: $\quad \eta(u, \lambda) \leftarrow \text{argmin}_{\eta(u, \lambda)} \mathcal{E}$
7: $\quad$ **if** $|\theta_{\text{old}}(u) - \theta(u)| < \tau_\theta$ and $|\eta_{\text{old}}(u, \lambda) - \eta(u, \lambda)| < \tau_\eta, \forall u$ and λ **then**
8: $\quad\quad$ **break**
9: $\quad$ **end if**
10: **end while**
11: **return** $\theta(u), \eta(u, \lambda)$

---

straightforward manner. To this end, we opt for an equivalent, yet simpler formulation of the cost functional, which expresses the data error term using the inverse function of the Fresnel transmission ratio in Eq. (11.5). Eventually, we reformulate the cost functional with a quadratic function of the zenith angle $\theta(u)$ in the data error term, whose minimum can be derived analytically.

With the refractive index $\eta(u, \lambda)$ fixed, we can invert the Fresnel transmission ratio function in Eq. (11.5) with respect to the zenith angle $\theta(u)$ as follows:

$$\sin \theta(u) \equiv \frac{\eta(u, \lambda)\sqrt{1 - R^2(u, \lambda)}}{\sqrt{\eta^2(u, \lambda) - 2R(u, \lambda)\eta(u, \lambda) + 1}}. \quad (11.17)$$

Note that the zenith angle is a geometric variable independent of the wavelength. However, the above equation provides wavelength-dependent estimates of the zenith angle. In practice, these estimates may not be the same across the spectrum due to measurement error and noise corruption. If the index of refraction is in hand, the value of $r(u, \lambda)$ computed from the ratio of maximal and minimal image irradiance can be used as an estimate for the function $R(u, \lambda)$. This yields a wavelength-dependent estimate of the zenith angle $\theta(u)$ at the current iteration, which is given by

$$\varphi(u, \lambda) = \arcsin\left(\frac{\eta(u, \lambda)\sqrt{1 - r^2(u, \lambda)}}{\sqrt{\eta^2(u, \lambda) - 2r(u, \lambda)\eta(u, \lambda) + 1}}\right). \quad (11.18)$$

In Eq. (11.18), we use the notation $\varphi(u, \lambda)$ for the wavelength-dependent estimate and distinguish it from the wavelength-independent zenith angle $\theta(u)$. We

11.3 Optimal Zenith Angle and Refractive Index

utilise this wavelength dependency and instead of directly minimising the original cost functional in Eq. (11.15), we seek to recover a zenith angle close to the wavelength-dependent estimates in Eq. (11.18) while satisfying the integrability constraint. Thus, we employ the alternative cost functional:

$$\mathcal{E}_1 = \int_S \int_W (\theta(u) - \varphi(u,\lambda))^2 \, d\lambda \, du$$
$$+ \beta(u) \int_S \left(\cos\alpha(u) \frac{\partial \theta(u)}{\partial y} - \sin\alpha(u) \frac{\partial \theta(u)}{\partial x} \right)^2 du, \quad (11.19)$$

as an alternative to that in Eq. (11.15).

Equation (11.19) poses the minimisation problem in a simpler setting. The merit of the alternative cost functional is that the Fresnel ratio error is quantified as a quadratic form of the function $\theta(u)$. This is important since this quadratic form is more tractable than the original error term, which contains a rational function in the expression for $R(u, \lambda)$. Moreover, we can rewrite Eq. (11.19) as follows:

$$\mathcal{E}_1 = \int_S f\left(u, \theta(u), \theta_x(u), \theta_y(u)\right) du, \quad (11.20)$$

by letting $f(\cdot)$ be given by

$$f\left(u, \theta(u), \theta_x(u), \theta_y(u)\right)$$
$$\triangleq \int_W (\theta(u) - \varphi(u,\lambda))^2 \, d\lambda + \beta(u)\left(\cos\alpha(u)\theta_y(u) - \sin\alpha(u)\theta_x(u)\right)^2.$$
$$(11.21)$$

The function $f(\cdot)$ is important, since it permits the use of calculus of variations to recover the minimiser of the functional in Eq. (11.20). We do this by noting that the minima must satisfy the following Euler–Lagrange equation:

$$\frac{\partial f}{\partial \theta} = \frac{\partial}{\partial x}\left(\frac{\partial f}{\partial \theta_x}\right) + \frac{\partial}{\partial y}\left(\frac{\partial f}{\partial \theta_y}\right). \quad (11.22)$$

By computing the derivatives of f so as to satisfy the Euler–Lagrange equation above, we arrive at the following differential equation:

$$\theta(u) \int_W d\lambda - \int_W \varphi(u,\lambda) \, d\lambda$$
$$= \beta(u) \times \left(\sin^2\alpha(u)\theta_{xx}(u) - \sin 2\alpha(u)\theta_{xy}(u) + \cos^2\alpha(u)\theta_{yy}(u) \right), \quad (11.23)$$

where $\theta_{xx}(u)$, $\theta_{yy}(u)$ and $\theta_{xy}(u)$ are the second-order and covariant derivatives of $\theta(u)$ with respect to the x- and y-axes of the coordinate system.

In the discrete case, where the imagery is acquired at K wavelength indexed bands $\lambda \in \{\lambda_1, \ldots, \lambda_K\}$, we have $\int_W d\lambda = K$. Therefore, $\theta(u)$ satisfies the differential equation

$$\theta(u) = \frac{1}{K} \int_W \varphi(u, \lambda) \, d\lambda$$
$$+ \frac{\beta(u)}{K} \left(\sin^2 \alpha(u) \theta_{xx}(u) - \sin 2\alpha(u) \theta_{xy}(u) + \cos^2 \alpha(u) \theta_{yy}(u) \right). \tag{11.24}$$

We note that Eq. (11.24) is a second-order partial differential equation with respect to $\theta(u)$. We further enforce the continuity and differentiability of the spatial domain by assuming that the function $\theta(u)$ is continuously twice differentiable, i.e. $\theta_{xy}(u) = \theta_{yx}(u)$. This assumption permits the decomposition of $\theta(u)$ into an orthogonal basis of integrable two-dimensional functions, in a similar manner to that in Frankot and Chellappa (1988). Since digital images have a limited band of spatial frequencies, the surface shading can be expressed as finite linear combinations of the real part of Fourier basis functions, which are cosine functions. This representation allows an analytical solution to the functional minimisation problem above. Moreover, we will show later that this representation also leads to a computationally efficient solution to Eq. (11.24).

Note that the function $\theta(u)$ can be viewed as a discrete function on a two-dimensional lattice. Let the size of the lattice be $X \times Y$, where X and Y are the image width and height, respectively. Based on the Nyquist–Shannon sampling theorem (Shannon 1949), the zenith angle $\theta(u)$ can be reconstructed using frequency components of up to one-half of the sampling frequency of the image. In image processing, these sampling frequencies υ, where $\upsilon = (\upsilon_x, \upsilon_y)$, are often chosen such that $\upsilon_x = \frac{2\pi i}{X}$, where $i = 0, 1, \ldots, X-1$, and $\upsilon_y = \frac{2\pi j}{Y}$, where $j = 0, 1, \ldots, Y-1$ (Gonzalez and Woods 2001). With these two-dimensional frequency components, the function $\theta(u)$ can be reconstructed with an orthogonal set of Fourier basis functions $e^{i(u^T \upsilon)} = e^{i(\upsilon_x x_u + \upsilon_y y_u)}$, where $\mathbf{i}$ is an imaginary number and $u = (x_u, y_u)$ is the pixel location. Formally, this is given by

$$\theta(u) = \sum_\upsilon \kappa_\upsilon e^{\mathbf{i}(u^T \upsilon)}. \tag{11.25}$$

Intuitively, Eq. (11.25) means that the shading of the surface can be decomposed into a linear combination of Fourier components with a range of frequencies matching that of the input imagery. In the equation, κ_υ is the coefficient (weight) of the Fourier basis function $e^{\mathbf{i}(u^T \upsilon)}$, which can be computed by making use of the equation

$$\kappa_\upsilon = \frac{1}{|\mathcal{S}|} \sum_u \theta(u) e^{-\mathbf{i}(u^T \upsilon)},$$

where $|\mathcal{S}|$ represents the number of image pixels.

11.3 Optimal Zenith Angle and Refractive Index

Similarly, the partial derivatives of $\theta(u)$ can also be expressed in terms of the Fourier basis as follows:

$$\theta_{xx}(u) = -\sum_{v} \kappa_v v_x^2 e^{i(u^T v)}, \tag{11.26}$$

$$\theta_{xy}(u) = -\sum_{v} \kappa_v v_x v_y e^{i(u^T v)}, \tag{11.27}$$

$$\theta_{yy}(u) = -\sum_{v} \kappa_v v_y^2 e^{i(u^T v)}. \tag{11.28}$$

Let $h(u) = \frac{1}{K} \int_W \varphi(u, \lambda) d\lambda$. By substituting Eqs. (11.25), (11.26), (11.27) and (11.28) into Eq. (11.24), we obtain

$$h(u) = \sum_{v} \kappa_v e^{i(u^T v)} \left(1 + \frac{\beta(u)}{K} \left(\sin^2 \alpha(u) v_x^2 - \sin 2\alpha(u) v_x v_y + \cos^2 \alpha(u) v_y^2 \right) \right). \tag{11.29}$$

Note that Eq. (11.29) applies to every image location u and every spatial frequency v. By making use of the expressions for $h(u)$ at every image location and frequency, we can construct a linear system with respect to the unknown vector $\mathbb{U} = [\kappa_v]^T$, which is, effectively, the concatenation of all the Fourier coefficients. The recovery of the coefficients κ_v can then be effected by solving the linear system $\mathbb{L}\mathbb{U} = \mathbb{H}$, with $\mathbb{H} = [h(u)]^T$, which is the vectorial concatenation of the known function values $h(u)$ for all the image locations u, and where $\mathbb{L}$ is a matrix with rows and columns indexed to the image pixels and spatial frequencies, respectively. In other words, the matrix element $\mathbb{L}_{u,v}$ corresponding to a given pixel u and a given frequency v is $\mathbb{L}_{u,v} = e^{i(u^T v)}(1 + \frac{\beta(u)}{K}(\sin^2 \alpha(u) v_x^2 - \sin 2\alpha(u) v_x v_y + \cos^2 \alpha(u) v_y^2))$. With a chosen Fourier basis and the azimuth angle $\alpha(u)$ obtained as described in Sect. 11.1.2, the matrix $\mathbb{L}$ can be computed in a straightforward manner. With the coefficients κ_v at hand, the zenith angle $\theta(u)$ can be recovered through the application of Eq. (11.25).

11.3.2 Refractive Index Computation

We now turn our attention to the estimation of the refractive index at each image location, making use of the Fresnel transmission ratio. To derive the refractive index directly from the Fresnel transmission ratio in Eq. (11.5), we are required to solve a quadratic equation involving the index of refraction $\eta(u, \lambda)$. This expression is given by

$$\left(\cos^2 \theta(u) - r^2(u, \lambda) \right) \times \eta^2(u, \lambda) + 2r(u, \lambda) \sin^2 \theta(u) \times \eta(u, \lambda) - \sin^2 \theta(u) = 0. \tag{11.30}$$

Algorithm 11.2 Refractive index selection

Require: $\eta(u, \lambda)$: Solutions to the refractive index at pixel u and wavelength λ
Ensure: $\eta^*(u, \lambda)$: The uniquely determined refractive index at pixel u and wavelength λ
1: **for all** wavelength λ **do**
2:    **for all** pixel u with a single physically plausible refractive index at wavelength λ **do**
3:       $determined(u, \lambda) \leftarrow$ **true**
4:    **end for**
5:    **while** there are pixels with two plausible refractive indices **do**
6:       **for all** pixel u with two solutions $\eta_1(u, \lambda)$ and $\eta_2(u, \lambda)$ **do**
7:          $\mathcal{N}(u) \leftarrow$ the spatial neighbourhood of u
8:          $\bar{\eta}(u, \lambda) \leftarrow Average_{v \in \mathcal{N}(u), determined(v, \lambda) = true} \eta(v, \lambda)$
9:          $\eta^*(u, \lambda) \leftarrow \eta_i(u, \lambda)$ which is closer to $\bar{\eta}(u, \lambda)$
10:         $determined(u, \lambda) \leftarrow$ **true**
11:      **end for**
12:    **end while**
13: **end for**
14: **return** $\eta^*(u, \lambda)$ $\forall$ pixel u and wavelength λ

In general, the quadratic equation above yields no more than two real-valued roots. The choice of refractive index value depends on the physical plausibility of these roots; i.e. the refractive index for dielectrics must be a real value greater than one. This choice is straightforward if only one of the roots is physically plausible.

In the case where the two roots are plausible, we can adopt an iterative approach which iteratively selects the root closer to the refractive index average at the same wavelength within the local spatial neighbourhood. This approach works under the assumption that there is a single solution to the refractive index at a number of pixels in the image. Initially, we label these pixels as having their refractive index uniquely determined. At each iteration, we assign the refractive index of pixels with two plausible solutions, making use of the regions whose refractive index is already determined. We do this by selecting the root which is in better accordance with the average of the refractive indices within its spatial neighbourhood. A pseudo-code of this iterative procedure is illustrated in Algorithm 11.2.

When neither root is a physically plausible solution to Eq. (11.30), we can opt for an approximation of the Fresnel transmission ratio that provides a single solution to the refractive index. To this end, we consider an approximating function that is a product of two separable factors, one of which contains solely the refractive index while the other depends on the zenith angle. By adopting a formulation similar to Schlick's approximation of the Fresnel reflection coefficient (Schlick 1994), we arrive at the expression

$$R^*(u, \lambda) = d + c\bigl(1 - \cos\theta(u)\bigr)^b, \tag{11.31}$$

where b, c and d are constants.

11.3 Optimal Zenith Angle and Refractive Index

Fig. 11.2 The Fresnel transmission ratio for several refractive indices and zenith angles in the range of $[0, \frac{\pi}{2}]$. (**a**) The true Fresnel transmission ratio function. (**b**) The approximating function

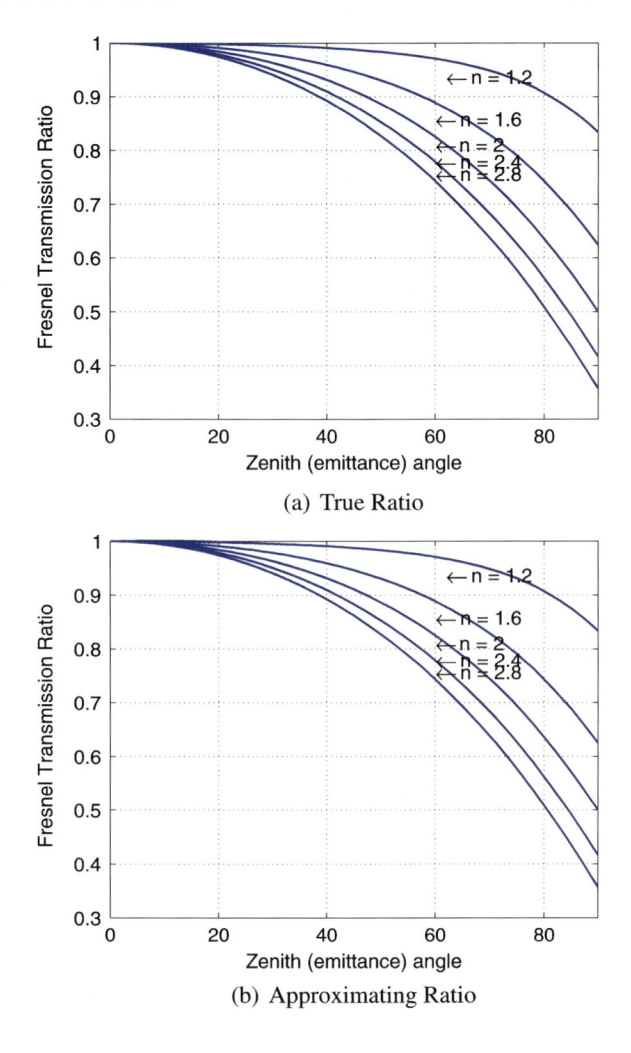

(a) True Ratio

(b) Approximating Ratio

Figure 11.2(a) shows the graph of the exact Fresnel transmission ratio for refractive index values ranging between 1.2 and 2.8, with an increment of 0.4. Empirical observations of this graph give rise to a power function with respect to the zenith angle. At the end points, $R = 1$ as $\theta(u) = 0$ and $R = \frac{1}{\eta(u,\lambda)}$ as $\theta(u) = \frac{\pi}{2}$. To satisfy these conditions, it is necessary that $c = \frac{1}{\eta(u,\lambda)} - 1$ and $d = 1$. Thus, we have

$$R^*(u, \lambda) = 1 + \left(\frac{1}{\eta(u, \lambda)} - 1\right)(1 - \cos\theta(u))^b. \tag{11.32}$$

In Eq. (11.32), b is a predetermined power that provides the best fit with respect to the true Fresnel transmission ratio over a range of parameter values. In this chapter, we consider material refractive indices between 1 and 3 and $\theta(u) \in [0, \frac{\pi}{2}]$. Using a one-dimensional search for the power b, we find that $b = 1.4$ minimises

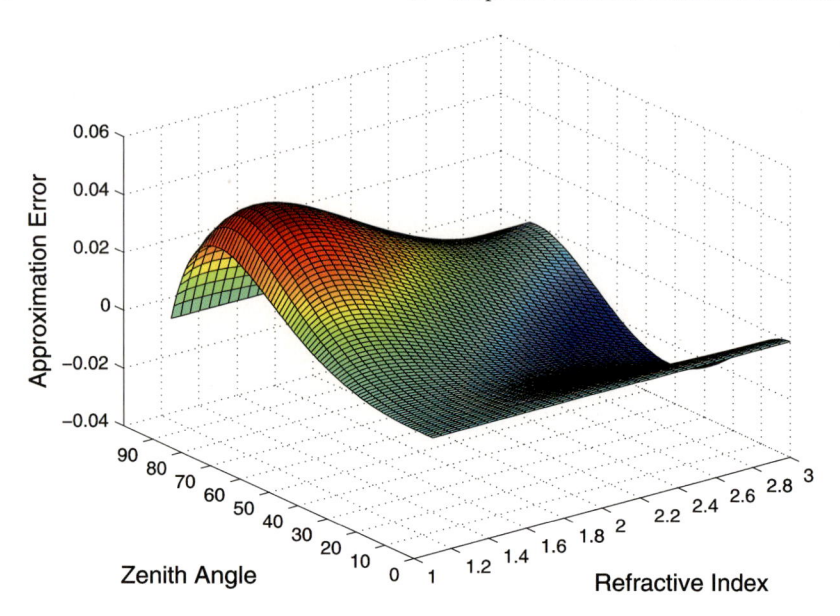

Fig. 11.3 The approximation error for the Fresnel transmission ratio plotted for refractive indices between 1 and 3 and zenith angles in the range of $[0, \frac{\pi}{2}]$

the L^1-error between the approximating and the true Fresnel transmission ratio. In Fig. 11.2(b), we have plotted the approximating Fresnel ratio for $b = 1.4$ and noted its similarity to the Fresnel ratio function.

To verify the approximation accuracy for the Fresnel transmission ratio, in Fig. 11.3 we plot the error function for $b = 1.4$, where it is represented as a surface with respect to both the zenith angle and refractive index. From the figure, we observe that the absolute error is below 0.04 for all the combinations of zenith angle and refractive index within the considered range. The error is largest near grazing zenith angles ($70° \leq \theta(u) < 90°$) for small refractive indices or at around $40°$ for refractive indices larger than 2.3. However, these are extreme cases and, generally, do not include the material and the geometry under study.

Using the approximated Fresnel ratio function, we arrive at a single approximate solution for the refractive index at each pixel and wavelength. Given the zenith angle $\theta(u)$ at the current iteration, the index of refraction is estimated to be

$$\eta(u, \lambda) = \frac{(1 - \cos\theta(u))^b}{(1 - \cos\theta(u))^b - 1 + r(u, \lambda)}. \tag{11.33}$$

Next, we apply the following strategy to ensure the physical plausibility of the approximating solution in Eq. (11.33). We note that the approximating refractive index is physically plausible, i.e. $\eta(u, \lambda) \geq 1$ if $(1 - \cos\theta(u))^b - 1 + r(u, \lambda) > 0$ and $r(u, \lambda) < 1$. In the case where $r(u, \lambda) = 1$, we can conclude that $\theta(u) = 0$, as can be observed in Fig. 11.2(a). However, in this case, the refractive index value can be arbitrary. Otherwise, when $r(u, \lambda) < 1$, i.e. $\theta(u) \neq 0$, we can guarantee the

11.3 Optimal Zenith Angle and Refractive Index

physical plausibility of the solution in Eq. (11.33) by scaling the zenith angles at all the image pixels such that $\cos\theta(u) < 1 - (1 - r(u, \lambda))^{\frac{1}{b}}$ for all u and λ.

Finally, having obtained a refractive index spectrum per surface location, we fit the spectrum to one of the dispersion equations in Eq. (11.13) or (11.14). As a result, we recover the dispersion coefficients in these equations and compute the closest approximation on the dispersion curve for the refractive index obtained earlier.

The Cauchy dispersion equation, i.e. Eq. (11.13), can be rewritten in the following matrix form:

$$\mathbf{n}(u) = \Lambda \mathbf{C}(u), \qquad (11.34)$$

where

$$\mathbf{n}(u) = \begin{bmatrix} \eta(u, \lambda_1) \\ \vdots \\ \eta(u, \lambda_K) \end{bmatrix}, \quad \mathbf{C}(u) = \begin{bmatrix} C_1(u) \\ \vdots \\ C_M(u) \end{bmatrix}, \quad \text{and}$$

$$\Lambda = \begin{bmatrix} 1 & \lambda_1^{-2} & \cdots & \lambda_1^{-2(M-1)} \\ 1 & \lambda_2^{-2} & \cdots & \lambda_2^{-2(M-1)} \\ & & \cdots & \\ 1 & \lambda_K^{-2} & \cdots & \lambda_K^{-2(M-1)} \end{bmatrix},$$

where M is the number of dispersion coefficients used.

Since $\mathbf{n}(u)$ and Λ are known, the system in Eq. (11.34) is over-determined if the number of dispersion coefficients is chosen such that $M \leq K$, where K is the number of wavelength indexed bands in the imagery. In this case, we can solve for the coefficients $\mathbf{C}(u)$ and compute the refractive index spectrum according to the dispersion equation.

For Sellmeier's equation, i.e. Eq. (11.14), the fitting task can be posed as a nonlinear least-squares optimisation problem. If the number of dispersion coefficients M is chosen such that $M \leq K$, then the optimisation problem above becomes well-constrained and the coefficients $\mathbf{C}(u)$ can be solved numerically by standard line-search or trust-region techniques.

Having obtained the azimuth and zenith angles of the surface normals, one can recover the shape under study by means of a surface integration method such as the one in Frankot and Chellappa (1988). On the other hand, with the dispersion coefficients in hand, the index of refraction can be recovered by the dispersion equation.

In Fig. 11.4, we present the depth maps recovered from the real-world imagery. These maps have been produced in such a way that the grey level corresponds to the surface height. To illustrate the recovery of shape, we employ Cauchy's dispersion equation as a constraint for the optimisation of the cost functional in Eq. (11.15). Here, Cauchy's dispersion equation consists of five terms and is of the eighth order. The fitting of refractive index spectra to this dispersion equation was performed via a constrained linear least-squares optimisation method (Coleman and Li 1996).

Fig. 11.4 *Top row*: input polarisation component images captured at 45°, rendered in trichromatic pseudo-colours. *Bottom row*: depth maps of the corresponding objects in the *top row*

Once the surface normal is recovered by making use of the polarisation information, we reconstruct the surface depth by means of normal field integration. To this end, we make use of the surface integration method introduced by Frankot and Chellappa (1988). Note that the three-dimensional structure of the reconstructed shapes is, in general, perceptually consistent with their original input images. We also note that the distortions across material boundaries are particularly visible in the reconstructed shapes of the bear and the dinosaur. This happens because the variation of polarisation across materials has been interpreted as a result of the variation in surface geometry.

In addition, we provide qualitative results for the refractive indices estimated from the real-world imagery. In Fig. 11.5, we plot the mean refractive index of two sample regions in each image, as a function of the wavelength. The top row of Fig. 11.5 shows the rectangular bounding boxes of these regions overlaying the 45° polarisation images presented in Fig. 11.4. The remaining rows show the mean estimated refractive index spectrum for each selected region as a result of fitting to the Cauchy dispersion equation. Note that the refractive index spectra are plotted in colours that match those of the bounding boxes of the selected regions in the input images. In the figure, the regions selected in the image of each object, except for the statue, are made of different materials. This results in the difference between the estimated refractive index spectra between the two regions in each of these images. Since the object in the second column is composed of almost the same material across its surface, the dispersion of the refractive index over the wavelength does not vary significantly across both of the selected regions. Furthermore, the refractive index values can provide a hint at the roughness of the surface. For example, the surface of region 1 in the first column consists of several ridges and valleys and is rougher than that of region 2. This observation is consistent with the ranges of refractive index values: between 1.45–1.55 for region 1 and 1.38–1.46 for region 2.

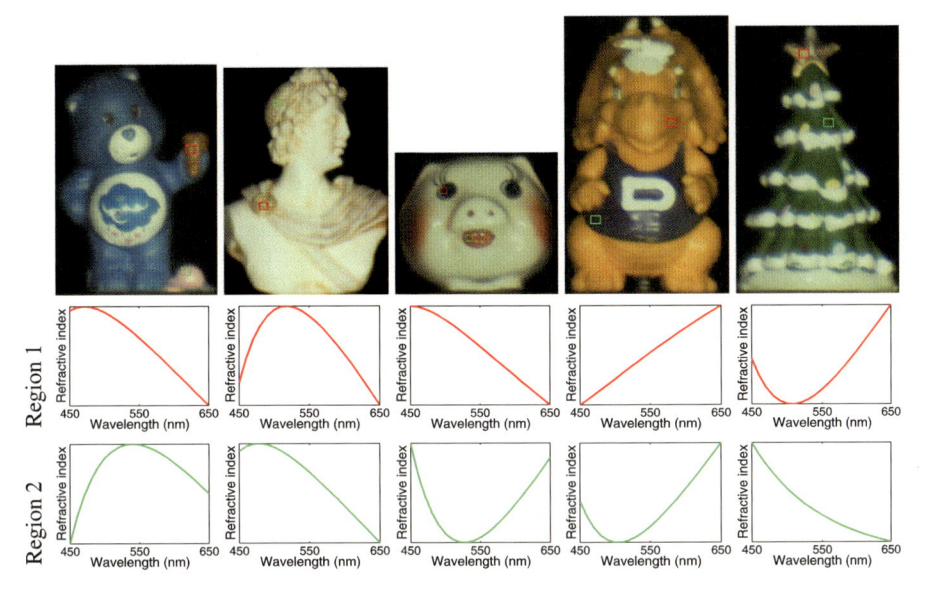

Fig. 11.5 Mean refractive index spectra of a number of selected regions in the input images. *First row*: the 45° polarisation component images rendered in pseudo-colour, with the selected regions indicated by rectangular bounding boxes. *Second and third rows*: the mean of the estimated refractive index spectra over the pixels in each selected region, fitted to the Cauchy dispersion equation. These spectra are plotted as *solid lines* with colours matching those of the region boundaries depicted in the *first row*

11.4 Notes

Since reflectance models, such as the Torrance–Sparrow model (Torrance and Sparrow 1967) and the Wolff model (Wolff 1994), draw on the Fresnel reflection theory, they share the common feature of considering the reflected light as a combination of polarisation components parallel to and perpendicular to the plane of reflection. Consequently, they can accommodate polarised light sources. One of the common features of these Fresnel-based models is that physical properties of the surface material and the geometry of the reflection process are expressed in a single equation with multiple degrees of freedom. As a result, the simultaneous recovery of photometric parameters and shape information becomes an under-constrained problem.

To tackle this ill-posedness, several attempts have been made to make use of polarisation images captured with varying viewpoint and light source direction. These methods either assume a known refractive index or a known surface orientation distribution, infer the surface orientation using single viewpoint polarisation images or, alternatively, require a complex instrumental set-up. Thus, although polarimetric imagery is an active topic of research, the recovery of object shape from a single view still remains a challenging task due to the presence of photometric artefacts and discontinuities on the object surface.

Polarisation has also proven to be an effective tool in revealing the material properties of surfaces from images. Early work by Wolff and Boult (Wolff 1989) showed how to classify image regions as belonging to metallic or dielectric materials. In this work, the Fresnel ratio was used to characterise the relative electrical conductivity of the surface as a discriminant feature for classification. Later, the same authors used a Fresnel reflectance model to predict the magnitude of polarisation at an arbitrary orientation in the image plane (Wolff and Boult 1991; Wolff 1990). Using this model, the Fresnel ratio can be estimated and used for the classification of dielectrics. More recently, Chen and Wolff (1998) employed the phase angle of polarisation for distinguishing conducting metals from dielectric materials. Their approach hinges on the physical observation that, upon reflection from a metal surface, the phase difference between polarisation components is altered (Born and Wolf 1999). It is worth noting that this is not the case for dielectrics.

References

Atkinson, G., & Hancock, E. R. (2005). Multi-view surface reconstruction using polarization. In *International conference on computer vision* (pp. 309–316).
Atkinson, G. A., & Hancock, E. R. (2006). Recovery of surface orientation from diffuse polarization. *IEEE Transactions on Image Processing, 15*(6), 1653–1664.
Atkinson, G. A., & Hancock, E. R. (2007). Shape estimation using polarization and shading from two views. *IEEE Transactions on Pattern Analysis and Machine Intelligence, 29*(11), 2001–2017.
Belhumeur, P. N., Kriegman, D. J., & Yuille, A. L. (1997). The bas-relief ambiguity. *Computer Vision and Pattern Recognition*, p. 1060.
Born, M., & Wolf, E. (1999). *Principles of optics: electromagnetic theory of propagation, interference and diffraction of light* (7th ed.). Cambridge: Cambridge University Press.
Boyd, S., & Vandenberghe, L. (2004). *Convex optimization*. Cambridge: Cambridge University Press.
Chen, H., & Wolff, L. B. (1998). Polarization phase-based method for material classification in computer vision. *International Journal of Computer Vision, 28*(1), 73–83.
Coleman, T. F., & Li, Y. (1996). A reflective newton method for minimizing a quadratic function subject to bounds on some of the variables. *SIAM Journal on Optimization, 6*(4), 1040–1058.
Drbohlav, O., & Sára, R. (2001). Unambigous determination of shape from photometric stereo with unknown light sources. In *International conference on computer vision* (pp. 581–586).
Frankot, R. T., & Chellappa, R. (1988). A method of enforcing integrability in shape from shading algorithms. *IEEE Transactions on Pattern Analysis and Machine Intelligence, 4*(10), 439–451.
Gonzalez, R. C., & Woods, R. E. (2001). *Digital image processing* (2nd ed.). Boston: Addison-Wesley Longman.
Kasarova, S. N., Sultanova, N. G., Ivanov, C. D., & Nikolo, I. D. (2007). Analysis of the dispersion of optical plastic materials. *Optical Materials, 29*, 1481–1490.
Miyazaki, D., Kagesawa, M., & Ikeuchi, K. (2004). Transparent surface modeling from a pair of polarization images. *IEEE Transactions on Pattern Analysis and Machine Intelligence, 26*(1), 73–82.
Miyazaki, D., Saito, M., Sato, Y., & Ikeuchi, K. (2002). Determining surface orientations of transparent objects based on polarization degrees in visible and infrared wavelengths. *Journal of the Optical Society of America A, 19*(4), 687–694.
Miyazaki, D., Tan, R. T., Hara, K., & Ikeuchi, K. (2003). Polarization-based inverse rendering from a single view. In *IEEE international conference on computer vision* (Vol. 2, p. 982).

References

Rahmann, S. (1999). Inferring 3D scene structure from a single polarization image. In *SPIE proceedings on polarization and color techniques in industrial inspection* (pp. 22–33).

Rahmann, S. (2000). Polarization images: a geometric interpretation for shape analysis. In *International conference on pattern recognition* (Vol. 3, pp. 538–542).

Rahmann, S., & Canterakis, N. (2001). Reconstruction of specular surfaces using polarization imaging. In *IEEE conference on computer vision and pattern recognition* (Vol. 1, pp. 149–155).

Saito, M., Sato, Y., Ikeuchi, K., & Kashiwagi, H. (1999). Measurement of surface orientations of transparent objects using polarization in highlight. *Journal of the Optical Society of America A*, *16*(9), 2286–2293.

Schlick, C. (1994). An inexpensive BRDF model for physically-based rendering. *Computer Graphics Forum*, *13*(3), 233–246.

Sellmeier, W. (1871). Zur Erklärung der abnormen Farbenfolge im Spectrum einiger Substanzen. *Annalen der Physik und Chemie*, *219*(6), 272–282.

Shannon, C. E. (1949). Communication in the presence of noise. *Proceedings of the Institute of Radio Engineers*, *37*(1), 10–21.

Thilak, V., Voelz, D. G., & Creusere, C. D. (2007). Polarization-based index of refraction and reflection angle estimation for remote sensing applications. *Applied Optics*, *46*(30), 7527–7536.

Torrance, K., & Sparrow, E. (1967). Theory for off-specular reflection from roughened surfaces. *Journal of the Optical Society of America*, *57*(9), 1105–1112.

Torrance, K. E., Sparrow, E. M., & Birkebak, R. C. (1966). Polarization, directional distribution, and off-specular peak phenomena in light reflected from roughened surfaces. *Journal of the Optical Society of America*, *56*, 916–924.

Wolff, L. B. (1989). Using polarization to separate reflection components. In *Computer vision and pattern recognition* (pp. 363–369).

Wolff, L. B. (1990). Polarization-based material classification from specular reflection. *IEEE Transactions on Pattern Analysis and Machine Intelligence*, *12*(11), 1059–1071.

Wolff, L. B. (1994). *Diffuse-reflectance model for smooth dielectric surfaces* (No. 11, pp. 2956–2968).

Wolff, L. B., & Boult, T. E. (1991). Constraining object features using a polarization reflectance model. *IEEE Transactions on Pattern Analysis and Machine Intelligence*, *13*(7), 635–657.

Zhu, Q., & Shi, J. (2006). Shape from shading: recognizing the mountains through a global view. In *Computer vision and pattern recognition* (pp. 1839–1846).

Index

A
Absorption detection, 118
  derivative analysis, 118
  scale space method, 120
    Fingerprint, 121
      maximum modulus wavelet transform, 124
    uni-modal segmentation, 125
Acousto optic tunable filter, 11
  sensitivity function, 12
Affine coordinate transformation, 112
Affine distortion correction, 115
Affine distortion matrix, *see* Distortion matrix
Albedo, 44, 46, 192
Applications
  biometrics, 103
  food sorting, 167
  mineral classification, 155
  pest detection, 4, 116
  photography
    material substitution, 3, 64
    re-illumination, 3
  skin recognition, 101, 147
Azimuth angle, 21

B
Band ratio, 82
Bas-relief ambiguity, 249
Bayesian winner-take-all, 155
Beckmann–Kirchhoff model, 39, 180
  diffuse scattering, 180
  specular spike, 181
  surface correlation length, 180
  surface height variation, 180
  surface roughness, 180
  surface slope, 180
  diffuse scattering, 41

  exponential correlation function, 41
  Fresnel correction, 42
  Gaussian correlation function, 41
  geometric attenuation, 41
  specular magnitude, 40
  specular spike, 40
  surface correlation length, 39
  surface height variation, 39
  surface roughness, 40
Bidirectional reflectance distribution function, *see* BRDF
Blind source separation, 233
BRDF, 21
  assumptions, 22
  relation to the image irradiance, 23

C
Calculus of variations, 176, 186
Calibration target, 84, 101, 184
Canonical correlation analysis, 159, 164
CCA, *see* Canonical correlation analysis
CCD
  noise calibration, 14
  photon flux, 14
  quantum efficiency, 14
  read noise, 14
  thermal noise, 14
Classifier fusion, 165
Colour matching functions, 24
  10-degree, 25
  2-degree, 25
Complete polarisation, 212
Continuum, 59, 125
  removal, 125
Contrast sensitivity function, 30
Convex/concave ambiguity, 193, 249

Cook–Torrance model, 47, 182
Coordinate descent, 251
Correlated colour temperature, 26, 55
    white balancing to D55, 31

D

Dark current image, 14
Deterministic annealing, 146
Dichromatic model, 43, 179, 227
    shading factor, 44
    specular coefficient, 44
Dichromatic plane, 58, 74, 227
Dielectric material, 44, 216, 219
Diffuse polarisation, 219
    phase, 222
    subsurface scattering, 221
Dispersion equation, 241, 249
    Cauchy's equation, 249, 259
    dispersion coefficients, 249
    Sellmeier's equation, 249
Distortion matrix
    affine, 112
    affine transformation, 113
    in the power spectrum, 113

E

Eikonal equation, 188
Electromagnetic wave, 210
    electric field, 211, 218, 221
    electric field amplitude, 221
    electric field vector, 210, 218, 220
    harmonic plane waves, 211
    magnetic field vector, 210
    plane of vibration, 211
    superposition of harmonic components, 212
Elliptic polarisation, 212
Entropy
    α-entropy, 161
    alpha divergence, 161
    alpha log-divergence, 162
    Jeffreys divergence, 163, 165
    Kullback–Leibler divergence, 162, 165
    Shanon's, 68
Entropy minimisation, *see* Specularity removal
Epanechnikov kernel, 245
    bandwidth, 246
Euclidean angle, 68
Euler–Lagrange equation, 186, 202, 253

F

Fast marching, 245
Finite difference, 186
Foreshortened surface area, 18

Fourier cosine transform, 111
Fourier kernel, 110
Fourier series, 254
Fresnel attenuation, *see* Fresnel reflection coefficients, 218
Fresnel ratio, *see* Fresnel reflection ratio, Fresnel transmission ratio
Fresnel reflectance model, 219
Fresnel reflection coefficients, 40, 45, 46, 180, 218
Fresnel reflection ratio, 218
Fresnel reflection theory, 176, 218, 221, 241, 242
Fresnel transmission coefficients, 221
Fresnel transmission ratio, 221, 242, 247
    approximation, 256

G

Gaussian sphere, 190
General irradiance equation, 185
General reflectance model, 177, 185
    diffuse reflection, 178
    relationship to specific models, 179
    specific model correspondence, 183
    specular reflection, 178
Gibbs distribution, 147
Green's theorem, 190

H

Harmonic kernel, 110
Helmholtz reciprocity principle, 23, 37
Huber's M-estimator, 78

I

ICA, *see* Independent component analysis
Illuminant direction
    slant angle, 196, 201
    tilt angle, 195, 197
Illuminant direction estimation, 193
    highlight, 204
    non-linear optimisation, 204
    occluding boundary, 198
        irradiance extrapolation, 200
    optimisation methods, 202
    shading statistics, 193
    shadow, 201
Illuminant direction estimator
    Brooks and Horns, 202
    disk method, 197
    Knill, 197
    Lee and Rosenfeld, 195
    Nillius and Eklundh, 200

Index

Illuminant direction estimator (cont.)
  Pentland, 194, 195
  Weinshall, 201
  Zheng and Chellappa, 196, 199
Illuminant position estimation, 203, 204
Illuminant spectrum estimation
  closed form solution, 58
  least-squares, 58
  Planckian locus, 57
  spectrum deconvolution, 59
Incident flux, 21
Incoherent electric field, 212
Independent component analysis, 233
Integral transform, 111
Irradiance equation, 175, 183, 192
Isotonic regression, 126
Isotropic surface, 23

K
Kernel weighting function, 245
Kirchhoff's wave scattering, 37

L
Lambertian model, 43
Lambert's cosine law, 43, 46, 223
Law of reflection, 217
Law of refraction, see Snell's law
Levenberg–Marquardt optimisation, 204, 215
Line search, 188
Liquid crystal tunable filter, 12
  sensitivity function, 13
Li's adaptive potential function, 78
Logarithmic differentiation, see Specularity removal
Lorentzian function, 131

M
Material dispersion, 249
Matrix factorisation
  constrained, 148
  non-negative, 148, 150
  non-negative with a sparsity constraint, 150
Max-cut problem, 245
Maximum entropy criterion, 146
Measurement noise, 81
Metamerism, 34
Mixture model, see Spectrum representation
Multiple-CCD multispectral cameras, 13
  typical spectral transmission, 13
Mutual information, 235

N
NDVI, 118
Normalised difference vegetation index, see NDVI

O
Off-specular reflection, 46, 222, 224
Oren–Nayar model, 45
  V-cavities, 45

P
Patch stitching, 245
PCA
  representation of the spectra, 92
  visualisation, 28
Perron–Frobenius theorem, 164
Photogrammetric variables, 178
Photometric calibration, 25
Photometric invariant recovery, 176, 189
Photometric process, 23
Photometric stereo, 191
  uncalibrated, 249
Photopic CIE luminosity curve, 30
Planckian black body equation, 56
Planckian locus, 27
Planck's law, 54
Plane of incidence, 219
Plane of reflection, 219, 222
Polarimeter, 209
Polarimetric BRDF, 246
Polarimetric image decomposition, 215
Polarimetric imaging, 209
Polarimetric planes, 226
Polarimetric reflection model, 222
  rough surface, 223
  specular line, 232
Polarisation
  Brewster's angle, 215
  Mapping degree of polarisation, 245
  Parallel component, 218, 220, 225
  Perpendicular component, 218, 220, 225
  Phase correspondence, 245
Polarisation camera, 209
Polariser angle, 214
Pool adjacent violators, 126
Pushbroom imager, 11

R
Reflectance map, 189
Reflectance model, 37, 177
  empirical, 37
  internal scattering, 46
  interreflection, 45
  masking, 45–47
  micro-facet slope, 47, 48
  micro-facet slope distribution, 48
    Beckmann, 48
    Gaussian, 48

Reflectance model (*cont.*)
  multi-layered, 48
  multiple reflections, 46
  phenomenological, 37, 43
  physics-based, 37, 39
  shadowing, 45–47
Reflection angle, 218
Reflection component separation, 227
  polarisation, 227
    diffuse radiance propagation, 229
    geometric method, 231
    non-negative least squares, 230
    probabilistic independence, 235
Reflection geometry, 38, 176
Refraction, 220
Refractive index, 40, 178, 186, 218, 221, 224, 247
  complex refractive index, 246
Refractive index recovery, 241, 246, 255
  cost functional, 250
Regulariser
  curvature consistency constraint, 79
  robust statistics, 78
  smoothness constraint, 71, 78
Renyi entropy, *see* Entropy
Robust regulariser, *see* Regulariser

S

SAM, *see* Spectral angle mapper
Savitzky–Golay noise filtering, 120, 128
Schlick's approximation, 256
Semi-definite programming, 245
SFF, *see* Spectral feature fitting
Shape and source from shading, 202
    iterative update, 203
    multi-level algorithms, 203
Shape from shading, 175, 188
    discretisation, 191
    gradient consistency constraint, 188
    iterative update, 191
    level sets method, 188
    loop integral constraint, 189
    perspective shape from shading, 188
    surface smoothness constraint, 188
    variational approach, 191
Shape recovery, 175, 241
    cost functional, 185, 250
    discrete approximation, 186
    functional optimisation, 186
    iterative update, 187
    update equation, 187
    variational approach, 183
Singular point, 244
Snell's law, 45, 180, 219, 242

Spatially varying illumination, 223, 224
  estimation, 228
Spectral angle, 154
Spectral angle mapper, 153
Spectral cross correlation, 109, 114
Spectral feature fitting, 154
  affinity of two spectra, 154
Spectrum
  Subdivision according to the CIE and the ISO, 9
Spectrum differentiation, 123
Spectrum representation
  mixture model, 90
  parametric, 130
    Lorentzian, 131
    Voight's, 131
  relationship to PCA, 92
  splines, *see* Spline
Specular polarisation, 217
  phase, 219
Specularity removal, 65, 227
  entropy minimisation, 68
  log-linear bias, 107
  logarithmic differentiation, 67
Spline
  definition, 93
  global interpolation, 99
  knot removal, 96
Standard illuminant, 53
  A, B and C series, 53
  D series, 54
  E and F series, 55
Staring array camera, 11
Stereographic projection, 190
Subsurface scattering, 220
Surface integrability, 186, 190, 241, 248
  global integrability constraint, 245
Surface integration, 260
Surface normal, 20
  azimuth angle, 219, 222
    disambiguation, 244
    estimation, 245
  zenith angle, 221, 247
    estimation, 246, 251

T

Tetracorder, 117
Torrance–Sparrow model, 46, 182, 223
  log-transformed, 203
  micro-facet, 46
  micro-facet distribution, 46
  reflection geometry, 46
  specular component, 182
Transmitted radiance sinusoid, 214

Transmitted radiance sinusoid (*cont.*)
    amplitude, 214
    degree of linear polarisation, 215
    fitting error, 245
    I_{max}, 214, 216, 224
    I_{min}, 214, 216, 224
    phase of polarisation, 214, 216, 244
    unpolarised intensity, 214
TRS, *see* Transmitted Radiance Sinusoid
Tukey's bi-weight function, 78

U
Unmixing, 141
    abundance fractions, 143
    affine hull, 152
    Craig's criterion, 152
    end member, 141
    linear, 148, 149
    luminance fractions, 144
    partial membership probability, 145
    regularised model, 151
    relationship to scene materials, 142
    simplex, 152
    simplical cone, 152
Unpolarised light, 212, 219

V
Variational calculus, *see* Calculus of variations
Vernold–Harvey model, 42, 181
Voight function, 131

W
Wave theory, 210
Wavelet transform, 124
White reference, 25
    halon G-50, 25
    spectralon, 25
Wien's radiation law, 56
Wolfe condition, 188
Wolff model, 44, 180
    diffuse albedo, 44
    subsurface scattering, 44

X
XYZ colour space, 27

Y
Young–Householder decomposition, 92

Z
Zenith angle, 21